Practical Methods of Optimization

Volume 2

Practical Methods of Optimization

Volume 2
Constrained Optimization

R. Fletcher
Department of Mathematics
University of Dundee, Scotland, U.K.

A Wiley–Interscience Publication

JOHN WILEY & SONS
Chichester · New York · Brisbane · Toronto

British Library Cataloguing in Publication Data:

Fletcher, R.
Practical methods of optimization.
Vol. 2: Constrained optimization
1. Mathematical optimization
I. Title
515 QA402.5 80-42063

ISBN 0 471 27828 9

Typeset by Preface Ltd, Salisbury, Wilts. and printed in the United States of America.

Contents

Preface

The general presentation of Volume 2 continues those aims which I set out in the preface to Volume 1 and I shall not repeat the comments here. The additional complication of constraints in an optimization problem does make the general level of the material more difficult and I have made some effort to present at least the basic ideas in a straightforward way. In some cases this has led me to first of all present some results in an intuitive way followed by a rigorous but more mathematical derivation. The chapters in Volume 2 number on from those in Volume 1 so that cross referencing between the volumes is unambiguous. The selection of topics for this volume also has required some thought. Most finite dimensional problems of a continuous nature have been included but I have generally kept away from problems of a discrete or combinatorial nature since they have an entirely different character and the choice of method can be very specialized. In this case the nearest thing to a general purpose method is the branch and bound method, and since this is a transformation to a sequence of continuous problems of the type covered in this volume, I have included a straightforward description of the technique.
A feature of this volume which I think is lacking in the literature is a treatment of non-differentiable optimization which is reasonably comprehensive and covers both theoretical and practical aspects adequately. I hope that the final chapter meets this need. Although only unconstrained problems of this type are discussed it is appropriate to do it in this volume since a considerable background of constrained optimization theory is required. The subject of geometric programming is also included in the book because I think that it is potentially valuable, and again I hope that this treatment will turn out to be more straightforward and appealing than others in the literature. The subject of nonlinear programming is covered in some detail but there are difficulties in that this is a very active research area. To some extent therefore the presentation mirrors my assessment and prejudice as to how things will turn out, in the absence of a generally agreed point of view. However I have also tried to present various alternative approaches and their merits and demerits. Linear constraint programming, on the other hand, is now well developed and here the difficulty is that there are two distinct points of view. One is the traditional approach in which algorithms are presented as generalizations of early linear programming methods which carry out pivoting in a tableau. The other is a more recent approach in terms of active set strategies: I regard this as more intuitive

and flexible and have therefore emphasized it, although both methods are presented and their relationship is explored.

The occurrence of errors in the text is vexatious for any author and my experience with Volume 1 and the greater complexity of Volume 2 assures me that a fair number of these will inevitably be present. I would be most grateful therefore to know of any serious errors and also to hear of any other observations which may be pertinent. A list of these for Volume 1 follows this preface. The fact that Volume 1 is relatively free from typographical errors is entirely due to my wife Mary who proofread the entire volume despite not understanding very much of it. Perhaps this was an advantage! She also prepared the subject index and I am very grateful for her assistance. I must also repeat my thanks to those people mentioned in Volume 1 who have helped me in many ways with the preparation of the book. In addition I am very grateful to R. S. Womersley for his constructive advice on Chapter 14. I would also like to thank Mrs V. Gandy, who typed much of the book, and Mrs C. Peters for their excellent typing. The preparation of this book has occupied me to a greater or lesser extent for the past seven years. The final impetus to complete Volume 2 came during a period of leave of absence at the Mathematics Department, University of Kentucky, Lexington, at the invitation of Professor R. Wets, and I am very grateful for this support and for the provision of typing and computing facilities.

Dundee, October 1980 **R. Fletcher**

Errata to Volume 1

p. 21, line 2 from bottom: . . . it follows that $(f^{(k)} - f^{(k+1)})/(-\mathbf{g}^{(k)\mathrm{T}}\boldsymbol{\delta}^{(k)}) \to 1$, contradicting (2.4.3). Hence $\mathbf{g}^{(k)} \to \mathbf{0}$. □

p. 22, equation (2.4.8): replace $\geqslant$ by $\leqslant$.

p. 27, equation (2.6.1): delete all subscript 2s.

p. 35, line 20: . . . $\|\mathbf{h}^{(k)}\| \to 0$. Thus . . . .

pp. 51 and 52, theorem 3.4.2: delete the word 'least' in the statement of the theorem and corollary.

p. 64, equation (4.1.9): insert – after first.

p. 78: run on line 16 after line 15 in (iv).

p. 83: It has been pointed out to me by Dr D. Sorenson that $\mathbf{G}^{(k)} + \nu\mathbf{I}$ positive semi-definite (together with (5.2.1) and feasibility) is also a sufficient (as well as necessary) condition for a global solution (not necessarily unique) to (5.1.2) using $\|\cdot\|_2$. This follows because

$$\tfrac{1}{2}(\boldsymbol{\delta} - \boldsymbol{\delta}^{(k)})^{\mathrm{T}}(\mathbf{G}^{(k)} + \nu\mathbf{I})(\boldsymbol{\delta} - \boldsymbol{\delta}^{(k)}) \geqslant 0$$

implies using (5.2.1) and (3.1.1) that

$$q^{(k)}(\boldsymbol{\delta}) - q^{(k)}(\boldsymbol{\delta}^{(k)}) \geqslant \tfrac{1}{2}\nu(\boldsymbol{\delta}^{(k)\mathrm{T}}\boldsymbol{\delta}^{(k)} - \boldsymbol{\delta}^{\mathrm{T}}\boldsymbol{\delta}) = 0$$

by the first order conditions. I am very grateful for this observation which nicely completes the theoretical properties of the system.

p. 107, line 4 from bottom: **J** in bold type not italics.

Chapter 7
Introduction

7.1 Preview

The motivation for studying constrained minimization has been discussed at some length in Chapter 1 of Volume 1 of this book. The mathematical background given there, and indeed many of the concepts which arise in unconstrained optimization, are important in the study of constrained optimization. In this volume, the selection of material from the extensive literature which exists has again been done with the main theme of practicality in mind. Thus topics such as reliability and effectiveness are uppermost; to some extent these are measured by convergence and rate of convergence results. These aspects are therefore studied in some detail, and together with the subject of optimality conditions they provide good material for an academic course. However the use of *experimentation* to validate the properties of an algorithm is still of paramount importance. In fact the study of constrained optimization is by no means as well advanced as for the unconstrained case. The writing of software is a much more complex task, and so comparative experimental results are much less widely available. Often there is even a lack of suitable test problems. Also many more special cases arise and the problem of assessing numerical evidence is more difficult. For all these reasons Volume 2 departs from the feature in Volume 1 of presenting detailed numerical evidence. Nonetheless important experimental results do exist in the literature and the selection of material is guided by such results. This lack of certainty also shows up in that the decision as to precisely what algorithm to recommend in any one case is often not clear. For this reason the availability of good well-documented library software is often poor. Thus I appreciate the fact that many algorithms are necessarily used which are not ideal and I have tried to make users aware of defects in these algorithms and to enable them to mitigate their worst effects.

The structure of most constrained optimization problems is essentially contained in the following:

$$\begin{aligned} &\text{minimize } f(\mathbf{x}) && \mathbf{x} \in \mathbb{R}^n \\ &\text{subject to } c_i(\mathbf{x}) = 0, && i \subset E \\ &\phantom{\text{subject to }} c_i(\mathbf{x}) \geqslant 0, && i \in I. \end{aligned} \tag{7.1.1}$$

As in Volume 1, $f(\mathbf{x})$ is the *objective function*, but there are additional *constraint functions* $c_i(\mathbf{x})$, $i = 1, 2, \ldots, p$. E is the index set of equations or equality constraints in the problem, I is the set of inequality constraints, and both these sets are finite. More general constraints can usually be put into this form: for example $c_i(\mathbf{x}) \leqslant b$ becomes $b - c_i(\mathbf{x}) \geqslant 0$. If any point $\mathbf{x}'$ satisfies all the constraints in (7.1.1) it is said to be a *feasible point* and the set of all such points is referred to as the *feasible region* R. As in Volume 1, maximization problems are easily handled by the transformation $\max f(\mathbf{x}) = -\min -f(\mathbf{x})$. Also, a local minimizer or solution (referred to by $\mathbf{x}^*$) is looked for, rather than a global minimizer, the computation of which can be difficult. It is possible to illustrate the effect of the constraints when $n = 2$ by drawing the zero contour of each constraint function. For an equality constraint, the line itself is the set of feasible points; for an inequality constraint the line marks the boundary of the feasible region and the infeasible side is conventionally shaded. This is shown in Figure 7.1.1. Case (i) has constraints $x_2 = x_1^2$, $\mathbf{x} \geqslant \mathbf{0}$, which can be written $c_1(\mathbf{x}) = x_2 - x_1^2$, $c_2(\mathbf{x}) = x_1$, $c_3(\mathbf{x}) = x_2$ and $E = \{1\}$, $I = \{2, 3\}$. Case (ii) has constraints $x_2 \geqslant x_1^2$, $x_1^2 + x_2^2 \leqslant 1$ which can be written as $c_1(\mathbf{x}) = x_2 - x_1^2$, $c_2(\mathbf{x}) = 1 - x_1^2 - x_2^2$, $I = \{1, 2\}$, and E empty. Formulation (7.1.1) covers most types of problem; however the condition that some variables x_i take only discrete values is not included. This type of condition is covered in *integer programming* which is largely beyond the scope of this book. However a useful general purpose algorithm is the branch and bound method which enables the problem to be reduced to a sequence of smooth problems and hence solved by other techniques given in this book. This is described in Section 13.1. Another type of condition which is not included in (7.1.1) is a constraint of the form $c_i(\mathbf{x}) > 0$; something more is said about this case in Section 7.2. In fact there is often some choice in how best to pose the problem in the first instance and a number of possibilities of this type are discussed in Section 7.2.

It is assumed in (7.1.1) that the functions $c_i(\mathbf{x})$ are continuous which implies that R is closed. It is also assumed that $f(\mathbf{x})$ is continuous for all $\mathbf{x} \in R$ and preferably for all $\mathbf{x} \in \mathbb{R}^n$. If in addition the feasible region is bounded ($\exists a > 0$ such that $\|\mathbf{x}\| \leqslant a \ \forall \ \mathbf{x} \in R$), then it follows that a solution $\mathbf{x}^*$ exists. If not then the problem may be *unbounded* ($f(\mathbf{x}) \to -\infty$) or may not have a minimizing point. The problem also has no solution when R is empty, that is when the constraints are inconsistent. In fact most practical methods require the stronger assumption that the objective and constraint functions are also smooth in that their first and often second continuous derivatives exist ($f, c_i \in \mathbb{C}^1$ or $\mathbb{C}^2$). The notation ∇f (= $\mathbf{g}$) and $\nabla^2 f$ (= $\mathbf{G}$) for

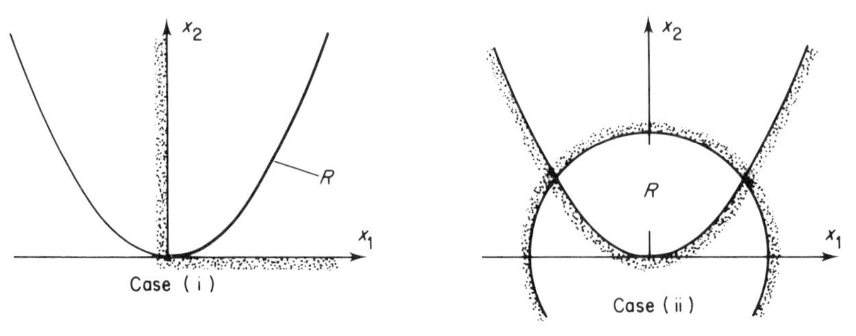

Figure 7.1.1 Examples of feasible regions

the gradient vector and Hessian matrix of f is described in Volume 1. The notation ∇c_i and $\nabla^2 c_i$ is used to denote the corresponding first and second derivatives of any constraint function c_i. The vector ∇c_i is also denoted by $\mathbf{a}_i$ and is referred to as the *normal vector* of the constraint c_i. Note that $\mathbf{a}_i$ refers to the ith vector in a set and not to the ith component of $\mathbf{a}$. These vectors are sometimes collected into columns of the *Jacobian matrix* $\mathbf{A}$ (although this rule is contradicted in the simplex method (Chapter 8) in which the normal vectors are the rows of $\mathbf{A}$). The vector $\mathbf{a}'_i$ (that is $\mathbf{a}_i(\mathbf{x})$ evaluated at $\mathbf{x} = \mathbf{x}'$) is the direction of greatest increase of $c_i(\mathbf{x})$ at $\mathbf{x}'$, and if $c_i = 0$ and $i \in I$ then the direction is on the feasible side of the constraint (see Figure 7.1.2) and is at right angles to the zero contour. Most of Volume 2 assumes the existence of these derivatives which can be used, for example, to characterize optimality conditions, as described in Chapter 9, which generalize the results for unconstrained minimization given in Volume 1, Section 2.1. This is not to say necessarily that user supplied formulae for these derivatives are required in any method. Mostly, however, formulae for first derivatives are required, and in some cases formulae for second derivatives also. Methods which require no derivative information have not been studied to any great extent and the obvious advice is to estimate these derivatives by finite differences (see Volume 1), although the resulting algorithm is likely to be less robust and effective when this is done. A different situation arises when the functions f and c_i do not have continuous derivatives, which is referred to as *non-differentiable* or *non-smooth optimization*. In this case methods for smooth problems are not appropriate and special attention must be given to the surfaces of non-differentiability. These behave somewhat like the boundary of a constraint in (7.1.1) and it is therefore appropriate to discuss the problem within the structure of this volume. This is done for unconstrained non-differentiable optimization in Chapter 14: in fact it is also possible to generalize these ideas to include both smooth and non-smooth constraint functions, but this is beyond the scope of the book, although some references are given.

Another important concept is that of an *active* or *binding constraint*. Active constraints at any point $\mathbf{x}'$ are defined by the index set

$$\mathscr{A}' = \mathscr{A}(\mathbf{x}') = \{i : c_i(\mathbf{x}') = 0\} \tag{7.1.2}$$

so that any constraint is active at $\mathbf{x}'$ if $\mathbf{x}'$ is on the boundary of its feasible region. If $\mathbf{x}'$ is feasible then $\mathscr{A}' \supset E$ clearly follows. In particular the set $\mathscr{A}^*$ of active constraints at the solution of (7.1.1) is of some importance. If this set is known then

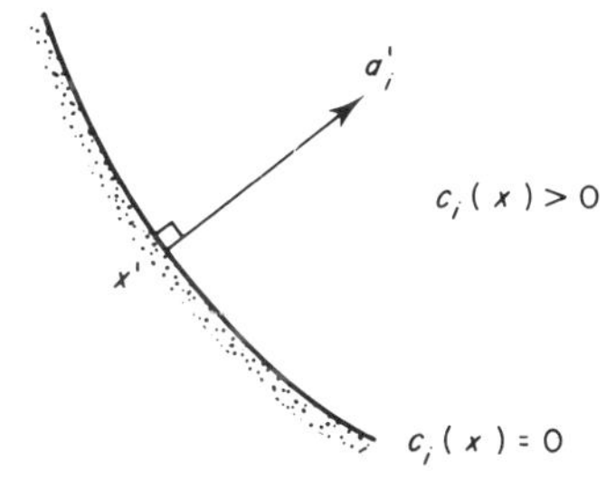

Figure 7.1.2 The normal vector

the remaining constraints can be ignored (locally) and the problem can be treated as an equality constraint problem with $E = \mathscr{A}^*$. Also constraints with $i \notin \mathscr{A}^*$ can be perturbed by small amounts without affecting the local solution whereas this is not usually true for an active constraint. An example is given by the problem: minimize $f(\mathbf{x}) = -x_1 - x_2$ subject to $x_2 \geqslant x_1^2$ and $x_1^2 + x_2^2 \leqslant 1$. Clearly from Figure 7.1.3 the solution is achieved at $\mathbf{x}^* = (1/\sqrt{2}, 1/\sqrt{2})^T$ when the contour of $f(\mathbf{x})$ is a tangent to the unit circle. Thus the active set in the notation of Figure 7.1.1(ii) is $\mathscr{A}^* = \{2\}$, and the circle constraint $c_2(\mathbf{x})$ is active. Likewise the parabola constraint $c_1(\mathbf{x})$ is inactive and can be perturbed or removed from the problem without changing $\mathbf{x}^*$. A further refinement of this definition to include strongly active and weakly active constraints is given in Figure 9.1.2.

Methods for the solution of (7.1.1) are usually iterative so that a sequence $\mathbf{x}^{(1)}, \mathbf{x}^{(2)}, \mathbf{x}^{(3)}, \ldots$, say, is generated from a given point $\mathbf{x}^{(1)}$, hopefully converging to $\mathbf{x}^*$. If $\mathbf{x}^*$ is a member of the sequence then the method is said to *terminate*. Some early methods for constrained optimization were developed in an ad hoc way and are not strongly supported theoretically. Because of this these methods are often unreliable and expensive for problems of any size and they are not described here. However a review of what has been attempted is given by Swann (1974). The subject of constrained optimization splits into two main parts, *linear constraint programming* and *nonlinear programming* which have quite different features. In linear constraint programming each constraint is a *linear function* $c_i(\mathbf{x}) = \mathbf{a}_i^T\mathbf{x} - b_i$. The boundary of the feasible region for any one such constraint is a hyperplane, and the normal vector ∇c_i is constant and is again the vector $\mathbf{a}_i$. Linear constraint problems can be handled by a combination of an elimination method and an active set method (see Section 7.2) and the iterates $\mathbf{x}^{(k)}$ are always feasible points. The simplest cases are when the objective function is either linear or quadratic (*linear programming* or *quadratic programming* – Chapters 8 and 10 respectively) in both of which cases algorithms which terminate can be determined. The application to a general objective function is given in Chapter 11, and in this case many of the possibilities for unconstrained optimization in Volume 1 carry over directly. For example there are analogues of Newton's method, quasi-Newton

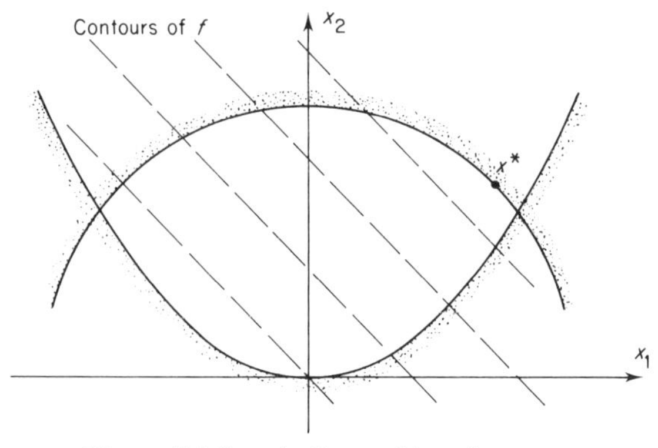

Figure 7.1.3 Active and inactive constraints

methods, the Gauss–Newton method, restricted step methods, and no-derivative methods which use finite difference approximations. Similar considerations hold in regard to using line searches and in regard to deciding what type of convergence test to use to terminate the iteration. Special cases of a linear constraint are the bounds $x_i \geqslant \ell_i$ or $x_i \leqslant u_i$ in which $\mathbf{a}_i$ is $\pm\mathbf{e}_i$, the ith coordinate vector, and it is particularly simple to handle such constraints. It is important that algorithms should take this into account.

The most difficult type of constrained minimization problem is nonlinear programming (Chapter 12) in which there exist some non-linear constraint functions in the problem. In this case a completely satisfactory general purpose method has yet to be agreed upon and the subject is one of intense research activity. Of course if the non-linear constraints can be rearranged so as to be eliminated directly then this should be done, but is not usually possible. Indirect elimination by solving a system of equations numerically is possible (Sections 7.2 and 12.4) but is not usually efficient and other difficulties exist; this idea is closely related to another approach known as a *feasible direction method.* A different approach is to attempt to transform the problem to one of unconstrained minimization by using *penalty functions* (Sections 12.1, 12.2, and 14.3). Efficiency depends on exactly how this is done, but it seems inevitable that some sort of penalty function must be used to get good global convergence properties. In many algorithms the iteration is determined by modelling the original problem in a suitable way. In particular a *linearization* of the constraint functions is often used. This is a first order Taylor series approximation about the current iterate $\mathbf{x}^{(k)}$:

$$c_i(\mathbf{x}^{(k)} + \boldsymbol{\delta}) \approx \ell_i^{(k)}(\boldsymbol{\delta}) = c_i^{(k)} + \mathbf{a}_i^{(k)\mathrm{T}}\boldsymbol{\delta}, \qquad i = 1, 2, \ldots, p. \tag{7.1.3}$$

The linearized function $\ell_i^{(k)}$ is defined in terms of the correction $\boldsymbol{\delta}$ to $\mathbf{x}^{(k)}$, the superscript k indicating that it is made on iteration k. This approximation enables linear constraint subproblems to be solved on each iteration. As in Volume 1, it is also possible to make a quadratic model of the objective function. However to take constraint curvature correctly into account it is appropriate to modify the quadratic term in a suitable way. Methods of this type are very important, although to obtain good global properties they must be incorporated with some type of penalty function (Sections 12.3 and 14.3). A special case of nonlinear programming is *geometric programming* in which the functions f and c_i have a polynomial type structure. It is possible to reduce this problem to a linear constraint problem which is more readily solved (Section 13.2). Often linear constraints and bounds arise in nonlinear programming problems: it is usually possible to take advantage of this fact to make the algorithm more efficient.

These algorithms, in particular those for nonlinear programming, depend on a study of optimality conditions for problem (7.1.1), and this theory is set out in Chapter 9. In Volume 1, I tried to write parts of the book in simple terms, avoiding the use of too much theory. To some extent I have done this here, for example in the presentation of linear and quadratic programming. However constrained minimization problems are much more complex than unconstrained problems and it is important for the user to have some grasp of this theory. This is especially true in regard to *Lagrange multipliers* and *first order conditions* and I have tried to give a simple semi-rigorous introduction in Section 9.1 showing how these multipliers arise and can be interpreted. A more rigorous presentation then follows. The same is true in regard to second order conditions in Section 9.3. Some simple notions of convexity and duality for smooth problems appear in Sections 9.4 and 9.5. The subject of *non-differentiable optimization* is arguably the most difficult that I have tried to cover in this book. Some presentations of this subject are extremely theoretical although I have tried to avoid this as much as possible. However at the expense of introducing a little more theory concerning optimality conditions for non-differentiable convex functions (Section 14.2), a reasonably elegant and not too difficult treatment can be given. A discussion of algorithms (Sections 14.4 and 14.5) is also given.

7.2 Elimination and Other Transformations

There is considerable scope for making transformations to a constrained minimization problem which reduce it to a form which is more readily solved. This can be advantageous and a number of such possibilities are discussed. However it is important to be aware from the outset that this procedure is not entirely without risk and that solutions of the original and transformed problems may not correspond on a one-to-one basis or that methods may not perform adequately on the transformed problem. A number of examples of this are given in this section and elsewhere in the book, and the user should be on his guard. The most simple possibility for equality constraints is to use the equations to eliminate some of the variables in the problem (*elimination*). If there are just m equations $\mathbf{c}(\mathbf{x}) = \mathbf{0}$ which can be rearranged directly to give

$$\mathbf{x}_1 = \boldsymbol{\phi}(\mathbf{x}_2) \tag{7.2.1}$$

where $\mathbf{x}_1$ and $\mathbf{x}_2$ are partitions of $\mathbf{x}$ in $\mathbb{R}^m$ and $\mathbb{R}^{n-m}$, then the original objective function $f(\mathbf{x}_1, \mathbf{x}_2)$ is replaced by

$$\psi(\mathbf{x}_2) = f(\boldsymbol{\phi}(\mathbf{x}_2), \mathbf{x}_2) \tag{7.2.2}$$

and $\psi(\mathbf{x}_2)$ is minimized over $\mathbf{x}_2$ without any constraints. A simple example is given in Question 7.3. Derivatives of ψ are readily obtained from those of f and $\mathbf{c}$ (see Question 7.5). Some care has to be taken to avoid an ill-conditioned rearrangement when forming (7.2.1), for instance with linear constraints it is advisable to use some sort of pivoting on the variables. In some cases the method may fail completely, as shown in Question 7.4. In fact it is possible to discuss elimination in more general terms, implicitly by first making a linear transformation of variables; this is described in Section 10.1. In cases where no direct rearrangement like (7.2.1) is available, it is possible to regard $\mathbf{c}(\mathbf{x}_1, \mathbf{x}_2) = \mathbf{0}$ as a system of non-linear equations which can be solved by the Newton–Raphson method (Volume 1, Section 6.2). In doing this $\mathbf{x}_2$ remains fixed and a vector $\mathbf{x}_1$ is determined which solves the equations. Thus $\mathbf{x}_1$ depends on $\mathbf{x}_2$ and so the process implicitly defines a function $\mathbf{x}_1 = \boldsymbol{\phi}(\mathbf{x}_2)$. This method is outlined in a more general form in Section 12.4;

however the process is not always the most efficient and there can be difficulties in getting the Newton–Raphson method to converge. An alternative transformation for the equality constraint problem is the *method of Lagrange multipliers* (Section 9.1) in which the system of non-linear equations (9.1.5) is solved which arises from the first order necessary conditions. Except in special cases this system must be solved numerically, which renders the method of little practical use. The method can also fail, not only when the solution of (9.1.5) corresponds to a constrained maximum point or saddle point, but also when the regularity condition (9.2.4) does not hold (see Question 9.14).

Elimination methods are not directly applicable to inequality constraint problems unless the set of active constraints $\mathscr{A}^*$ is known. However it is possible to use a trial and error sort of method in which a guess $\mathscr{A}$ is made at the set of active constraints, and constraints in $\mathscr{A}$ are then treated as equalities, neglecting the remaining inequality constraints. The resulting equality constraint problem is then solved by elimination or by the method of Lagrange multipliers, giving a solution $\hat{\mathbf{x}}$. It is necessary to check that $\hat{\mathbf{x}}$ is feasible with respect to the constraints which have been ignored. If not, one of these is added to the active set and the above process is repeated. If $\hat{\mathbf{x}}$ is feasible then it is also necessary to check that the first order conditions are satisfied. To do this requires the calculation of the corresponding Lagrange multiplier vector $\hat{\boldsymbol{\lambda}}$. Since $\lambda_i = \partial f/\partial c_i$ to first order measures the effect of perturbations in the c_i on f, it is necessary for an inequality constraint $c_i(\mathbf{x}) \geqslant 0$ that $\lambda_i \geqslant 0$ at the solution, for otherwise a feasible perturbation would reduce f. Thus if there are any $\hat{\lambda}_i < 0$, one such constraint must be removed from the active set and the process repeated again. On the other hand, if $\hat{\boldsymbol{\lambda}} \geqslant \mathbf{0}$ then the required solution is located. Methods of this type can be used in an informal way on small problems. However they are most useful in solving all types of linear constraint problem when systematic procedures can be devised. Such methods include the *simplex method* for linear programming and the *active set method* for all types of linear constraint programming. So-called *exchange algorithms* for best linear L_1 and L_∞ data fitting are also examples of this type of procedure. Systematic procedures using active set methods for non-linear constraints based on solving the equality constraint problems by implicit elimination can also be devised (Section 12.4) but there are some difficulties which are not readily overcome. It is difficult to handle constraints of the form $c_i(\mathbf{x}) > 0$ in an active set method because the feasible region is not closed and the constraint cannot be active at a solution. However it can be useful to include them in the problem via the transformation $c_i(\mathbf{x}) \geqslant \epsilon > 0$, possibly solving a sequence of problems in which $\epsilon \downarrow 0$ if the constraints happen to be active. The reason for doing this might be to prevent or dissuade $f(\mathbf{x})$ being evaluated at an infeasible point at which it is not defined (for example the problem: $\min x \log_e x$ subject to $x > 0$). It may not be satisfactory just to ignore the constraints because the problem may then become unbounded or have a global solution with $c_i(\mathbf{x}) \leqslant 0$, which is of no interest.

Some other transformations are worthy of note which relate equality and inequality constraint problems. For example a constraint $c_i(\mathbf{x}) = 0$ can be equivalently replaced by two opposite inequality constraints $c_i(\mathbf{x}) \geqslant 0$ and $-c_i(\mathbf{x}) \geqslant 0$.

This enables (7.1.1) to be reduced to an inequality constraint problem. However there are some practical disadvantages due to degeneracy and other reasons and the idea is best avoided, although it can occasionally be useful. The alternative possibility is to write $c_i(\mathbf{x}) \geqslant 0$ as the equality constraint $\min(c_i(\mathbf{x}), 0) = 0$. Unfortunately this function is not a $\mathbb{C}^1$ function and so is usually excluded on this count. Another possibility is to replace a constraint $c_i(\mathbf{x}) \geqslant 0$ by adding an extra variable, z say, giving an equality constraint $c_i(\mathbf{x}) = z$ and a bound $z \geqslant 0$. The variable z is referred to as a *slack variable* since it measures the slack in the inequality constraint. This transformation is most useful in the simplex method for linear programming which requires all general inequality constraints to be handled in this way, but is not necessary in active set methods which treat inequalities of any type directly. Furthermore, following an idea introduced later in this section, it is possible to do away with the need for the bound $z \geqslant 0$. This is done by adding a *quadratic slack variable* y and replacing $c_i(\mathbf{x}) \geqslant 0$ by the (non-linear) equality constraint $c_i(\mathbf{x}) = y^2$. This removes the need to treat inequality constraints directly. However this transformation does cause some distortion as explained below and in this case it may be somewhat dangerous, in particular because of the following feature. Let for example $c(\mathbf{x}) \geqslant 0$ be the only constraint and let $\mathbf{x}'$ be such that $c' = 0$ and $\mathbf{g}' = \mathbf{a}'\lambda'$ where $\lambda' < 0$. Then $\mathbf{x}'$ and λ' do not satisfy first order conditions for a solution. Yet the vector $\mathbf{x}'$ augmented by $y' = 0$ does satisfy first order conditions in the transformed equality constraint problem with the same λ'. Thus the transformation does not seem to be able to distinguish whether or not constraints are active on the basis of first order information. I have also heard bad reports of quadratic slacks in practice which might well be accountable for in this way.

Many other useful transformations arise in constrained optimization and are used in subsequent chapters. Perhaps the most well-known idea is the use of *penalty functions* for nonlinear programming. The idea is to transform the problem to one of unconstrained optimization by adding to the objective function a penalty term which weights constraint violations. In *sequential penalty functions* $\mathbf{x}^*$ is found as the limit of the minimizing points of a sequence of penalty functions, as some controlling parameter is changed. More recently the value has been realized of an *exact penalty function* which has $\mathbf{x}^*$ as its local minimizer. These transformations are described in some detail in Sections 12.1, 12.2, 12.5, and 14.3. Other transformations of some importance are those arising in duality (Section 9.5), integer programming (Section 13.1), and geometric programming (Section 13.2), amongst others.

It is also possible to make transformations of variables in an attempt to simplify the problem. For example the bound $x_i \geqslant 0$ can be removed by defining a new variable y_i which replaces x_i, such that $x_i = y_i^2$. Then for any y_i in $(-\infty, \infty)$ it follows that $x_i \geqslant 0$ so the bound does not need to be explicitly enforced. Another similar transformation for $\ell_i \leqslant x_i \leqslant u_i$ is to let y_i satisfy $x_i = \ell_i + (u_i - \ell_i)\sin^2 y_i$. For strict constraints $x_i > 0$ it is possible to use $x_i = e^{y_i}$. The advantage of these transformations is that they do extend the range of problems which can be handled by an unconstrained minimization routine. This is not to say that minimization with simple bounds $\ell_i \leqslant x_i \leqslant u_i$ is at all difficult; in fact the opposite is true and it is probably more efficient to treat the problem directly. It is simply that subroutines which minimize functions subject only to bounds are much less readily available to the user at present. These ideas can also be used to transform inequality constraints to equalities (see above in regard to quadratic slacks), although this possibility should be viewed with some suspicion for the reason given above. These transformations do cause some distortion which often may not be favourable. For example the problem $\min x^2$ subject to $x \geqslant 0$, after transforming $x = y^2$, becomes $\min y^4$. This has a singular Hessian at the solution which causes any standard minimization method based on a quadratic model to converge slowly. Another example is the convex programming problem $\min(x - 1)^2$ subject to $x \geqslant 0$. Although the transformation is well behaved at the solution $x^* = 1$, it induces a stationary point with a non-positive-definite Hessian matrix at $x = 0$ and both these features could possibly cause difficulties (see also Question 7.6). Thus although such transformations can be useful, the user should be aware that they are not entirely risk free.

Another transformation which enables $|x_i|$ functions to be handled is to replace the variable x_i by two non-negative variables x_i^+ and x_i^- representing the positive and negative parts of x_i (that is $\max(x_i, 0)$ and $\max(-x_i, 0)$). The conditions $x_i^+ \geqslant 0$ and $x_i^- \geqslant 0$ are explicitly included in the problem; also whenever x_i appears in the problem it is replaced by $x_i^+ - x_i^-$ and similarly $|x_i|$ is replaced by $x_i^+ + x_i^-$ (see Question 8.12). This latter replacement is only valid if one of x_i^+ or x_i^- is zero, which can sometimes be guaranteed, for example in otherwise linear problems when both x_i^+ and x_i^- together cannot be basic (Chapter 8). This transformation can also be used to handled unbounded variables in a linear programming problem. An alternative technique for handling $|x_i|$ terms is described in Section 8.4. These ideas can be extended to functions $|c_i(x)|$ by adding extra variables y_i and the equality constraint $c_i(\mathbf{x}) = y_i$ (see Question 8.11), thus enabling L_1 approximation problems to be handled by smooth techniques. Similar ideas for minimizing max functions or L_∞ functions can be tackled by introducing an extra variable v as described in Section 14.1. However all these techniques are really attempting to solve non-differentiable optimization problems as smooth problems. In the current state of the art this can be useful, but when software becomes readily available for some of the better more direct methods described in Sections 14.4 and 14.5, these should be preferred.

Finally the very important transformation of *scaling* either the constraints or the variables in the problem is discussed. Scaling of a constraint set is achieved by multiplying each constraint function by a constant chosen so that the value of each constraint function, evaluated for typical values of $\mathbf{x}$, is of the same order of magnitude. This can be important in that this scales the Lagrange multipliers (inversely) and so can make more reliable the test on the magnitude of a multiplier which is used in some algorithms. A well-scaled matrix is also important in some linear algebra routines when pivoting tests are made. Moreover when using penalty functions which involve quantities like $\mathbf{c}^T\mathbf{c}$ or $\|\mathbf{c}\|_1$, it is important that constraints are scaled. In a similar way scaling of the variables can sometimes be important. This again arises when pivoting tests on the variables are made, or when implicitly using some norm of the variables, for example in restricted step methods or methods with a bias towards steepest descent (see Volume 1). In practice variables are usually scaled by multiplying each one by a suitable constant. However a non-linear scaling which can be useful for variables $x_i > 0$ is to use the transformation $x_i = e^{y_i}$. Then variables of magnitudes $10^{-6}, 10^{-3}, 10^0, 10^3, \ldots$, say, which typically can occur in kinetics problems, are transformed into logarithmic variables with magnitudes which are well scaled.

Questions for Chapter 7

1. Calculate the Jacobian matrix $\nabla \boldsymbol{\ell}^T$ of the linear system $\boldsymbol{\ell}(\mathbf{x}) = \mathbf{A}^T\mathbf{x} - \mathbf{b}$. If $\boldsymbol{\ell}$ is obtained by linearizing a non-linear system as in (7.1.3) show that both systems have the same Jacobian matrix.
2. Obtain the gradient vector and the Hessian matrix of the functions $f(\mathbf{x}) + h(\mathbf{c}(\mathbf{x}))$ and $f(\mathbf{x}) + \boldsymbol{\lambda}^T\mathbf{c}(\mathbf{x})$. In the latter case treat the cases both where $\boldsymbol{\lambda}$ is a constant vector and where it is a function $\boldsymbol{\lambda}(\mathbf{x})$.
3. Find the solution of the problem minimize $-x - y$ subject to $x^2 + y^2 = 1$ by graphical means and also by eliminating x, and show that the same solution is obtained. Discuss what happens, however, if the square root which is required is chosen to have a negative sign.
4. Solve the problem minimize $x^2 + y^2$ subject to $(x - 1)^3 = y^2$ both graphically and also by eliminating y. In the latter case show that the resulting function of x has no minimizer and explain this apparent contradiction. What happens if the problem is solved by eliminating x?
5. Consider finding derivatives of the functions $\boldsymbol{\phi}(\mathbf{x}_2)$ and $\psi(\mathbf{x}_2)$ defined in (7.2.1) and (7.2.2). Define partitions

$$\nabla = \begin{pmatrix} \nabla_1 \\ \nabla_2 \end{pmatrix}, \qquad \mathbf{g} = \begin{pmatrix} \mathbf{g}_1 \\ \mathbf{g}_2 \end{pmatrix}, \qquad \mathbf{A} = \begin{bmatrix} \mathbf{A}_1 \\ \mathbf{A}_2 \end{bmatrix}$$

   and show by using the chain rule that $\nabla_2 \boldsymbol{\phi}^T = -\mathbf{A}_2\mathbf{A}_1^{-1}$ and hence that $\nabla_2 \psi = \mathbf{g}_2 - \mathbf{A}_2\mathbf{A}_1^{-1}\mathbf{g}_1$. Second derivatives of ψ are most conveniently obtained as in (12.4.6) and (12.4.7) by setting $\mathbf{V}^T = [\mathbf{0} : \mathbf{I}]$ and hence $\mathbf{Z}^T = [-\mathbf{A}_2\mathbf{A}_1^{-1} : \mathbf{I}]$ and $\mathbf{S}^T = [\mathbf{A}_1^{-1} : \mathbf{0}]$.
6. Consider the problem minimize $f(x_1, x_2)$ subject to $x_1 \geqslant 0, x_2 \geqslant 0$ when the transformation $x = y^2$ is used. Show that $\mathbf{x}' = \mathbf{0}$ is a stationary point of the transformed function, but is not minimal if any $g_i' \leqslant 0$. If $\mathbf{g}' = \mathbf{0}$ then second order information in the original problem would usually enable the question of whether $\mathbf{x}'$ is a minimizer to be determined. Show that in the transformed problem this cannot be done on the basis of second order information.

Chapter 8
Linear programming

8.1 Structure

The most simple type of constrained optimization problem is obtained when the functions $f(\mathbf{x})$ and $c_i(\mathbf{x})$ in (7.1.1) are all linear functions of $\mathbf{x}$. The resulting problem is known as a *linear programming* (LP) problem. Such problems have been studied since the earliest days of electronic computers, and the subject is often expressed in a quasi-economic terminology which to some extent obscures the basic numerical processes which are involved. This presentation aims to make these processes clear, whilst retaining some of the traditional nomenclature which is widely used. One main feature of the traditional approach is that linear programming is expressed in the *standard form*

$$\begin{aligned}&\underset{\mathbf{x}}{\text{minimize}}\ f(\mathbf{x}) \triangleq \mathbf{c}^{\mathrm{T}}\mathbf{x}\\&\text{subject to } \mathbf{A}\mathbf{x}=\mathbf{b}, \qquad \mathbf{x}\geqslant \mathbf{0},\end{aligned} \tag{8.1.1}$$

where $\mathbf{A}$ is an $m \times n$ matrix, and $m \leqslant n$ (usually $<$). The symbol $\triangleq$ means 'defined by'. Thus the allowable constraints on the variables are either linear equations or non-negativity bounds. The coefficients $\mathbf{c}$ in the linear objective function are often referred to as *costs*. An example with four variables ($n = 4$) and two equations ($m = 2$) is

$$\begin{aligned}&\text{minimize}\ x_1+2x_2+3x_3+4x_4\\&\text{subject to}\ x_1+\ x_2+\ x_3+\ x_4=1\\&\qquad\qquad x_1\qquad\ +\ x_3-3x_4=\tfrac{1}{2},\\&\quad x_1\geqslant 0, x_2\geqslant 0, x_3\geqslant 0, x_4\geqslant 0.\end{aligned} \tag{8.1.2}$$

More general LP problems can be reduced to standard form without undue difficulty, albeit with some possible loss of efficiency. For instance a general linear inequality $\mathbf{a}^{\mathrm{T}}\mathbf{x} \leqslant b$ can be transformed using a slack variable z (see Section 7.2) to the equation $\mathbf{a}^{\mathrm{T}}\mathbf{x} + z = b$ and the bound $z \geqslant 0$. Alternatively the dual transformation can sometimes be used advantageously to obtain a standard form and this is described in more detail in Section 9.5. More general bounds $x_i \geqslant \ell_i$ can be dealt with by a shift of origin, and if no bound exists at all on x_i in the original problem, then the standard form can be reached by introducing non-negative variables x_i^+

and x_i^-, as described in Section 7.2. In fact very little is lost in complexity if the bounds in (8.1.1) are expressed as

$$\boldsymbol{\ell} \leqslant \mathbf{x} \leqslant \mathbf{u}, \tag{8.1.3}$$

as Question 8.8 illustrates. The merit of using other possible standard forms is discussed in more detail in Section 8.3. However for the most part this text will concentrate on the solution of problems which are already in the standard form (8.1.1).

It is important to realize that a problem in standard form may have no solution, either because there is no feasible point (the problem is *infeasible*), or because $f(\mathbf{x}) \to -\infty$ for $\mathbf{x}$ in the feasible region (the problem is *unbounded*). However it is shown that there is no difficulty in detecting these situations, and so the text concentrates on the usual case in which a solution exists (possibly not unique). It is also convenient to assume that the equations are independent, so that they have no trivial linear combination. In theory this situation can always be achieved, either by removing dependent equations or by adding artificial variables (see Section 8.4 and Question 8.21), although in practice there may be numerical difficulties if this dependence is not recognized.

If (8.1.1) is considered in more detail, it can be seen that if $m = n$, then the equations $\mathbf{Ax} = \mathbf{b}$ determine a unique solution, and the objective function $\mathbf{c}^{\mathrm{T}}\mathbf{x}$ and the bounds $\mathbf{x} \geqslant \mathbf{0}$ play no part. In most cases however $m < n$, so that the system $\mathbf{Ax} = \mathbf{b}$ is underdetermined and $n - m$ degrees of freedom remain. In particular the system can determine only m variables, given values for the remaining $n - m$ variables. For example the equations $\mathbf{Ax} = \mathbf{b}$ in (8.1.2) can be rearranged as

$$\begin{aligned} x_1 &= \tfrac{1}{2} - x_3 + 3x_4 \\ x_2 &= \tfrac{1}{2} \qquad\quad - 4x_4 \end{aligned} \tag{8.1.4}$$

which determines x_1 and x_2 given values for x_3 and x_4, or alternatively as

$$\begin{aligned} x_1 &= \tfrac{7}{8} - \tfrac{3}{4}x_2 - x_3 \\ x_4 &= \tfrac{1}{8} - \tfrac{1}{4}x_2 \end{aligned} \tag{8.1.5}$$

which determines x_1 and x_4 from x_2 and x_3, and in other ways as well. It is important to consider what values these remaining $n - m$ variables can take in the standard form problem. The objective function $\mathbf{c}^{\mathrm{T}}\mathbf{x}$ is linear and so contains no curvature which can give rise to a minimizing point. Hence such a point must be created by the conditions $x_i \geqslant 0$ becoming active on the boundary of the feasible region. For example if (8.1.5) is used to eliminate the variables x_1 and x_4 from the problem (8.1.2), then the objective function can be expressed as

$$f = x_1 + 2x_2 + 3x_3 + 4x_4 = \tfrac{11}{8} + \tfrac{1}{4}x_2 + 2x_3. \tag{8.1.6}$$

Clearly this function has no minimum value unless the conditions $x_2 \geqslant 0, x_3 \geqslant 0$ are imposed, in which case the minimum occurs when $x_2 = x_3 = 0$. An illustration is given in Figure 8.1.1 for the more simple conditions $x_1 + 2x_2 = 1$ and $x_1 \geqslant 0$, $x_2 \geqslant 0$. The feasible region is the line joining the points $\mathbf{a} = (0, \tfrac{1}{2})^{\mathrm{T}}$ and $\mathbf{b} = (1, 0)^{\mathrm{T}}$. When the objective function $f(\mathbf{x})$ is linear the solution must occur at either $\mathbf{a}$ or $\mathbf{b}$

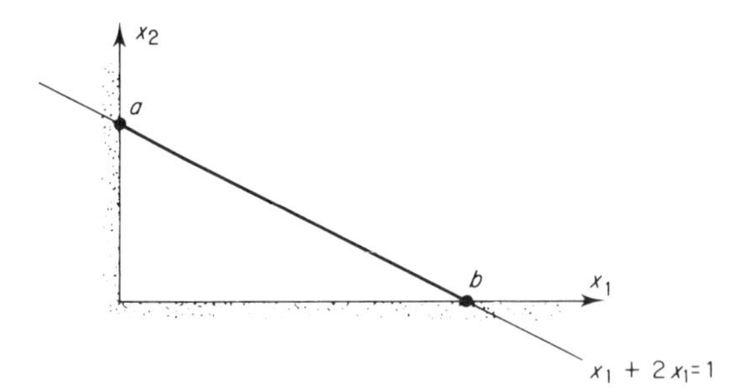

Figure 8.1.1 Constraints for a simple LP problem

with either $x_1 = 0$ or $x_2 = 0$ (try different linear functions, for example $f = x_1 + x_2$ or $f = x_1 + 3x_2$). If however $f(\mathbf{x}) = x_1 + 2x_2$ then any point on the line segment is a solution and this includes both **a** and **b**. This corresponds to the existence of a non-unique solution.

To summarize therefore, a solution of an LP problem in standard form always exists at one particular *extreme point* or *vertex* of the feasible region, with at least $n - m$ variables having *zero value*, and the remaining m variables being uniquely determined by the equations $\mathbf{Ax} = \mathbf{b}$ and taking non-negative values. This result is fundamental to the development of LP methods, and can be established rigorously using the notions of convexity (Section 9.4). The proof is sketched out in some detail in Questions 9.20 to 9.22.

The main difficulty in linear programming is to find which $n - m$ variables take zero value at the solution. The earliest method for solving this problem is the *simplex method*, which tries different sets of possibilities in a systematic way. This method is described in Section 8.2 and is still predominant today, albeit often in more sophisticated forms. Different variations of the method exist, depending upon exactly which intermediate quantities are computed. The earliest *tableau form* became superseded by the more efficient *revised simplex method*, both of which are described in Section 8.2. More recently methods based on using matrix factorizations have been suggested in order to control round-off errors more effectively. For large sparse LP problems, *product form methods* have enabled problems of up to 10^5 variables to be solved in practice. Both these developments are described in Section 8.5. An apparently different approach to LP is the active set method described in Section 8.3 which, however, turns out to be equivalent to the simplex method with slack variables, although different intermediate matrices are stored. The problem of calculating initial feasible points for LP and other linear constraint problems is described in Section 8.4. All these methods have one possible situation in which they can fail to solve a problem which has a well-defined solution. This is referred to as degeneracy and is described in Section 8.6.

8.2 The Simplex Method

The simplex method for solving an LP problem in standard form generates a sequence of feasible points $\mathbf{x}^{(1)}, \mathbf{x}^{(2)}, \ldots$ which terminates at a solution. Since there exists an extreme point at which the solution occurs, each iterate $\mathbf{x}^{(k)}$ is an extreme point. Thus $n - m$ of the variables have zero value at $\mathbf{x}^{(k)}$ and are referred to as *nonbasic variables* (the index set $N^{(k)}$), and the remaining m variables have a non-negative value (usually positive) and are referred to as *basic variables* (the index set $B^{(k)}$). The simplex method makes systematic changes to these sets after each iteration, in order to find the choice which gives the optimal solution. The superscript (k) is often omitted for clarity. At each iteration it is convenient to assume that the variables are permuted so that the basic variables are the first m elements of $\mathbf{x}$. Then $\mathbf{x}^{\mathrm{T}} = (\mathbf{x}_B^{\mathrm{T}}, \mathbf{x}_N^{\mathrm{T}})$ can be written, where $\mathbf{x}_B$ and $\mathbf{x}_N$ refer collectively to the basic and nonbasic variables respectively. The matrix $\mathbf{A}$ in (8.1.1) can also be partitioned similarly into $\mathbf{A} = [\mathbf{A}_B : \mathbf{A}_N]$ where the *basis matrix* $\mathbf{A}_B$ is $m \times m$ and $\mathbf{A}_N$ is $m \times (n - m)$. The equations $\mathbf{Ax} = \mathbf{b}$ can thus be written

$$[\mathbf{A}_B : \mathbf{A}_N]\begin{pmatrix}\mathbf{x}_B \\ \mathbf{x}_N\end{pmatrix} = \mathbf{A}_B\mathbf{x}_B + \mathbf{A}_N\mathbf{x}_N = \mathbf{b}. \tag{8.2.1}$$

At an extreme point it is always possible to find a partitioning into B and N such that $\mathbf{A}_B$ is non-singular (assuming that $\mathbf{A}$ has full rank; see Question 9.20). Also since $\mathbf{x}_N^{(k)} = \mathbf{0}$ it is possible to write

$$\mathbf{x}^{(k)} = \begin{pmatrix}\mathbf{x}_B \\ \mathbf{x}_N\end{pmatrix}^{(k)} = \begin{pmatrix}\hat{\mathbf{b}} \\ \mathbf{0}\end{pmatrix} \tag{8.2.2}$$

where $\hat{\mathbf{b}} = \mathbf{A}_B^{-1}\mathbf{b}$. Since the basic variables must take non-negative values it is required that $\hat{\mathbf{b}} \geqslant \mathbf{0}$. The partitioning $B^{(k)}$ and $N^{(k)}$ and the extreme point $\mathbf{x}^{(k)}$ with the above properties ($\mathbf{x}_B^{(k)} = \hat{\mathbf{b}} \geqslant \mathbf{0}$, $\mathbf{x}_N^{(k)} = \mathbf{0}$, $\mathbf{A}_B$ non-singular) is referred to as a *basic feasible solution* (*b.f.s.*). An example is provided by the LP problem (8.1.2) with the choice $B = \{1, 2\}$ and $N = \{3, 4\}$ (that is the variables x_1 and x_2 are basic, and x_3 and x_4 are nonbasic). Then

$$\mathbf{A}_B = \begin{bmatrix}1 & 1 \\ 1 & 0\end{bmatrix}, \qquad \mathbf{A}_N = \begin{bmatrix}1 & 1 \\ 1 & -3\end{bmatrix},$$

and

$$\hat{\mathbf{b}} = \mathbf{A}_B^{-1}\mathbf{b} = \begin{bmatrix}0 & 1 \\ 1 & -1\end{bmatrix}\begin{pmatrix}1 \\ \frac{1}{2}\end{pmatrix} = \begin{pmatrix}\frac{1}{2} \\ \frac{1}{2}\end{pmatrix} \geqslant \mathbf{0}.$$

Since $\mathbf{A}_B$ is non-singular and $\hat{\mathbf{b}} \geqslant \mathbf{0}$, this choice of B and N determines a basic feasible solution. It is also of interest to know the value ($\hat{f}$ say) of the objective function at a basic feasible solution. Partitioning $\mathbf{c}^{\mathrm{T}} = (\mathbf{c}_B^{\mathrm{T}}, \mathbf{c}_N^{\mathrm{T}})$ and using (8.2.2), it follows that

$$\hat{f} = \mathbf{c}^{\mathrm{T}}\mathbf{x}^{(k)} = \mathbf{c}_B^{\mathrm{T}}\hat{\mathbf{b}}.$$

In the example above, $\mathbf{c}_B = (1, 2)^{\mathrm{T}}$ so that the objective function has the value $\hat{f} = 1\frac{1}{2}$. The nomenclature 'basic feasible solution' can be somewhat confusing (basic feasible point would be better) since 'solution' refers to a solution of the constraint equations $\mathbf{Ax} = \mathbf{b}$ and $\mathbf{x} \geqslant \mathbf{0}$, and not to the overall solution of (8.1.1), which is therefore referred to as an *optimal* b.f.s. It should be noticed that not every possible choice of B and N gives rise to a b.f.s. For instance in (8.1.2), $B = \{1, 3\}$ gives a singular matrix $\mathbf{A}_B$, and $B = \{2, 4\}$ gives a vector $\hat{\mathbf{b}}$ for which $\hat{b}_4 < 0$. (Note that indices such as $\hat{b}_4$ refer to the variable to which the element corresponds, and not to the position in the vector $\hat{\mathbf{b}}$.) Consequently the determination of an initial b.f.s. $B^{(1)}$, $N^{(1)}$ and $\mathbf{x}^{(1)}$ is a non-trivial task. Suitable methods exist which incorporate the logic of the simplex method, and these are discussed in Section 8.4.

It is a simple matter to discover whether a basic feasible solution is optimal. Equation (8.2.1) is used to eliminate the basic variables from the objective function, yielding a reduced objective function $f(\mathbf{x}_N)$ of the nonbasic variables only, whose coefficients give this information directly. For example in the LP problem (8.1.2) with the choice $B = \{1, 2\}$, $N = \{3, 4\}$, (8.2.1) can be rearranged as shown in (8.1.4) and substituted into the objective function in (8.1.1) to give

$$f(x_3, x_4) = 1\tfrac{1}{2} + 2x_3 - x_4. \tag{8.2.3}$$

Now at the b.f.s. x_3 and x_4 take the values $x_3 = x_4 = 0$, but in general may satisfy $x_3 \geqslant 0$ and $x_4 \geqslant 0$. Thus only an *increase* in the value of any nonbasic variable is allowed. Furthermore f can be decreased from its value of $\hat{f} = 1\frac{1}{2}$ by increasing x_4, since x_4 has a negative coefficient in (8.2.3), and so it follows that this b.f.s. is not optimal. On the other hand, if the coefficients of the nonbasic variables in the reduced objective function are *all non-negative* as in (8.1.6), then the corresponding b.f.s. is optimal since there is no feasible change to the nonbasic variables which will reduce $f(\mathbf{x}_N)$.

In general terms the elimination of the basic variables uses the rearrangement of (8.2.1) given by

$$\mathbf{x}_B = \mathbf{A}_B^{-1}(\mathbf{b} - \mathbf{A}_N \mathbf{x}_N) = \hat{\mathbf{b}} - \mathbf{A}_B^{-1}\mathbf{A}_N \mathbf{x}_N. \tag{8.2.4}$$

The reduced objective function can then be written as

$$\begin{aligned} f(\mathbf{x}_N) &= \mathbf{c}_B^T \mathbf{x}_B + \mathbf{c}_N^T \mathbf{x}_N \\ &= \mathbf{c}_B^T(\hat{\mathbf{b}} - \mathbf{A}_B^{-1}\mathbf{A}_N \mathbf{x}_N) + \mathbf{c}_N^T \mathbf{x}_N \\ &= \hat{f} + \hat{\mathbf{c}}_N^T \mathbf{x}_N \end{aligned} \tag{8.2.5}$$

say, where the coefficients $\hat{\mathbf{c}}_N$ are defined by

$$\hat{\mathbf{c}}_N = \mathbf{c}_N - \mathbf{A}_N^T \boldsymbol{\pi}, \tag{8.2.6}$$

and where

$$\boldsymbol{\pi} = \mathbf{A}_B^{-T}\mathbf{c}_B. \tag{8.2.7}$$

In the example above $\boldsymbol{\pi} = (2, -1)^T$ and $\hat{\mathbf{c}}_N = (2, -1)^T$. The coefficients $\hat{\mathbf{c}}_N$ are known as the *reduced costs* at the basic feasible solution. By virtue of the discussion in the previous paragraph, the basic feasible solution is optimal if the reduced costs satisfy the test

$$\hat{\mathbf{c}}_N \geqslant \mathbf{0}. \tag{8.2.8}$$

The first part of a simplex method iteration therefore is to determine the reduced costs. If (8.2.8) holds then the basic feasible solution is optimal and the method terminates. If (8.2.8) also contains some element $\hat{c}_q = 0$, then a similar argument shows that x_q can be increased with f staying constant, so that the solution is *non-unique* (assuming that a non-zero step is permitted – for example if $\hat{\mathbf{b}} > \mathbf{0}$ – see next paragraph and Question 8.2).

Usually (8.2.8) is not satisfied, in which case the simplex method proceeds to find a new basic feasible solution which has a lower value of the objective function. Firstly a variable x_q, $q \in N$, is chosen for which $\hat{c}_q < 0$, and $f(\mathbf{x}_N)$ is decreased by increasing x_q whilst the other nonbasic variables retain their zero value. Usually the most negative $\hat{c}_q$ is chosen, that is

$$\hat{c}_q = \min_{i \in N} \hat{c}_i$$

although other selections have been investigated (Goldfarb and Reid, 1977). As x_q increases, the values of the basic variables $\mathbf{x}_B$ change as indicated by (8.2.4), in order to keep the system $\mathbf{A}\mathbf{x} = \mathbf{b}$ satisfied. Usually the need to keep $\mathbf{x}_B \geqslant \mathbf{0}$ to maintain feasibility limits the amount by which x_q can be increased. In the above example $\hat{c}_4$ is the only negative reduced cost, so x_4 is chosen to be increased, whilst x_3 retains its zero value. The effect on x_1 and x_2 is indicated by (8.1.4), and x_1 increases like $\frac{1}{2} + 3x_4$ whilst x_2 decreases like $\frac{1}{2} - 4x_4$. Thus x_2 becomes zero when x_4 reaches the value $\frac{1}{8}$, and no further increase in x_4 is permitted.

In general, because x_q is the only nonbasic variable which changes, it follows from (8.2.4) that the effect on $\mathbf{x}_B$ is given by

$$\begin{aligned} \mathbf{x}_B &= \hat{\mathbf{b}} - \mathbf{A}_B^{-1}\mathbf{a}_q x_q \\ &= \hat{\mathbf{b}} + \mathbf{d}x_q \end{aligned} \tag{8.2.9}$$

say, where $\mathbf{a}_q$ is the column of $\mathbf{A}$ corresponding to variable q, and where

$$\mathbf{d} = -\mathbf{A}_B^{-1}\mathbf{a}_q \tag{8.2.10}$$

can be thought of as the derivative of the basic variables with respect to changes in x_q. In particular if any d_i has a *negative* value then an increase in x_q causes a reduction in the value of the basic variable x_i. From (8.2.9), x_i becomes zero when $x_q = \hat{b}_i/-d_i$. The amount by which x_q can be increased is limited by the *first* basic variable, x_p say, to become zero, and the index p is therefore that for which

$$\frac{\hat{b}_p}{-d_p} = \min_{\substack{i \in B \\ d_i < 0}} \frac{\hat{b}_i}{-d_i}. \tag{8.2.11}$$

It may be, however, that there are no indices $i \in B$ such that $d_i < 0$. In this case $f(\mathbf{x})$ can be decreased without limit, and this is the manner in which an unbounded solution is indicated. In practical terms this usually implies a mistake in setting up the problem in which some restriction on the variables has been omitted.

In geometric terms, the increase of x_q and the corresponding change to $\mathbf{x}_B$ in (8.2.9) causes a move from the extreme point $\mathbf{x}^{(k)}$ along an *edge* of the feasible region. The termination of this move because $x_p, p \in B$, becomes zero indicates that a new extreme point of the feasible region is reached. Correspondingly there are again $n - m$ variables with zero value (x_p and the x_i, $i \in N^{(k)}$, $i \neq q$), and these are the nonbasic variables at the new b.f.s. Thus the new sets $N^{(k+1)}$ and $B^{(k+1)}$ are obtained by replacing q by p in $N^{(k)}$, and p by q in $B^{(k)}$, respectively. The form of the iteration ensures that $\hat{\mathbf{b}} \geqslant \mathbf{0}$ at the new b.f.s. and it is possible to show also that $\mathbf{A}_B$ is non-singular (see (8.2.15)), so that the conditions for a b.f.s. remain satisfied. In the example above the nonbasic variable x_4 is increased and the basic variable x_2 becomes zero. Thus 4 and 2 are interchanged between the sets $B = \{1, 2\}$ and $N = \{3,4\}$ giving rise to a new basic feasible solution determined by $B = \{1,4\}$ and $N = \{2, 3\}$. It is shown by virtue of (8.1.5) and (8.1.6) that the resulting b.f.s. is in fact optimal.

With the determination of a new b.f.s., the description of an iteration of the simplex method is complete. The method repeats the sequence of calculations until an optimal b.f.s. is recognized. Usually each iteration reduces $f(\mathbf{x})$, and since the number of vertices is finite, it follows that the iteration must terminate. There is one case, however, in which this may not be true, and this is when $\hat{b}_i = 0$ and $d_i < 0$ occurs. Then no increase in x_q is permitted, and although a new partitioning B and N can be made as before, no decrease in $f(\mathbf{x})$ is made. In fact it is possible for the algorithm to *cycle* by returning to a previous set B and N and so fail to terminate even though a well-determined solution does exist. The possibility that $\hat{b}_i = 0$ is known as *degeneracy* and the whole subject is discussed in more detail in Section 8.6.

For small illustrative problems ($n = 2$ or 3) it is straightforward to follow the above scheme of computation. That is given B and N, then

$$\mathbf{A}_B^{-1} = \frac{\text{adjoint}(\mathbf{A})}{\det(\mathbf{A})} \tag{8.2.12}$$

is calculated, and hence $\hat{\mathbf{b}} = \mathbf{A}_B^{-1}\mathbf{b}$. Then (8.2.7) and (8.2.6) enable the reduced costs $\hat{\mathbf{c}}_N$ to be obtained. Either the algorithm terminates by (8.2.8) or the least $\hat{c}_q$ determines the index q. Computation of $\mathbf{d}$ by (8.2.10) is followed by the test (8.2.11) to determine p. Finally p and q are interchanged giving a new B and N with which to repeat the process.

For larger problems some more efficient scheme is desirable which avoids the computation of $\mathbf{A}_B^{-1}$ from (8.2.12) on each iteration. The earliest method was the *tableau form* of the simplex method. In this method the data is arranged in a tableau:

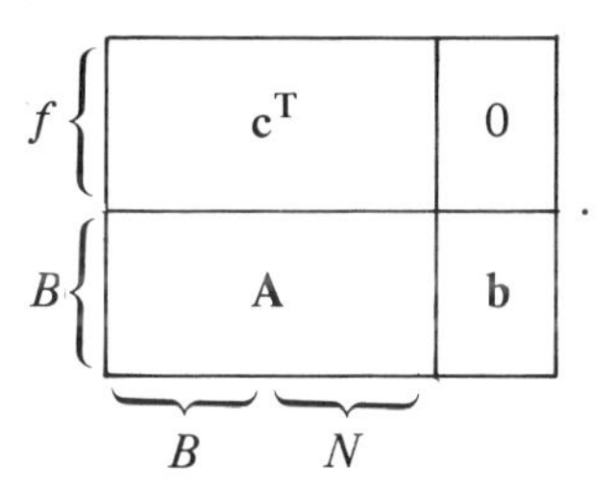

Then by making *row operations* on the tableau (adding or subtracting multiples of one row (not the f row) from another, or by scaling any row (not the f row)), it is possible to reduce the tableau to the form:

$$\begin{array}{c} f\left\{ \right. \\ B\left\{ \right. \end{array} \begin{array}{|c|c|c|} \hline \mathbf{0}^{\mathrm{T}} & \hat{\mathbf{c}}_N^{\mathrm{T}} & -\hat{f} \\ \hline \mathbf{I} & \hat{\mathbf{A}}_N & \hat{\mathbf{b}} \\ \hline \end{array} \quad (8.2.13)$$

$$\underbrace{\quad}_{B}\ \underbrace{\quad}_{N}$$

where $\hat{\mathbf{A}}_N = \mathbf{A}_B^{-1}\mathbf{A}_N$, which represents the *canonical* or *reduced form* of the LP problem. Essentially the process is that used in obtaining (8.1.4) and (8.2.3) from (8.1.2): the operations on the B rows are equivalent to premultiplying by $\mathbf{A}_B^{-1}$, and those on the f row are equivalent to premultiplying by $\mathbf{c}_B^{\mathrm{T}}$ in the new B rows and subtracting this row from the f row. In fact an initial canonical form (8.2.13) is usually available directly as a result of the operations to find the initial b.f.s. (Section 8.4). Tests (8.2.8) and (8.2.11) can be carried out directly on information in the tableau (note that $\mathbf{d} = -\hat{\mathbf{a}}_q$) so that the indices p and q can be determined. The new tableau is obtained by first scaling row p so that the element in column q is unity. Multiples of row p are then subtracted from other rows so that the remaining elements in column q become zero, and column q becomes a column of the unit matrix. The previous column (p) with this property becomes full during these operations. Then columns p and q of (8.2.13) are interchanged, giving a new tableau in canonical form. In fact this interchange need not take place, in which case the columns of the unit matrix in (8.2.13) no longer occur in the first n columns of the tableau. The iteration is then repeated on the new tableau. An example of the tableau form applied to (8.1.2) is given in Table 8.2.1. In general the quantity $\hat{a}_{pq}$ $(= -d_p)$ must be non-zero (see (8.2.11)) and plays the role of *pivot* in these row operations, in a similar way to the pivot in Gaussian elimination. In Table 8.2.1 the pivot element is circled.

It became apparent that the tableau form is inefficient in that it updates the whole tableau $\hat{\mathbf{A}}_N$ at each iteration. The earlier part of this section shows that the simplex method can be carried out with an explicit knowledge of $\mathbf{A}_B^{-1}$, and it is possible to carry out updating operations on this matrix. Since $\mathbf{A}_B^{-1}$ is often smaller than $\hat{\mathbf{A}}_N$, the resulting method, known as the *revised simplex method*, is usually more efficient. The effect of a basis change is to replace the column $\mathbf{a}_p$ by $\mathbf{a}_q$ in $\mathbf{A}_B^{(k)}$, which can be written as the rank one change

$$\mathbf{A}_B^{(k+1)} = \mathbf{A}_B + (\mathbf{a}_q - \mathbf{a}_p)\mathbf{e}_p^{\mathrm{T}} \quad (8.2.14)$$

(suppressing superscript (k)). By using the Sherman–Morrison formula (Volume 1, Question 3.13) it follows that

$$[\mathbf{A}_B^{(k+1)}]^{-1} = [\mathbf{A}_B]^{-1} - \frac{(\mathbf{d} + \mathbf{e}_p)\mathbf{v}_p^{\mathrm{T}}}{d_p} \quad (8.2.15)$$

Table 8.2.1 The tableau form applied to problem (8.1.2)

Data:						
f	1	2	3	4	0	
x_1	1	1	1	1	1	
x_2	1	0	1	-3	$\frac{1}{2}$	
	x_1	x_2	x_3	x_4		

$k = 1$:						
f	0	0	2	-1	$-1\frac{1}{2}$	(see (8.2.3))
x_1	1	0	1	-3	$\frac{1}{2}$	(see (8.1.4))
x_2	0	1	0	(4)	$\frac{1}{2}$	
	x_1	x_2	x_3	x_4		

f	0	$\frac{1}{4}$	2	0	$-1\frac{3}{8}$	(see (8.1.6))
x_1	1	$\frac{3}{4}$	1	0	$\frac{7}{8}$	(see (8.1.5))
x_2	0	$\frac{1}{4}$	0	1	$\frac{1}{8}$	
	x_1	x_2	x_3	x_4		

$k = 2$:					
f	0	0	2	$\frac{1}{4}$	$-1\frac{3}{8}$
x_1	1	0	1	$\frac{3}{4}$	$\frac{7}{8}$
x_4	0	1	0	$\frac{1}{4}$	$\frac{1}{8}$
	x_1	x_4	x_3	x_2	

where $\mathbf{v}_p^T$ is the row of $\mathbf{A}_B^{-1}$ corresponding to x_p. The fact that $d_p \neq 0$ is implied by (8.2.11) ensures that $\mathbf{A}_B^{(k+1)}$ is non-singular. It is also possible to update the vectors $\boldsymbol{\pi}$ and $\hat{\mathbf{b}}$ by using (8.2.15).

A disadvantage of representing $\mathbf{A}_B^{-1}$ in the revised simplex method is that it is not adequate for solving large problems. Here the matrix $\mathbf{A}$ is sparse, and the problem is only manageable if this sparsity can be taken into account. Storing $\mathbf{A}_B^{-1}$ which is usually a full matrix is no longer possible. Also it has more recently been realized that representing $\mathbf{A}_B^{-1}$ can sometimes cause numerical problems due to magnification of round-off errors. Thus there is currently much research into other representations of $\mathbf{A}_B$ in the form of matrix factorizations which are easily invertible, numerically stable, and which enable sparsity to be exploited. These developments are described further in Section 8.5.

8.3 Other LP Techniques

It has already been remarked that general LP problems can be reduced to the standard form (8.1.1) by adding extra variables. This section reviews other ways in which a general LP problem can be tackled. One possibility is to use *duality* results to transform the problem. These are dealt with in more detail in Section 9.5 where three different LP duals of differing complexity are given. The resulting dual problem is then solved by the simplex method, possibly after adding extra variables to give a standard form. The aim in using a dual is to generate a transformed problem which has a more favourable structure, in particular one with fewer rows in the constraint matrix. Another type of transformation which can be used is known as *decomposition* in which a large structured problem is decomposed into a more simple master problem which is defined in terms of several smaller subproblems. The aim is to solve a sequence of smaller problems in such a way as to determine the solution of the original large problem. An example of this type of transformation is given in more detail in Section 8.5.

A different technique, known as *parametric programming*, concerns finding the solution of an LP problem which is a perturbation of another problem for which the solution is available. This may occur, for example, when the effect of changes in the data ($\mathbf{A}$, $\mathbf{b}$, or $\mathbf{c}$) is considered, or when an extra constraint or variable is added to the problem. The effect of changes in the data gives useful information regarding the *sensitivity* of various features in the model, and may often be more important than the solution itself. The rate of change of the values of the function and the basic variables with respect to $\mathbf{b}$ or $\mathbf{c}$ can be determined directly from $\hat{\mathbf{b}} = \mathbf{A}_B^{-1}\mathbf{b}$ and $\hat{f} = \mathbf{c}_B^T\hat{\mathbf{b}}$ (see Question 8.14). To find the effect of small finite non-zero changes, the aim is to retain as much information as possible from the unperturbed solution when re-solving the problem. If only $\mathbf{b}$ or $\mathbf{c}$ is changed then a representation of $\mathbf{A}_B^{-1}$ is directly available. If $\mathbf{A}_B$ is changed then the partition into B and N can still be valuable. The only difficulty is that changes to $\mathbf{b}$ or $\mathbf{A}_B$ might result in the condition $\hat{\mathbf{b}} \geqslant \mathbf{0}$ being violated. It is important to be able to restore feasibility without restarting ab initio, and this is discussed further in Section 8.4. Once feasibility is obtained then simplex steps are taken to achieve optimality. Another type of parametric programming arises when extra constraints are added (see Section 13.1 for example) which cut out the solution. The same approach of taking simplex steps to restore first feasibility and then optimality can be used. An alternative possibility is to use the dual formulation, when extra constraints in the primal just correspond to extra variables in the dual. The previous solution is then still dual feasible, and all that is required is to restore optimality.

Another way of treating the general LP problem

$$\begin{aligned} \underset{\mathbf{x}}{\text{minimize}}\ & f(\mathbf{x}) \triangleq \mathbf{c}^T\mathbf{x} \\ \text{subject to}\ & \mathbf{a}_i^T\mathbf{x} = b_i, \quad i \in E \\ & \mathbf{a}_i^T\mathbf{x} \geqslant b_i, \quad i \in I \end{aligned} \tag{8.3.1}$$

for finite index sets E and I, is the *active set method*, which devolves from ideas used in more general linear constraint problems (Chapters 10 and 11), and which is perhaps more natural than the traditional approach via (8.1.1). Similar introductory considerations apply as in Section 8.1, and in particular the solution usually occurs at an extreme point or vertex of the feasible region. (When this situation does not hold, for example in $\min x_1 + x_2$ subject to $x_1 + x_2 = 1$, then it is possible to create a vertex by removing undetermined variables, or equivalently by having *pseudo-constraints* $x_i = 0$ in the active set – see Section 8.4.) Each iterate $\mathbf{x}^{(k)}$ method is therefore a feasible vertex defined by the equations

$$\mathbf{a}_i^{\mathrm{T}}\mathbf{x} = b_i, \qquad i \in \mathscr{A} \tag{8.3.2}$$

where $\mathscr{A}$ is the *active set* of n constraint indices, and where the vectors $\mathbf{a}_i$ are independent. Except in degenerate cases (see Section 8.6), $\mathscr{A}$ is the set of active constraints $\mathscr{A}(\mathbf{x}^{(k)})$ in (7.1.2).

Each iteration in the active set method consists of a move from one vertex to another along a common edge. At $\mathbf{x}^{(k)}$ the Lagrange multiplier vector

$$\boldsymbol{\lambda} = \mathbf{A}^{-1}\nabla f = \mathbf{A}^{-1}\mathbf{c} \tag{8.3.3}$$

is evaluated, where $\mathbf{A}$ now denotes the $n \times n$ matrix with columns $\mathbf{a}_i$, $i \in \mathscr{A}$. Let the columns of $\mathbf{A}^{-\mathrm{T}}$ be written $\mathbf{s}_1, \mathbf{s}_2, \ldots, \mathbf{s}_n$. Then the construction (9.1.14) shows that the vectors $\mathbf{s}_i$, $i \in \mathscr{A} \cap I$ are the directions of all feasible edges at $\mathbf{x}^{(k)}$, and the multipliers λ_i are the slopes of $f(\mathbf{x})$ along these edges. Thus if

$$\lambda_i \geqslant 0, \qquad i \in \mathscr{A} \cap I \tag{8.3.4}$$

then no feasible descent directions exist, so $\mathbf{x}^{(k)}$ is optimal and the iteration terminates. Otherwise the most negative λ_i, $i \in \mathscr{A} \cap I$ (λ_q say) is chosen, and a search is made along the downhill edge

$$\mathbf{x} = \mathbf{x}^{(k)} + \alpha\mathbf{s}_q, \qquad \alpha \geqslant 0. \tag{8.3.5}$$

The ith constraint has the value $c_i^{(k)} = \mathbf{a}_i^{\mathrm{T}}\mathbf{x}^{(k)} - b_i$ at $\mathbf{x}^{(k)}$ and the value $c_i(\mathbf{x}) = c_i^{(k)} + \alpha\mathbf{a}_i^{\mathrm{T}}\mathbf{s}_q$ at points along the edge. The search is terminated by the first inactive constraint (p say) to become active. Candidates for p are therefore indices $i \notin \mathscr{A}$ with $\mathbf{a}_i^{\mathrm{T}}\mathbf{s}_q < 0$ and any such constraint function becomes zero when

$$\alpha = \frac{b_i - \mathbf{a}_i^{\mathrm{T}}\mathbf{x}^{(k)}}{\mathbf{a}_i^{\mathrm{T}}\mathbf{s}_q}.$$

Thus the index p and the corresponding value of α are defined by

$$\alpha = \min_{\substack{i:\, i \notin \mathscr{A}, \\ \mathbf{a}_i^{\mathrm{T}}\mathbf{s}_q < 0}} \frac{b_i - \mathbf{a}_i^{\mathrm{T}}\mathbf{x}^{(k)}}{\mathbf{a}_i^{\mathrm{T}}\mathbf{s}_q}. \tag{8.3.6}$$

The corresponding point defined by (8.3.5) is the new vertex $\mathbf{x}^{(k+1)}$, and the new active set is obtained by replacing q by p in $\mathscr{A}$. The iteration is then repeated from this new vertex until termination occurs. Only one column of $\mathbf{A}$ is changed at each iteration, so $\mathbf{A}^{-1}$ can be readily updated as in (8.2.15).

In fact the active set method is closely related to the simplex method, when the slack variables are added to (8.3.1). It is not difficult (see Question 8.17) to show that the Lagrange multiplier vector in (8.3.3) is the reduced costs vector in (8.2.6), so that the optimality tests (8.3.4) and (8.2.8) are identical. Likewise the index q which is determined at a non-optimal solution is the same in both cases. Moreover the tests (8.3.6) and (8.2.11) which determine p are also equivalent, so the same interchanges in active constraints (nonbasic slack variables) are made. In fact the only difference between the two methods lies in how the inverse matrix information is represented, because the active set method updates the $n \times n$ matrix $\mathbf{A}^{-1}$ whereas the revised simplex method updates the $m \times m$ matrix $\mathbf{A}_B^{-1}$. Thus for problems in standard form (8.1.1) which only have a few equations ($m \ll n$), a smaller matrix is updated by the revised simplex method, which is therefore preferable. On the other hand, if the problem is like (8.3.1) and there are $m \gg n$ constraints (mostly inequalities), then the simplex method requires many slack variables to be added, and the active set method updates the smaller inverse matrix and is therefore preferable.

In both cases the inefficiency of the worse method is caused by the fact that columns of the unit matrix occur in the matrix ($\mathbf{A}$ or $\mathbf{A}_B$) whose inverse is being updated. For example if bounds $\mathbf{x} \geqslant \mathbf{0}$ (or more generally $\boldsymbol{\ell} \leqslant \mathbf{x} \leqslant \mathbf{u}$) are recognized in the active set method, then the matrix $\mathbf{A}$ can be partitioned (after a suitable permutation of variables) into the form

$$\mathbf{A} = \overset{\quad n-p \quad\; p}{\begin{bmatrix} \mathbf{A}_1 & \mathbf{0} \\ \mathbf{A}_2 & \mathbf{I} \end{bmatrix}} \begin{matrix} n-p \\ p \end{matrix}$$

where p is the number of active bounds. Then only $\mathbf{A}_1^{-1}$ need be recurred, and $\mathbf{A}^{-1}$ can be recovered from the expression

$$\mathbf{A}^{-1} = \begin{bmatrix} \mathbf{A}_1^{-1} & \mathbf{0} \\ -\mathbf{A}_2\mathbf{A}_1^{-1} & \mathbf{I} \end{bmatrix}.$$

In this way the active set method becomes as efficient as possible, if sparsity in the remaining constraint normals can be ignored. However a similar modification to the revised simplex method can be made when basic slack variables are present. Then $\mathbf{A}_B$ also contains columns of the unit matrix, which enables a smaller submatrix to be recurred. If this is done then the methods are entirely equivalent. However as a personal choice I prefer the description of linear programming afforded by the active set method, since it is more natural and direct, and does not rely on introducing extra variables to solve general LP problems.

8.4 Feasible Points for Linear Constraints

In this section the *method of artificial variables* is described for finding an initial basic feasible solution for use in the simplex method. The underlying idea is of wide generality and it is shown that it is possible to determine a similar but more flexible and efficient method for use in a wide variety of situations. These include the active set methods for minimization subject to linear constraints described in Sections 8.3 (linear objective function), 10.3 (quadratic objective), and 11.2 (general objective). The resulting methods are also very suitable for restoring feasibility when doing parametric programming. The idea of an L_1 exact penalty function for nonlinear constraints (Section 14.3) is also seen to be a generalization of these ideas.

In the first instance the problem is considered of finding a basic feasible solution to the constraints

$$\mathbf{A}\mathbf{x} = \mathbf{b}, \qquad \mathbf{x} \geqslant \mathbf{0} \tag{8.4.1}$$

which arise in the standard form of linear programming (8.1.1). Extra variables $\mathbf{r}$ (often referred to as *artificial variables*) are introduced into the problem, such that

$$\mathbf{r} = \mathbf{b} - \mathbf{A}\mathbf{x} \tag{8.4.2}$$

These variables can be interpreted as *residuals* of the equations $\mathbf{A}\mathbf{x} = \mathbf{b}$. In the method of artificial variables it is first ensured that $\mathbf{b} \geqslant \mathbf{0}$ by first reversing the sign of any row of $\mathbf{A}$ and $\mathbf{b}$ for which $b_i < 0$. Then the auxiliary problem

$$\begin{aligned} &\underset{\mathbf{x},\mathbf{r}}{\text{minimize}} \ \sum_i r_i \\ &\text{subject to } \mathbf{A}\mathbf{x} + \mathbf{r} = \mathbf{b}, \qquad \mathbf{x} \geqslant \mathbf{0}, \ \mathbf{r} \geqslant \mathbf{0} \end{aligned} \tag{8.4.3}$$

is solved in an attempt to reduce the residuals to zero. Let $\mathbf{x}'$ and $\mathbf{r}'$ solve (8.4.3); clearly if $\mathbf{r}' = \mathbf{0}$ then $\mathbf{x}'$ is feasible in (8.4.1) as required, whereas if $\mathbf{r}' \neq \mathbf{0}$ then there is *no feasible point* for (8.4.1). (This latter observation is true since if $\mathbf{x}'$ is feasible in (8.4.1) then $\mathbf{x}', \mathbf{0}$ is a feasible point of (8.4.3) with $\Sigma r_i = 0$ which contradicts the optimality of $\mathbf{r}' \neq \mathbf{0}$.) Also (8.4.3) is a linear program in standard form with variables $\mathbf{x}, \mathbf{r}$ for which the coefficient matrix is $[\mathbf{A} : \mathbf{I}]$ and the costs are $\mathbf{0}, \mathbf{e}$ where $\mathbf{e}$ is a vector of ones. An initial basic feasible solution is given by having the $\mathbf{r}$ variables basic and the $\mathbf{x}$ variables nonbasic so that $\mathbf{r}^{(1)} = \mathbf{b}$ and $\mathbf{x}^{(1)} = \mathbf{0}$. The initial basis matrix is simply $\mathbf{A}_B = \mathbf{I}$. Thus the simplex method itself can be used to solve (8.4.3) directly from this b.f.s.

A feature of the calculation is that if the artificial variable r_i becomes nonbasic then the constraint $(\mathbf{A}\mathbf{x} - \mathbf{b})_i = 0$ becomes satisfied and need never be relaxed, that is r_i need never become basic again. Thus once an artificial variable becomes nonbasic, it is removed from the computation entirely, and some effort may be saved. If a feasible point exists ($\mathbf{r}' = \mathbf{0}$) the optimum basic feasible solution to (8.4.3) has no basic r_i variables (assuming no degeneracy) so that all the artificial variables must have been removed. Therefore the remaining variables $\mathbf{x}^T = (\mathbf{x}_B^T, \mathbf{x}_N^T)$ are partitioned to give a basic feasible solution to (8.4.1). The same tableau $\hat{\mathbf{A}}_N$ or inverse basis matrix $\mathbf{A}_B^{-1}$ is directly available to be used in the solution of the main LP (8.1.1). Because the solution of (8.4.3) is a preliminary to the solution of (8.1.1), the two are sometimes referred to as *phase I* and *phase II* of the simplex method respectively. In some degenerate cases (for example when the constraint matrix $\mathbf{A}$ is rank deficient) it is possible that some artificial variables may remain in the basis after (8.4.3) is solved. Since $\mathbf{r} = \mathbf{0}$ nothing is lost by going on to solve the main problem with the artificial variables remaining in the basis, essentially just as a means of enabling the basic variables to be eliminated (see Question 8.21).

An example of the method to find a basic feasible solution to the constraints in (8.1.2) is described. Since $\mathbf{b} \geqslant \mathbf{0}$ no sign change is required. Artificial variables r_1 and r_2 (x_5 and x_6 say) are added to give the phase I problem

$$\begin{aligned}
&\underset{\mathbf{x}}{\text{minimize}}\ x_5 + x_6 \\
&\text{subject to}\ x_1 + x_2 + x_3 + x_4 + x_5 = 1 \\
&\qquad\qquad x_1 + x_3 - 3x_4 + x_6 = \tfrac{1}{2} \\
&\qquad x_i \geqslant 0, \quad i = 1, 2, \ldots, 6.
\end{aligned}$$

A basic feasible solution for this problem is $B = \{5, 6\}$, $N = \{1, 2, 3, 4\}$, with $\mathbf{A}_B = \mathbf{I}$ by construction. A calculation shows that $\hat{\mathbf{c}}_N = (-2, -1, -2, 2)^\mathrm{T}$ so that x_1 $(q = 1)$ is chosen to be increased. Then $\mathbf{d} = (-1, -1)^\mathrm{T}$ so both x_5 and x_6 are reduced, but x_6 reaches zero first. Thus x_1 is made basic and x_6 nonbasic. Since x_6 is artificial it is removed from the computation, and a smaller problem is solved with $B = \{1, 5\}$, $N = \{2, 3, 4\}$. In this case the calculation shows that x_4 is increased and x_5 goes to zero. When x_4 is made basic and x_5 removed, there remains the partition $B = \{1, 4\}$, $N = \{2, 3\}$. Both the artificial variables have been removed and so phase I is complete. This basic feasible solution is carried forward into phase II and by chance is found to be optimal (see Section 8.2), so no further basis changes are made.

The way in which an artificial variable is introduced is very similar to that in which a slack variable is introduced (see Sections 8.1 and 7.2). In fact if a slack variable has already been added into an inequality $\mathbf{a}_i^\mathrm{T}\mathbf{x} \leqslant b_i$, and if $b_i \geqslant 0$, then it is possible to use a slack variable in the initial basis and not add an artificial variable as well. For example consider the system

$$\begin{aligned}
x_1 + x_2 + x_3 + x_4 &\leqslant 1 \\
x_1 + x_3 - 3x_4 &= \tfrac{1}{2}, \qquad \mathbf{x} \geqslant \mathbf{0}.
\end{aligned}$$

Adding a slack variable x_5 gives

$$\begin{aligned}
x_1 + x_2 + x_3 + x_4 + x_5 &= 1 \\
x_1 + x_3 - 3x_4 &= \tfrac{1}{2}, \qquad \mathbf{x} \geqslant \mathbf{0}
\end{aligned} \tag{8.4.4}$$

which is in standard form. Since $b_1 = 1 \geqslant 0$ the slack variable x_5 can be used in the initial basis, and so an artificial variable (x_6 say) need only be added into the second equation. The resulting cost function is the sum of the *artificial variables only*, so in this case the phase I problem is

$$\begin{aligned}
&\text{minimize}\ x_6 \\
&\text{subject to}\ x_1 + x_2 + x_3 + x_4 + x_5 = 1 \\
&\qquad\qquad x_1 + x_3 - 3x_4 + x_6 = \tfrac{1}{2}, \qquad \mathbf{x} \geqslant \mathbf{0}
\end{aligned}$$

and the initial partition is $B = \{5, 6\}$, $N = \{1, 2, 3, 4\}$. As before artificial variables are removed from the problem when they become nonbasic, but slack variables remain in the problem throughout. In this example it happens that x_6 is removed on the first iteration and the resulting partition $B = \{1, 5\}$, $N = \{2, 3, 4\}$ gives a basic feasible solution for (8.4.4), to be used in phase II. In general if $b_i < 0$ after adding a slack variable to an inequality $\mathbf{a}_i^\mathrm{T}\mathbf{x} \leqslant b_i$, then it is also necessary to add an artificial variable to that equation. This is cumbersome and can be avoided by adopting the more general framework which follows, and which does not require any sign changes to ensure $\mathbf{b} \geqslant \mathbf{0}$.

At the expense of a small addition to the logic in phase I, a much more flexible technique for obtaining feasibility can be developed (Wolfe, 1965). This is to allow negative variables in the basis and to extend the cost function to give the auxiliary problem on the kth step as

$$\text{minimize} \quad \sum_{i:x_i^{(k)}<0} -x_i + \sum_{i:r_i^{(k)}<0} -r_i + \sum_{i:r_i^{(k)}>0} r_i \tag{8.4.5}$$

$$\text{subject to } \mathbf{Ax} + \mathbf{r} = \mathbf{b}, \qquad x_i \geqslant 0 \quad \forall i : x_i^{(k)} \geqslant 0.$$

In this problem it is understood that an element of the cost vector is taken as -1 if $x_i^{(k)} < 0$ and 0 if $x_i^{(k)} \geqslant 0$, and likewise as -1 if $r_i^{(k)} < 0$ and $+1$ if $r_i^{(k)} > 0$ so that the costs may change from one iteration to another. *Any* set of basic variables can be chosen initially (assuming $\mathbf{A}_B$ is non-singular) without the need to have $\hat{\mathbf{b}} \geqslant \mathbf{0}$, and simplex steps are made as before, except that (8.2.11) must be modified in an obvious way to allow for a negative variable being increased to zero. Once x_i becomes non-negative then the condition $x_i \geqslant 0$ is subsequently enforced and artificial variables are removed once they become zero. There is no need with (8.4.5) to have both artificial and slack variables in any one equation. An example is given in Question 8.22. In fact, because any set of basic variables can be chosen, the artificial variable is added only as a means to enable an easily invertible matrix $\mathbf{A}_B$ to be chosen. Thus if such a matrix is already available then artificial variables are not required at all.

Another consequent advantage of (8.4.5) occurs in *parametric programming* (Section 8.3) when the solution from a previous problem (B and $\mathbf{A}_B^{-1}$) is used to start phase I for the new problem. It is likely that the old basis is close to a feasible basis if the changes are small. However it is quite possible that the perturbation causes feasibility ($\hat{\mathbf{b}} \geqslant \mathbf{0}$) to be lost, and it is much more efficient to use (8.4.5) in conjunction with the old basis, rather than starting the phase I procedure with a basis of artificial variables. Also a better feasible point is likely to be obtained. For example in (8.4.4) if b_2 is perturbed to $-\frac{1}{2}$ and the same partition $B = \{1, 5\}$, $N = \{2, 3, 4\}$ is used, then $\hat{\mathbf{b}} = (-\frac{1}{2}, \frac{3}{2})^{\mathrm{T}}$ can be calculated. Thus the cost function in (8.4.5) is $-x_1$ and it follows that $\hat{\mathbf{c}}_N = (0, 1, -3)^{\mathrm{T}}$ so that x_4 is chosen to be increased. Then $\mathbf{d} = (3, -4)^{\mathrm{T}}$ so the increase in x_4 causes x_1 to increase towards zero and x_5 to decrease towards zero. x_1 reaches zero first when $x_4 = \frac{1}{6}$, so the new partition is $B = \{4, 5\}$, $N = \{1, 2, 3\}$ with $\hat{\mathbf{b}} = (\frac{1}{6}, \frac{5}{6})^{\mathrm{T}}$. Since there are now no negative variables, feasibility is restored and phase II can commence.

Use of the auxiliary problem (8.4.5) is closely related to a technique for solving linear L_1 approximation problems in which $\| \mathbf{r} \|_1$ $(= \Sigma_i | r_i |)$ is minimized, where $\mathbf{r}$ is defined by (8.4.2). The $| r_i |$ terms are handled by introducing positive and negative variables r_i^+ and r_i^-, where $r_i^+ \geqslant 0$, $r_i^- \geqslant 0$, and by replacing $r_i = r_i^+ - r_i^-$ and $| r_i | = r_i^+ + r_i^-$ as described in Section 7.2. Barrodale and Roberts (1973) suggest a modification for improving the efficiency of the simplex method so that fewer iterations are required to solve this particular type of problem. The idea is to allow the simplex iteration to pass through a number of vertices at one time so as to minimize the total cost function. The same type of modification can be used advantageously with (8.4.5). Let the nonbasic variable x_q (either an element of $\mathbf{x}$ or $\mathbf{r}$) for which $\hat{c}_q < 0$ be chosen to be increased as in Section 8.2. If a variable $x_i < 0$ is increased to zero, then the corresponding element of $\mathbf{c}_B$ is changed from -1 to 0 and $\hat{c}_q$ is easily re-computed. If $\hat{c}_q < 0$ still holds then x_q can continue to be increased without making a pivot step to change $\mathbf{A}_B^{-1}$ or $\hat{\mathbf{A}}_N$. Only if $\hat{c}_q \geqslant 0$ is a pivot step made. Likewise if a variable $r_i < 0$ (or >0) becomes zero then the corresponding element of $\mathbf{c}_B$ is changed from -1 to $+1$ (or $+1$ to -1) and $\hat{c}_q$ is re-computed and tested in the same way. In this modification an artificial variable need not be removed immediately it becomes zero and $\mathbf{x}$ variables may be allowed to go negative if the cost function is reduced. For best efficiency it is valuable to scale the rows of $\mathbf{Ax} = \mathbf{b}$, or equivalently to scale the elements of the cost function in (8.4.5) to balance the contribution from each term (to the same order of magnitude).

It is possible to suggest an entirely equivalent technique for calculating a feasible point for use with the active set method for linear programming (Section 8.3), or with other active set methods for linear constraints (Sections 10.3 and 11.2). The technique is described here in the case that E is empty, although it is readily modified to the more general case. The phase I auxiliary problem (Fletcher, 1970) which is solved at $\mathbf{x}^{(k)}$ is

$$\begin{aligned} &\text{minimize} \quad \sum_{i \in V^{(k)}} (b_i - \mathbf{a}_i^{\mathrm{T}}\mathbf{x}) \\ &\text{subject to } \mathbf{a}_i^{\mathrm{T}}\mathbf{x} \geqslant b_i, \qquad i \notin V^{(k)} \end{aligned} \tag{8.4.6}$$

where $V^{(k)} = V(\mathbf{x}^{(k)})$ is the set of violated (that is infeasible) constraints at $\mathbf{x}^{(k)}$. Thus the cost function in (8.4.6) is the sum of moduli of violated constraints. The iteration commences by making any convenient arbitrary choice of an active set $\mathscr{A}$ of n constraints, and this determines an initial vertex $\mathbf{x}^{(1)}$. At $\mathbf{x}^{(k)}$ the gradient of the cost function is $-\sum_{i \in V^{(k)}} \mathbf{a}_i$, and the method proceeds as in (8.3.3) to use this vector to calculate multipliers and hence to determine an edge $\mathbf{s}_q$ along which the cost function has negative slope. A search along $\mathbf{x}^{(k)} + \alpha\mathbf{s}_q$ is made to minimize the sum of constraint violations $\sum_{i \in V(\mathbf{x})}(b_i - \mathbf{a}_i^{\mathrm{T}}\mathbf{x})$. This search is always terminated by an inactive constraint (p say) becoming active, and p then replaces q in the active set. The method terminates when $\mathbf{x}^{(k)}$ is found to be a feasible point, and the resulting active set and the matrix $\mathbf{A}^{-1}$ in (8.3.3) are passed forward to the phase II problem (for example (8.3.1)).

To create the initial vertex $\mathbf{x}^{(1)}$ easily in the absence of any parametric programming information, it is also possible to have something similar to the artificial variable method. In fact artificial constraints or *pseudoconstraints* $x_i = 0$, $i = 1, 2, \ldots, n$, are added to the problem, if they are not already present, and form the initial active set (Fletcher and Jackson, 1974). Pseudoconstraints with a non-zero Lagrange multiplier are always eligible for removal from the active set, and they do not contribute to the cost function in the line search. The value of this device is that the initial matrix used in (8.3.3), etc., is $\mathbf{A} = \mathbf{I}$, which is immediately available. Once a pseudoconstraint becomes inactive it is removed from the problem; this usually ensures that the pseudoconstraints are rapidly removed from the problem in favour of actual constraints. However pseudoconstraints with a zero multiplier are allowed to remain in the active set at a solution. This can be advantageous in phase I when the feasible region has no true vertex, other than at large values of **x** which might cause round-off difficulties. In phase II it also solves the problem of eliminating redundant variables (for example as in min $x_1 + x_2$ subject to $x_1 + x_2 = 1$) in a systematic way.

One final idea related to feasible points is worthy of mention. It is applicable to both a standard form problem (8.1.1) or to the more general form (8.3.1) used in the active set method. The feasible point which is found by the methods described in this section is essentially arbitrary, and may be far from the solution. It can be advantageous to bias the phase I cost function in (8.4.5) or (8.4.6) by adding to it a multiple $\nu f(\mathbf{x})$ of the phase II cost function. For all ν sufficiently small the solution of phase I is also that of phase II so the possibility exists of combining both these stages into one. The limit $\nu \to 0$ corresponds to the phase I–phase II iteration above, and is sometimes referred to (after multiplying through by $M = 1/\nu \gg 0$) as the *big M method*. However the method is usually more efficient if larger values of ν are taken, although if ν is too large then a feasible point is not obtained. In fact it is not difficult to see that the method has become one for minimizing an L_1 *exact penalty function* for the LP problem (8.3.1), as described in Section 14.3. Thus the result of theorem 14.3.1 indicates that the bound $\nu \leqslant 1/\|\boldsymbol{\lambda}^*\|_\infty$ limits the choice of ν where $\boldsymbol{\lambda}^*$ is the optimum Lagrange multiplier vector for (8.3.1).

This type of idea is of particular advantage for parametric programming in QP problems (Chapter 10), because the active set $\mathscr{A}$ from a previously solved QP problem may have less than n elements and so cannot be used as the initial vertex in the phase I problem (8.4.6). However if $\nu f(\mathbf{x})$ (now a quadratic function) is added to the phase I cost function in (8.4.6), then a QP-like problem results and the previous active set and the associated inverse information can be carried forward and used directly as initial data for this problem. No doubt a similar idea could also be used in more general linear constraint problems (Chapter 11), although I do not think that this possibility has been investigated.

8.5 Stable and Large-scale Linear Programming

A motivation for considering alternatives to the tableau method or the revised simplex method is that the direct representation of $\mathbf{A}_B^{-1}$ or its implicit occurrence in the matrix $\hat{\mathbf{A}} = \mathbf{A}_B^{-1}\mathbf{A}_N$ can lead to difficulties over round-off error. If an intermediate b.f.s. is such that $\mathbf{A}_B$ is nearly singular then $\mathbf{A}_B^{-1}$ becomes large and representing this matrix directly introduces large round-off errors. If on a later iteration $\mathbf{A}_B$ is well conditioned then $\mathbf{A}_B^{-1}$ is not large but the previous large round-off errors remain and their relative effect is large. This problem is avoided if a stable factorization of $\mathbf{A}_B$ is used which does not represent the inverses of potentially nearly singular matrices. Typical examples of this for solving equations $\mathbf{Ax} = \mathbf{b}$ are the **LU** factors (with pivoting) or the **QR** factors (**L** is lower triangular, **U**, **R** are upper triangular, and **Q** is orthogonal). Some generalizations of these ideas in linear programming are described in this section.

Also in this section, methods are discussed which are suitable for solving large sparse LP problems because they have some features in common. In terms of (8.1.1) this usually refers to problems in which the number of equations m is greater than a few hundred, and in which the coefficient matrix $\mathbf{A}$ is sparse, that is it contains a high proportion of zero elements. The standard techniques of Section 8.2 (or 8.3) fail because the matrices such as $\mathbf{A}_B^{-1}$ or $\hat{\mathbf{A}}_N$ do not retain the sparsity of $\mathbf{A}$ and must be stored in full, which is usually beyond the capacity of the rapid access computer store. By taking the sparsity structure of $\mathbf{A}$ into account it is possible to solve large sparse problems. There are two main types of method, one of which is a *general sparse matrix method* which allows for any sparsity structure and attempts to update an invertible representation of $\mathbf{A}_B$ which maintains a large proportion of the sparsity. The other type is the *decomposition method* which attempts to reduce the problem to a number of much smaller LP problems which can be solved conventionally. Decomposition methods usually require a particular structure in the matrix $\mathbf{A}$ so are not of general applicability. One practical example of this type of method is given, and other possibilities are referenced. It is not possible to say that any one method for large-scale LP is uniformly best because comparisons are strongly dependent on problem size and structure, but my interpretation of current thinking is that even those problems which are suitable for solving by decomposition methods can also be handled efficiently by general sparse matrix methods, so a good implementation of the latter type of method is of most importance.

The earliest attempt to take sparsity into account is the *product form method.* Equation (8.2.15) for updating the matrix $\mathbf{A}_B^{-1}$ can be written

$$[\mathbf{A}_B^{(k+1)}]^{-1} = \mathbf{M}_k[\mathbf{A}_B^{(k)}]^{-1} \tag{8.5.1}$$

where $\mathbf{M}_k$ is the matrix

$$\mathbf{M}_k = \begin{bmatrix} 1 & & & -d_1/d_p & & & \\ & \ddots & & -d_2/d_p & & & \\ & & 1 & \vdots & & & \\ & & & -d_p^{-1} & & & \\ & & & \vdots & 1 & & \\ & & & \vdots & & \ddots & \\ & & & -d_n/d_p & & & 1 \end{bmatrix} \tag{8.5.2}$$

$\uparrow$ column p

where $\mathbf{d}$ is defined in (8.2.10). Thus if $\mathbf{A}_B^{(1)} = \mathbf{I}$, then $[\mathbf{A}_B^{(k)}]^{-1}$ can be represented by the product

$$[\mathbf{A}_B^{(k)}]^{-1} = \mathbf{M}_{k-1} \cdots \mathbf{M}_2\mathbf{M}_1. \tag{8.5.3}$$

In the product form method the non-trivial column of each $\mathbf{M}_j, j = 1, 2, \ldots, k-1$, is stored in a computer file in packed form, that is just the non-zero elements and their row index. These vectors were known as *eta vectors* in early implementations. Then the representation (8.5.3) is used whenever operations with $\mathbf{A}_B^{-1}$ are required. As k increases the file of eta vectors grows longer and operating with it becomes more expensive. Thus ultimately it becomes cheaper to *reinvert* the basis matrix by restarting with a unit matrix and introducing the basic columns to give new eta vectors. This process of reinversion is common to most product form methods, and the choice of iteration on which it occurs is usually made on grounds of efficiency, perhaps using readings from a computer clock.

Product form methods are successful because the number of non-zeros required to represent $\mathbf{A}_B^{-1}$ as a product can be smaller than the requirement for storing the usually full matrix $\mathbf{A}_B^{-1}$ itself. However there are reasons why (8.5.3), known as the *Gauss–Jordan product form*, is not the most efficient. Consider the Gauss–Jordan method for eliminating variables in a set of linear equations with coefficient matrix $\mathbf{A}$ (ignoring pivoting). This can be written as the reduction of $\mathbf{A}$ $(= \mathbf{A}^{(1)})$ to a unit matrix $\mathbf{I}$ $(= \mathbf{A}^{(n+1)})$ by premultiplication by a sequence of Gauss–Jordan elementary matrices

$$\mathbf{I} = \mathbf{M}_n \cdots \mathbf{M}_2 \mathbf{M}_1 \mathbf{A} \tag{8.5.4}$$

where $\mathbf{M}_p$ has the form (8.5.2) with $d_j = -a_{jp}^{(p)}$ for all j. Another method for solving linear equations is Gaussian elimination, which is equivalent to the factorization $\mathbf{A} = \mathbf{LU}$ (again ignoring pivoting), where $\mathbf{L}$ is a unit lower triangular matrix and $\mathbf{U}$ is upper triangular. This can also be written in product form with

$$\begin{aligned} \mathbf{L} &= \mathbf{L}_1 \mathbf{L}_2 \cdots \mathbf{L}_{n-1} \\ \mathbf{U} &= \mathbf{U}_n \mathbf{U}_{n-1} \cdots \mathbf{U}_1 \end{aligned} \tag{8.5.5}$$

where $\mathbf{L}_i$ and $\mathbf{U}_i$ are the elementary matrices

$$\mathbf{L}_i = \begin{bmatrix} 1 & & & & & \\ & \ddots & & & & \\ & & 1 & & & \\ & & & 1 & & \\ & & & \ell_{i+1,i} & 1 & \\ & & & \vdots & & \ddots & \\ & & & \ell_{n,i} & & & 1 \end{bmatrix} \tag{8.5.6}$$

$\uparrow$ column i

$$\mathbf{U}_i = \begin{bmatrix} 1 & & & & & \\ & \ddots & & & & \\ & & 1 & & & \\ & & & u_{ii} & \cdots & u_{in} \\ & & & & 1 & \\ & & & & & \ddots & \\ & & & & & & 1 \end{bmatrix} \leftarrow \text{row } i \tag{8.5.7}$$

and where ℓ_{ij} and u_{ij} are the elements of the factors $\mathbf{L}$ and $\mathbf{U}$ of $\mathbf{A}$. It has been observed (see various chapters in Reid (1971), for example, and Question 8.23) that an **LU** factorization of a sparse matrix **A** retains much of the sparsity of **A** itself; typically the number of non-zero elements to be stored might be no more than double the number of non-zeros in **A**, especially when there is some freedom to do pivoting (as in reinversion).

It is now possible to explain why the Gauss–Jordan product form is not very efficient. It can be shown (see Question 8.24) that the non-trivial elements below the diagonal in the Gauss–Jordan elementary matrices $\mathbf{M}_k$ are the corresponding elements of **L** (with opposite sign), whereas the elements on and above the diagonal are the elements of $\mathbf{U}^{-1}$. Since $\mathbf{U}^{-1}$ is generally full it follows that only a limited amount of sparsity is retained in the factors $\mathbf{M}_k$. A simple example of this result is given in Question 8.23. This analysis also shows that use of a Gauss–Jordan product form does not solve the problem of numerical stability. If $\mathbf{A}_B$ is nearly singular then **U** can also be nearly singular and $\mathbf{U}^{-1}$ is then large. Thus the potential difficulty due to the presence of large round-off errors is not solved.

In view of this situation more recent research has concentrated on trying to achieve the advantages of both stability and sparsity possessed by an **LU** decomposition. The method of Bartels and Golub (Bartels, 1971) is of this type and uses a combination of Gaussian elimination and row and column interchanges. Assume without loss of generality that the initial basis matrix has triangular factors $\mathbf{A}_B^{(1)} = \mathbf{LU}$. Then on the kth iteration $\mathbf{A}_B^{(k)}$ has the invertible representation defined by the expression

$$(\mathbf{L}_r\mathbf{P}_r)(\mathbf{L}_{r-1}\mathbf{P}_{r-1})\cdots(\mathbf{L}_1\mathbf{P}_1)\mathbf{L}^{-1}\mathbf{A}_B^{(k)}\mathbf{Q}^{(k)} = \mathbf{U}^{(k)} \tag{8.5.8}$$

where $\mathbf{U}^{(k)}$ is upper triangular, $\mathbf{Q}^{(k)}$ is a permutation matrix, and $\mathbf{L}_i$ and $\mathbf{P}_i$ are certain elementary lower triangular and permutation matrices for $i = 1, 2, \ldots, r$. (Note that r refers to $r^{(k)}$, $\mathbf{U}^{(1)} = \mathbf{U}$, and $\mathbf{Q}^{(1)} = \mathbf{I}$.) An operation with $\mathbf{A}_B^{-1}$ is readily accomplished since from (8.5.8) it only requires permutations and operations with triangular matrices.

After a simplex iteration the new matrix $\mathbf{A}_B^{(k+1)}$ is defined by (8.2.14), and since only one column is changed, it is possible to write

$$(\mathbf{L}_r\mathbf{P}_r)(\mathbf{L}_{r-1}\mathbf{P}_{r-1})\cdots(\mathbf{L}_1\mathbf{P}_1)\mathbf{L}^{-1}\mathbf{A}_B^{(k+1)}\mathbf{Q}^{(k)} = \mathbf{S} \tag{8.5.9}$$

where **S** differs in only one column (t say) from $\mathbf{U}^{(k)}$ and is therefore the *spike matrix* illustrated in Figure 8.5.1. This is no longer upper triangular and is not readily returned to triangular form. However if the columns t to n of **S** are cyclically permuted in the order $t \leftarrow t+1 \leftarrow \cdots \leftarrow n \leftarrow t$ then an upper Hessenberg matrix **H** is obtained (see Figure 8.5.1). If this permutation is incorporated into $\mathbf{Q}^{(k)}$ giving $\mathbf{Q}^{(k+1)}$ then (8.5.9) can be rewritten as

$$(\mathbf{L}_r\mathbf{P}_r)(\mathbf{L}_{r-1}\mathbf{P}_{r-1})\cdots(\mathbf{L}_1\mathbf{P}_1)\mathbf{L}^{-1}\mathbf{A}_B^{(k+1)}\mathbf{Q}^{(k+1)} = \mathbf{H}. \tag{8.5.10}$$

It is now possible to reduce **H** to upper triangular form $\mathbf{U}^{(k+1)}$ in a stable manner by a sequence of elementary operations. Firstly a row interchange may be required to ensure that $|H_{tt}| > |H_{t+1,t}|$ and this operation is represented by $\mathbf{P}_{r+1}$. Then the off-diagonal element in column t is eliminated by premultiplying by

$$\mathbf{L}_{r+1} = \begin{bmatrix} 1 & & & & & & \\ & \ddots & & & & & \\ & & 1 & & & & \\ & & & 1 & & & \\ & & & -\ell_{t+1,\,t} & 1 & & \\ & & & & & \ddots & \\ & & & & & & 1 \end{bmatrix}$$

where $\ell_{t+1,\,t} = H_{t+1,\,t}/H_{tt}$ (or its inverse) and is bounded by 1 in modulus. By repeating these operations for off-diagonals $t+2, \ldots, n$, a reduction of the form

$$(\mathbf{L}_{r(k+1)}\mathbf{P}_{r(k+1)}) \cdots (\mathbf{L}_{r+1}\mathbf{P}_{r+1})\mathbf{H} = \mathbf{U}^{(k+1)}$$

is obtained. Incorporating this with (8.5.10) shows that $\mathbf{A}_B^{(k+1)}$ has become expressed in the form (8.5.8).

The Bartels–Golub algorithm was originally proposed as a stable form of the simplex method, but because of its close relation to an **LU** factorization, it is well suited to an algorithm which attempts to maintain sparse factors in compact form. Reid (1975) describes such an algorithm which builds up a file of the row operations which premultiply $\mathbf{A}_B^{(k)}$ in (8.5.8). The matrix $\mathbf{U}^{(k)}$ is also stored in compact form in random access storage. Reid also describes a further modification in which additional row and column permutations are made to shorten the length of the spike before reducing to Hessenberg form.

Another idea for stable LP is to update **QR** factors of $\mathbf{A}_B$ (or $\mathbf{A}_B^{\mathrm{T}}$), where **Q** is orthogonal and **R** is upper triangular (Gill and Murray, 1973). Since $\mathbf{A}_B$ is part of the problem data and is always available, it is also attractive to exploit the possibility of not storing **Q**, by using the representation

$$\mathbf{A}_B^{-1} = \mathbf{R}^{-1}\mathbf{Q}^{\mathrm{T}} = \mathbf{R}^{-1}\mathbf{R}^{-\mathrm{T}}\mathbf{A}_B^{\mathrm{T}}.$$

Operations with $\mathbf{R}^{-1}$ or $\mathbf{R}^{-\mathrm{T}}$ are carried out by backward or forward substitution. Also, since $\mathbf{A}_B^{\mathrm{T}}\mathbf{A}_B = \mathbf{R}^{\mathrm{T}}\mathbf{R}$, when $\mathbf{A}_B$ is updated by the rank one change (8.2.14), a rank two change in $\mathbf{A}_B^{\mathrm{T}}\mathbf{A}_B$ is made and hence **R** can be updated efficiently by using the methods described by Fletcher and Powell (1975) or Gill and Murray (1978a). A form of this algorithm suitable for large-scale LP in which the factor is updated

$$\begin{bmatrix} \times & \times & \times & \times & \times & \times & \times & \times \\ & \times & \times & \times & \times & \times & \times & \times \\ & & \times & \times & \times & \times & \times & \times \\ & & \times & \times & \times & \times & \times & \times \\ & & \times & & \times & \times & \times & \times \\ & & \times & & & \times & \times & \times \\ & & \times & & & & \times & \times \\ & & \times & & & & & \times \end{bmatrix} \qquad \begin{bmatrix} \times & \times & \times & \times & \times & \times & \times & \times \\ & \times & \times & \times & \times & \times & \times & \times \\ & & \times & \times & \times & \times & \times & \times \\ & & \times & \times & \times & \times & \times & \times \\ & & & \times & \times & \times & \times & \times \\ & & & & \times & \times & \times & \times \\ & & & & & \times & \times & \times \\ & & & & & & \times & \times \end{bmatrix}$$

Figure 8.5.1 Spike and Hessenberg matrices in the Bartels–Golub method

in product form is given by Saunders (1972). Many other algorithms for stable or sparse LP have been suggested which often incorporate similar ideas of using triangular or orthogonal factorizations together with permutations to promote a favourable matrix structure. A number of recent developments are described by Gill and Murray (1974b). Comparisons between the algorithms are difficult to make, in particular because they depend critically on the size and structure of the LP problem itself, but some results are given by Reid (1975).

A different approach to sparsity is to attempt to find a partial *decomposition* of the problem into smaller subproblems, each of which can be solved repeatedly under the control of some master problem. The decomposition method of Dantzig and Wolfe (1960) illustrates the principles well, and is described here although some other possibilities are described by Gill and Murray (1974b). The idea is of limited applicability since it depends on the main problem having a favourable structure, and even then it is not clear that the resulting method is better than using a general sparse matrix LP package. A well-known situation which might be expected to favour decomposition is when the LP problem models a system composed of a number of quasi-independent subsystems. Each subsystem has its own equations and variables, but there are a few additional equations which interrelate all the variables and may represent, for example, the distribution of shared resources throughout the system. Let each subsystem (for $i = 1, 2, \ldots, r$) be expressed in terms of n_i variables $\mathbf{x}_i$ and be subject to the conditions

$$\mathbf{C}_i \mathbf{x}_i = \mathbf{b}_i, \qquad \mathbf{x}_i \geqslant \mathbf{0}, \tag{8.5.11}$$

where $\mathbf{C}_i$ and $\mathbf{b}_i$ have m_i rows. Also let there be m_0 additional conditions

$$\sum_{i=1}^{r} \mathbf{B}_i \mathbf{x}_i = \mathbf{b}_0 \tag{8.5.12}$$

so that $\mathbf{b}_0$ and the matrices $\mathbf{B}_i$ have m_0 rows. Then the overall problem can be expressed in terms of an LP problem of the form (8.1.1) with

$$n = \sum_{i=1}^{r} n_i, \qquad m = \sum_{i=0}^{r} m_i,$$

$$\mathbf{x}^{\mathrm{T}} = (\mathbf{x}_1^{\mathrm{T}}, \mathbf{x}_2^{\mathrm{T}}, \ldots, \mathbf{x}_r^{\mathrm{T}}), \qquad \mathbf{b}^{\mathrm{T}} = (\mathbf{b}_0^{\mathrm{T}}, \mathbf{b}_1^{\mathrm{T}}, \ldots, \mathbf{b}_r^{\mathrm{T}}),$$

and the coefficient matrix has the structure

$$\mathbf{A} = \begin{bmatrix} \mathbf{B}_1 & \mathbf{B}_2 & \mathbf{B}_3 & \cdots & \mathbf{B}_r \\ \mathbf{C}_1 & & & & \\ & \mathbf{C}_2 & & & \\ & & \mathbf{C}_3 & & \\ & & & \ddots & \\ & & & & \mathbf{C}_r \end{bmatrix} . \tag{8.5.13}$$

The Dantzig–Wolfe decomposition method is applicable to this problem and uses the following idea. Assume for each i that (8.5.11) has p_i extreme points which are columns of the matrix $\mathbf{E}_i$, and that any feasible $\mathbf{x}_i$ can be expanded as a convex combination

$$\mathbf{x}_i = \mathbf{E}_i\boldsymbol{\theta}_i, \qquad \mathbf{e}^{\mathrm{T}}\boldsymbol{\theta}_i = 1, \qquad \boldsymbol{\theta}_i \geqslant \mathbf{0} \tag{8.5.14}$$

of these extreme points (see Section 9.4), where $\mathbf{e}$ denotes a vector of ones. This assumption is valid if the feasible region of (8.5.11) is bounded for each i, although in fact the assumption of boundedness is unnecessary. Then the main problem can be transformed to the equivalent LP problem

$$\begin{aligned} &\underset{\boldsymbol{\theta}}{\text{minimize}}\ \mathbf{f}^{\mathrm{T}}\boldsymbol{\theta} \\ &\text{subject to}\ \mathbf{G}\boldsymbol{\theta} = \mathbf{h}, \qquad \boldsymbol{\theta} \geqslant \mathbf{0} \end{aligned} \tag{8.5.15}$$

where

$$\boldsymbol{\theta} = \begin{pmatrix} \boldsymbol{\theta}_1 \\ \boldsymbol{\theta}_2 \\ \vdots \\ \boldsymbol{\theta}_r \end{pmatrix}, \qquad \mathbf{f} = \begin{pmatrix} \mathbf{f}_1 \\ \mathbf{f}_2 \\ \vdots \\ \mathbf{f}_r \end{pmatrix}, \qquad \mathbf{h} = \begin{pmatrix} \mathbf{b}_0 \\ 1 \\ 1 \\ \vdots \\ 1 \end{pmatrix},$$

$$\mathbf{G} = \begin{bmatrix} \mathbf{G}_1 & \mathbf{G}_2 & \mathbf{G}_3 & \cdots & \mathbf{G}_r \\ \mathbf{e}^{\mathrm{T}} & & & & \\ & \mathbf{e}^{\mathrm{T}} & & & \\ & & \mathbf{e}^{\mathrm{T}} & & \\ & & & \ddots & \\ & & & & \mathbf{e}^{\mathrm{T}} \end{bmatrix}$$

and where $\mathbf{G}_i = \mathbf{B}_i\mathbf{E}_i$ and $\mathbf{f}_i^{\mathrm{T}} = \mathbf{c}_i^{\mathrm{T}}\mathbf{E}_i$. This problem has many fewer equations $(m_0 + r)$ but very many more variables (Σp_i). However it turns out that it is possible to solve (8.5.15) by the *revised* simplex method without explicitly evaluating all the matrices $\mathbf{E}_i$.

Consider therefore an iteration of the revised simplex method for (8.5.15). There is no difficulty in having available the basis matrix $\mathbf{G}_B$, which has $m_0 + r$ columns (one from each subsystem). Each column is calculated from one extreme point. As usual this matrix determines the values $\hat{\mathbf{h}} = \mathbf{G}_B^{-1}\mathbf{h}$ of the basic variables $\boldsymbol{\theta}_B$, and $\hat{\mathbf{h}} \geqslant \mathbf{0}$. However the reduced costs of the nonbasic variables defined by

$$\hat{\mathbf{f}}_N = \mathbf{f}_N - \mathbf{G}_N^{\mathrm{T}}\boldsymbol{\pi}$$

where $\boldsymbol{\pi} = \mathbf{G}_B^{-\mathrm{T}}\mathbf{f}_B$ are not available since this would require all the remaining extreme points to be calculated. Nonetheless the smallest reduced cost can be calculated indirectly. Let $j \in N$ index a variable θ_j in the subvector $\boldsymbol{\theta}_i$, and let $\boldsymbol{\pi}$ be partitioned into $\boldsymbol{\pi}^{\mathrm{T}} = (\mathbf{u}^{\mathrm{T}}, \mathbf{v}^{\mathrm{T}})$ where $\mathbf{v}$ has r elements. Then by definition of $\mathbf{f}_i$ and $\mathbf{G}_i$, the corresponding reduced cost $\hat{f}_j$ can be written

$$\hat{f}_j = f_j - \mathbf{g}_j^{\mathrm{T}}\mathbf{u} - v_i = (\mathbf{c}_i^{\mathrm{T}} - \mathbf{u}^{\mathrm{T}}\mathbf{B}_i)\boldsymbol{\xi}_j - v_i \tag{8.5.16}$$

where $\mathbf{g}_j$ is the corresponding column of $\mathbf{G}_i$ and $\boldsymbol{\xi}_j$ is the corresponding extreme point of (8.5.11). Consider finding the smallest $\hat{f}_j$ for all indices j in the subset i; this is equivalent to finding the extreme point of (8.5.11) which minimizes (8.5.16). Since the solution of an LP problem occurs at an extreme point, this is equivalent to solving the LP

$$\begin{aligned} &\underset{\mathbf{x}_i}{\text{minimize}} \quad (\mathbf{c}_i - \mathbf{B}_i^{\mathrm{T}}\mathbf{u})^{\mathrm{T}}\mathbf{x}_i - v_i \\ &\text{subject to } \mathbf{C}_i\mathbf{x}_i = \mathbf{b}_i, \qquad \mathbf{x}_i \geqslant \mathbf{0}. \end{aligned} \tag{8.5.17}$$

Thus the smallest reduced cost for *all* j is found by solving (8.5.17) for all $i = 1, 2, \ldots, r$ so as to give the extreme point ($\boldsymbol{\xi}_q$ say) which for all i gives the smallest cost function value in (8.5.17). Thus the smallest reduced cost $\hat{f}_q$ is determined and the simplex method can proceed by increasing the variable θ_q.

If the method does not terminate, the next step is to find the basic variable θ_p which first becomes zero. Column q of $\mathbf{G}$ is readily computed using $\mathbf{g}_q = \mathbf{B}_i^{\mathrm{T}}\boldsymbol{\xi}_q$, where i now refers to the subsystem which includes the variable q. Thus $\mathbf{d}$ in (8.2.10) can be calculated so that the solution of (8.2.11) determines θ_p in the usual way. The indices p and q are interchanged, and the same column of $\mathbf{G}$ enables the matrix $\mathbf{G}_B$ to be updated using (8.2.15), which completes the description of the algorithm.

One possible situation which has not been considered is that the solution to (8.5.17) may be unbounded. This can be handled with a small addition to the algorithm as described by Hadley (1962). Another feature of note is that the sub-problem constraints (8.5.11) can be any convex polyhedron and the same development applies. A numerical example of the method in which this feature occurs is given by Hadley (1962).

As a final remark on large-scale LP, it is observed that the idea which occurs in the Dantzig–Wolfe method of solving an LP without explicitly generating all the columns in the coefficient matrix, is in fact of much wider applicability. It is possible to model some applications by LP problems in which the number of variables is extremely large, and to utilize some other feature to find which reduced cost is least. An interesting example of this is the ship scheduling problem; see Appelgren (1971), for example.

8.6 Degeneracy

It is observed in Section 8.2 that the simplex method converges when each iteration is known to reduce the objective function. Failure to terminate can arise when an iteration occurs on which $\hat{b}_p = 0$ and $d_p < 0$. Then no change to the variable x_q is permitted, and hence no decrease in f. Although a new partitioning B and N can be made, it is possible for the algorithm to return to a previous set B and N and hence to *cycle* infinitely without locating the solution. The situation in which $\hat{b}_i = 0$ occurs is known as *degeneracy*. The same can happen in the active set method when

$\mathbf{a}_i^T\mathbf{x}^{(k)} = b_i$ for some indices $i \notin \mathscr{A}$. If $\mathbf{a}_i^T\mathbf{s}_q < 0$ then these constraints allow no progress along the search direction $\mathbf{s}_q$. Although changes in the active set $\mathscr{A}$ can be made, the possibility of cycling again arises. This is to be expected since the algorithms are equivalent as described in Question 8.17, and so the examples of degeneracy, etc., described below apply equally to both the simplex and active set methods. Degeneracy can arise in other active set methods (Sections 10.3 and 11.2) and the remarks in this section also apply there.

An example in which cycling occurs has been given by Beale (see Hadley, 1962) and is described in Question 8.25. For cycling to occur, a number of elements $\hat{b}_i = 0$ must occur, and the way in which ties are broken in (8.2.11) is crucial. Beale's example uses the plausible rule that p is chosen as the first such p which occurs in the tableau. (This is not the same as the first in numerical order but is a well-defined rule.) In practice, however, degeneracy and cycling are rarely observed and it has become fashionable to assert that the problem can be neglected, or can be solved by making small perturbations to the data. On the other hand, I doubt that many LP codes monitor whether degeneracy or near degeneracy arises, so it may be that the problem arises more frequently and goes undetected. Graves and Brown (1979) indicate that on some large sparse LP systems, between 50 and 90 per cent of iterations are degenerate. I have also come across degeneracy on two or three occasions in quadratic programming problems.

It is possible to state a rule (Charnes, 1952) for resolving ties in (8.2.11) which does not cause cycling, and hence which in theory resolves the degeneracy problem. (In practice this may not be the case because of round-off error difficulties, a subject which is considered further at the end of this section.) Consider perturbing the right hand side vector $\mathbf{b}$ to give $\mathbf{b}(\epsilon)$ where

$$\mathbf{b}(\epsilon) = \mathbf{b} + \sum_{j=1}^{n} \epsilon^j \mathbf{a}_j. \tag{8.6.1}$$

Multiplying through by $\mathbf{A}_B^{-1}$ and denoting $\hat{\mathbf{A}} = \mathbf{A}_B^{-1}\mathbf{A}$ gives

$$\hat{\mathbf{b}}(\epsilon) = \hat{\mathbf{b}} + \sum_{j=1}^{n} \epsilon^j \hat{\mathbf{a}}_j \tag{8.6.2}$$

where $\hat{\mathbf{a}}_j$ is the jth column of $\hat{\mathbf{A}}$. This perturbation breaks up the degenerate situation, giving instead a closely grouped set of non-degenerate extreme points. The tie break rule (which does not require ϵ to be known) comes about by finding how the simplex method solves this perturbed problem for sufficiently small ϵ. Each element in (8.6.1) or (8.6.2) is a polynomial of degree n in ϵ. The following notation is useful. If

$$u(\epsilon) = u_0 + u_1\epsilon + \cdots + u_n\epsilon^n$$

is any such polynomial, $v(\epsilon)$ similarly, then $u(\epsilon) > v(\epsilon)$ for all sufficiently small $\epsilon > 0$ if and only if

$$
\begin{array}{llll}
 & u_0 > v_0 & & \\
\text{or} & u_0 = v_0 & \text{and} \quad u_1 > v_1 & \\
\text{or} & u_0 = v_0, & u_1 = v_1 \quad \text{and} & u_2 > v_2, \\
\vdots & & & \\
\text{or} & u_0 = v_0, & u_1 = v_1 \cdots u_{n-1} = v_{n-1} \quad \text{and} & u_n > v_n,
\end{array}
\tag{8.6.3}
$$

and $u(\epsilon) = v(\epsilon)$ iff $u_i = v_i \; \forall \, i = 0, 1, \ldots, n$. Dantzig, Orden, and Wolfe (1955) let (8.6.3) define a *lexicographic ordering* $\mathbf{u} \succ \mathbf{v}$ of the vectors $\mathbf{u}$ and $\mathbf{v}$ whose elements are the coefficients of the polynomial and show that the perturbation method can be interpreted in this way. I shall use the notation $u(\epsilon) \succ v(\epsilon)$ to mean that $u(\epsilon) > v(\epsilon)$ for all sufficiently small $\epsilon > 0$.

Assume for the moment that the initial vertex satisfies $\hat{\mathbf{b}}^{(1)}(\epsilon) \succ \mathbf{0}$ (that is $\hat{b}_i^{(1)}(\epsilon) \succ 0 \; \forall \, i \in B^{(1)}$). The rule for updating the tableau is

$$
\begin{aligned}
\hat{b}_i^{(k+1)} &= \hat{b}_i^{(k)} - \frac{\hat{a}_{iq}}{\hat{a}_{pq}} \hat{b}_p^{(k)}, \qquad i \neq p \\
\hat{b}_p^{(k+1)} &= \frac{1}{\hat{a}_{pq}} \hat{b}_p^{(k)}.
\end{aligned}
\tag{8.6.4}
$$

Also $\hat{\mathbf{a}}_q = -\mathbf{d}$ by definition, so $\hat{a}_{pq} = -d_p > 0$ from (8.2.11). Thus if $\hat{\mathbf{b}}^{(k)}(\epsilon) \succ \mathbf{0}$, then both $\hat{b}_p^{(k+1)}(\epsilon) \succ 0$ and $\hat{b}_i^{(k+1)}(\epsilon) \succ 0$ for all $i : d_i \geqslant 0, \, i \neq p$. Now let (8.2.11) be solved in the lexicographic sense

$$
\frac{\hat{b}_p(\epsilon)}{-d_p} = \min_{\substack{i \in B \\ d_i < 0}} \frac{\hat{b}_i(\epsilon)}{-d_i}.
\tag{8.6.5}
$$

It follows using $\hat{a}_{iq} = -d_i$ that

$$
\hat{b}_i(\epsilon) \succ \frac{\hat{a}_{iq}}{\hat{a}_{pq}} \hat{b}_p^{(k)}(\epsilon).
$$

(Equality is excluded because this would imply two equal rows in the tableau, which contradicts the independence assumption on $\mathbf{A}$.) Hence from (8.6.4), $\hat{b}_i^{(k+1)}(\epsilon) \succ 0$ for $i : d_i < 0$. Thus $\hat{\mathbf{b}}^{(k)}(\epsilon) \succ \mathbf{0}$ implies $\hat{\mathbf{b}}^{(k+1)}(\epsilon) \succ \mathbf{0}$, and so inductively the result holds for all k. Thus the points obtained when the simplex method is applied to the perturbed problem for sufficiently small ϵ are both feasible and non-degenerate. It therefore follows that $f^{(k)}(\epsilon) \succ f^{(k+1)}(\epsilon)$, and so the algorithm must terminate at the solution. There is no difficulty in ensuring that $\hat{\mathbf{b}}^{(1)}(\epsilon) \succ \mathbf{0}$. If the initial b.f.s. is non-degenerate then this condition is implied by $\hat{\mathbf{b}}^{(1)} > \mathbf{0}$. If not, then there always exists a unit matrix in the tableau, and the columns of the tableau should be rearranged so that these columns occur first, which again implies that $\hat{\mathbf{b}}^{(1)}(\epsilon) \succ \mathbf{0}$.

The required tie break rule in (8.2.11) is therefore the one obtained by solving (8.6.5) in the lexicographic sense of (8.6.3). Thus in finding $\min \hat{b}_i(\epsilon)/-d_i$ $\forall \, i \in B, d_i < 0$, the zero order terms $\hat{b}_i/-d_i$ are compared first. If there are any ties then the first order terms $\hat{a}_{i1}/-d_i$ are used to break the ties. If any ties still remain, then the second order terms $\hat{a}_{i2}/-d_i$ are used, and so on. The rule clearly depends on the column ordering and any initial ordering for which $\hat{\mathbf{b}}^{(1)}(\epsilon) \succ \mathbf{0}$ may be used (see Question 8.26). However once chosen this ordering must remain fixed, and the rearrangement of the tableau into B and N columns in (8.2.13) must be avoided (or accounted for). The method assumes that the columns stay in a fixed order and that the columns of $\mathbf{I}$ may occur anywhere in the tableau. A similar method (Wolfe, 1963b) is to introduce the virtual perturbations only as they are required, and is particularly suitable for practical use.

An equivalent rule for breaking ties in the active set method can be determined by virtue of Question 8.17. The equivalent tableau is given by

$$[\hat{\mathbf{b}} : \hat{\mathbf{A}}] = \left[\begin{array}{c|ccc} \mathbf{x}^{(k)} & \mathbf{I} & & -\mathbf{A}_1^{-1} \\ \mathbf{z}_2^{(k)} & & \mathbf{I} & -\mathbf{A}_2^{\mathrm{T}}\mathbf{A}_1^{-1} \end{array}\right] \qquad (8.6.6)$$
$$\qquad\qquad\qquad \underbrace{}_{x} \quad \underbrace{}_{z_2} \quad \underbrace{\phantom{-A_2^TA_1^{-1}}}_{z_1}$$

and if degeneracy occurs in some inactive slacks z_i then ties are resolved by examining the corresponding ith rows of $[\mathbf{I} : -\mathbf{A}_2^{\mathrm{T}}\mathbf{A}_1^{-1}]$ in accordance with some fixed ordering of the slack variables.

An alternative way of resolving degeneracy exists which avoids iterating the full LP a number of times as in the perturbation method. Instead a smaller non-degenerate LP is solved to determine a search direction $\mathbf{s}$ in the nonbasic variables which moves away from the degenerate vertex. Let $\hat{\mathbf{A}}_D$ denote the partition of the matrix $\hat{\mathbf{A}}_N$ whose rows correspond to degenerate basic variables. Consider changing each x_i, $i \in N$, by an amount αs_i, so that $\mathbf{s}$ is the required search direction. Since $\mathbf{s}$ is required to reduce the objective function, it is normalized to have negative slope by $\mathbf{c}_N^{\mathrm{T}}\mathbf{s} = -1$. Feasibility of $\mathbf{x}_N$ requires that $\mathbf{s} \geqslant \mathbf{0}$, and feasibility of the degenerate variables requires that $-\hat{\mathbf{A}}_D\mathbf{s} \geqslant \mathbf{0}$. Any vector $\mathbf{s}$ which is feasible in the above conditions can be chosen, so the freedom is taken up by minimizing $\|\mathbf{s}\|_1$. Thus $\mathbf{s}$ solves the problem

$$\begin{aligned} &\underset{\mathbf{s}}{\text{minimize}} \quad \mathbf{e}^{\mathrm{T}}\mathbf{s} \\ &\text{subject to } -\hat{\mathbf{c}}_N^{\mathrm{T}}\mathbf{s} = 1 \\ &\qquad\qquad -\hat{\mathbf{A}}_D\mathbf{s} \geqslant \mathbf{0}, \qquad \mathbf{s} \geqslant \mathbf{0}, \end{aligned} \qquad (8.6.7)$$

where $\mathbf{e}$ is a vector of ones. If no solution exists then the degenerate vertex is optimal. This problem is also degenerate (at $\mathbf{s} = \mathbf{0}$), but its dual problem

$$\begin{aligned} &\underset{\lambda_0, \boldsymbol{\lambda}}{\text{maximize}} \quad \lambda_0 \\ &\text{subject to } -\hat{\mathbf{c}}_N\lambda_0 - \hat{\mathbf{A}}_D^{\mathrm{T}}\boldsymbol{\lambda} \leqslant \mathbf{e}, \qquad \boldsymbol{\lambda} \geqslant \mathbf{0} \end{aligned} \qquad (8.6.8)$$

(see Section 9.5) is generally not. This problem has only $d + 1$ variables where d is the number of degenerate variables, and can usually be solved readily by the active set method (see Question 8.27).

A similar approach can be taken to resolving degeneracy in the active set method.

In this case a feasible direction s requires that $\mathbf{a}_i^T \mathbf{s} \geqslant \mathbf{0}$ for all $i \in \mathscr{A}^{(k)} = \{i : \mathbf{a}_i^T \mathbf{x}^{(k)} = b_i\}$. A downhill direction is obtained by normalizing $\mathbf{c}^T \mathbf{s} = -1$ ($\mathbf{c} = \nabla f$; see (8.3.1)) and the remaining freedom is taken up by minimizing $\| \mathbf{s} \|_1$, giving the problem

$$\begin{aligned} & \underset{\mathbf{s}}{\text{minimize}} \quad \| \mathbf{s} \|_1 \\ & \text{subject to } \mathbf{c}^T \mathbf{s} = -1 \\ & \qquad \mathbf{a}_i^T \mathbf{s} \geqslant 0, \qquad i \in \mathscr{A}^{(k)}. \end{aligned} \tag{8.6.9}$$

Again if no solution exists then $\mathbf{x}^{(k)}$ is optimal. Also (8.6.9) is degenerate, but its dual problem

$$\begin{aligned} & \underset{\mathbf{v}, \lambda_0, \boldsymbol{\lambda}}{\text{minimize}} \quad \lambda_0 \\ & \text{subject to } \mathbf{v} = \mathbf{c}\lambda_0 + \mathbf{A}\boldsymbol{\lambda} \\ & \qquad \| \mathbf{v} \|_\infty \leqslant 1, \qquad \boldsymbol{\lambda} \geqslant \mathbf{0} \end{aligned} \tag{8.6.10}$$

(Watson, 1978) is generally not, and can be solved. It may be that solving problems like (8.6.8) or (8.6.10) offers a more rapid way of resolving a degenerate vertex and so this type of possibility should be kept in mind as an alternative to the perturbation method. However it is not easy to incorporate these subproblems into an LP package, especially if degeneracy in these subproblems is allowed for (introducing a recursive aspect), which presumably can occur in theory.

In practice, however, cycling is not the only difficulty caused by degeneracy, and may not even be the most important. I have also observed difficulties due to round-off errors which can cause an algorithm to break down. The situation is most easily explained in terms of the active set method and is illustrated in Figure 8.6.1. There are three constraints in $\mathbb{R}^3$ which have a common line and are therefore degenerate. Initially constraint 1 is in $\mathscr{A}$, and the search along $\mathbf{s}^{(1)}$ approaches the degenerate vertex $\mathbf{x}^{(2)}$. The tie concerning which constraint to include in $\mathscr{A}$ is resolved in favour of constraint 2. The next search is along $\mathbf{s}^{(2)}$ with constraints 1

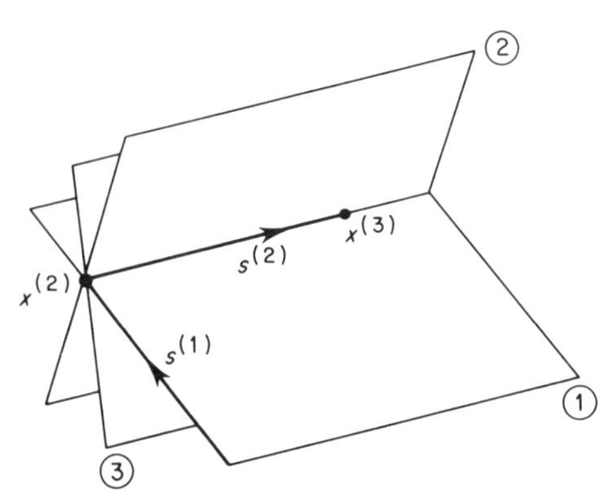

Figure 8.6.1 A practical difficulty caused by degeneracy

and 2 active; all appears normal and cycling does not occur at $\mathbf{x}^{(2)}$. What can happen, however, is that due to round-off error, the slack in constraint 3 is calculated as $\sim\epsilon$ (in place of zero) and the component of $-\mathbf{a}_3$ along $\mathbf{s}^{(2)}$ is also $\sim\epsilon$ in place of zero. Thus constraint 3 can become active at a spurious vertex $\mathbf{x}^{(3)}$ with a value of $\alpha \sim 1$ in (8.3.6). The algorithm thus breaks down because $\mathbf{x}^{(3)}$ is not a true vertex, and the matrix of active constraints $\mathbf{A}$ in (8.3.3) is singular. An equivalent situation arises in the simplex method when true values of $\hat{b}_i$ and $-d_i$ are zero, but are perturbed to $\sim\epsilon$ by round-off error. Then the solution of (8.2.11) can give a spurious b.f.s. in which the matrix $\mathbf{A}_B$ is singular. I have observed this type of behaviour on two or three occasions in practice when applying a quadratic programming package, and the only remedy has been to remove degenerate constraints from the problem. It may not be easy for the user to do this (or even to recognize that it is necessary) although it is certainly important for him or her to be aware of the possibility of failure on this account. Clearly further research is required to determine a suitable modification to linear constraint algorithms which is simple and yet effective in controlling all the ill effects of a degenerate problem.

Questions for Chapter 8

1. Consider the linear program

$$\begin{aligned} &\text{minimize} \quad -4x_1 - 2x_2 - 2x_3 \\ &\text{subject to} \quad 3x_1 + x_2 + x_3 = 12 \\ &\qquad\qquad\quad x_1 - x_2 + x_3 = -8, \qquad \mathbf{x} \geqslant \mathbf{0}. \end{aligned}$$

Verify that the choice $B \equiv \{1, 2\}$, $N \equiv \{3\}$ gives a basic feasible solution, and hence solve the problem.

2. Consider the LP problem of Question 1 and replace the objective function by the function $-4x_1 - 2x_3$. Show that the basic feasible solutions derived from $B = \{1, 2\}, N = \{3\}$ and $B = \{2, 3\}, N = \{1\}$ are both optimal, and that any convex combination of these solutions is also a solution to the problem.

   In general describe how the fact that an LP problem has a non-unique solution can be recognized, and explain why.

3. Sketch the set of feasible solutions of the following inequalities:

$$\begin{aligned} x_1 \geqslant 0, \qquad & x_2 \geqslant 0 \\ x_1 + 2x_2 & \leqslant 4 \\ -x_1 + x_2 & \leqslant 1 \\ x_1 + x_2 & \leqslant 3. \end{aligned}$$

At which points of this set does the function $x_1 - 2x_2$ take (a) its maximum and (b) its minimum value?

4. Consider the problem

$$\begin{aligned} &\text{maximize} \quad x_2 \\ &\text{subject to} \quad 2x_1 + 3x_2 \leqslant 9 \\ &\qquad\qquad\quad |x_1 - 2| \leqslant 1, \qquad \mathbf{x} \geqslant \mathbf{0}. \end{aligned}$$

(i) Solve the problem graphically.
(ii) Formulate the problem as a standard LP problem.

5. Use the tableau form of the simplex method to solve the LP problem

$$\begin{aligned} \text{minimize} \quad & 5x_1 - 8x_2 - 3x_3 \\ \text{subject to} \quad & 2x_1 + 5x_2 - x_3 \leqslant 1 \\ & -3x_1 - 8x_2 + 2x_3 \leqslant 4 \\ & -2x_1 - 12x_2 + 3x_3 \leqslant 9, \qquad \mathbf{x} \geqslant \mathbf{0} \end{aligned}$$

after reducing the problem to standard form.

6. A manufacturer uses resources of material (m) and labour (ℓ) to make up to four possible items (a to d). The requirements for these, and the resulting profits are given by

item a requires $4m + 2\ell$ resources and yields £5/item profit,
item b requires $m + 5\ell$ resources and yields £8/item profit,
item c requires $2m + \ell$ resources and yields £3/item profit,
item d requires $2m + 3\ell$ resources and yields £4/item profit.

There are available up to 30 units of material and 50 units of labour per day. Assuming that these resources are fully used and neglecting integrality constraints, show that the manufacturing schedule for maximum profit is an LP problem in standard form. Show that the policy of manufacturing only the two highest profit items yields a b.f.s. which is not optimal. Find the optimal schedule. Evaluate also the schedule in which equal amounts of each item are manufactured and show there is an under-use of one resource. Compare the profit in each of these cases.

7. Use the simplex method to show that for all μ in the range $-3 < \mu < -\frac{1}{2}$, the solution of the problem

$$\begin{aligned} \text{minimize} \quad & \mu x_1 - x_2 \\ \text{subject to} \quad & x_1 + 2x_2 \leqslant 4 \\ & 6x_1 + 2x_2 \leqslant 9, \qquad \mathbf{x} \geqslant \mathbf{0} \end{aligned}$$

occurs at the same point $\mathbf{x}^*$ and find x_1^* and x_2^*.

8. Consider modifying the revised simplex method to solve a more general standard form of LP,

$$\begin{aligned} \text{minimize} \quad & \mathbf{c}^T\mathbf{x} \\ \text{subject to} \quad & \mathbf{Ax} = \mathbf{b}, \qquad \boldsymbol{\ell} \leqslant \mathbf{x} \leqslant \mathbf{u}. \end{aligned}$$

Show that the following changes to the simplex method are required. Nonbasic variables take the value $x_i^{(k)} = \ell_i$ or u_i, and $\hat{\mathbf{b}} = \mathbf{A}_B^{-1}(\mathbf{b} - \mathbf{A}_N \mathbf{x}_N^{(k)})$ is used to calculate $\hat{\mathbf{b}}$. The optimality test (8.2.8) is changed to require that the set

$$\{i : i \in N, \hat{c}_i < 0 \text{ if } x_i = \ell_i, \hat{c}_i > 0 \text{ if } x_i = u_i\}$$

is empty; if not $\hat{c}_q$ is chosen to min $|\hat{c}_i|$ for i in this set. The choice of p in (8.2.11) is determined *either* when a variable in B reaches its upper or lower bound *or* when the variable x_q reaches its other bound. In the latter case no changes to B or N need to be made. Write down the resulting formula for the choice of p.

9. Using the method of Question 8, verify that $B \equiv \{1, 2\}, N \equiv \{3, 4\}, \mathbf{x}_N = \mathbf{0}$ gives a b.f.s. for the LP problem

$$\begin{array}{ll} \text{minimize} & x_1 \qquad\quad - 3x_3 \\ \text{subject to} & x_1 + x_2 - \;\; x_3 + 2x_4 = 6 \\ & 2x_1 \qquad\;\; - \;\; x_3 - 2x_4 = 2 \end{array}$$
$$0 \leqslant x_1 \leqslant 4, \; 0 \leqslant x_2 \leqslant 10, \; 0 \leqslant x_3 \leqslant 4, \; 0 \leqslant x_4 \leqslant 2$$

and hence solve the problem.

10. By drawing a diagram of the feasible region, solve the problems

$$\begin{array}{ll} \text{minimize} & \qquad x_2 \\ \text{subject to} & x_1 + x_2 = 1 \\ & x_1 - x_2 \leqslant 2 \\ & x_1 \qquad \geqslant 0 \end{array}$$

(a) when $x_2 \geqslant 0$ also applies and (b) when x_2 is unrestricted. Verify that the solutions are correct by posing each problem in standard form, determining the basic variables from the diagram, and then checking feasibility and optimality of the b.f.s. In case (b) introduce variables x_2^+ and x_2^- as described in Section 7.2.

11. A linear data fitting problem can be stated as finding the best solution for $\mathbf{x}$ of the system of equations $\mathbf{Ax} = \mathbf{b}$ in which $\mathbf{A}$ has more rows than columns (overdetermined). In some circumstances it can be valuable to find that solution which minimizes $\Sigma_i \mid r_i \mid$ where $\mathbf{r} = \mathbf{Ax} - \mathbf{b}$, rather than the more usual Σr_i^2 (least squares). Using the idea of replacing r_i by two variables r_i^+ and r_i^-, one of which will always be zero, show that the problem can be stated as an LP (not in standard form).

12. Consider the problem

$$\begin{array}{ll} \text{minimize} & |x| + |y| + |z| \\ \text{subject to} & x - 2y = 3 \\ & -y + \;\; z \leqslant 1 \\ & x, y, z \text{ unrestricted in sign.} \end{array}$$

Define suitable non-negative variables x^+, x^-, y^+, y^-, etc., and write down an LP problem in standard form. Verify that the variables x^+ and z^+ provide an initial basic feasible solution, and hence solve the LP problem. What are the optimum function value and variables in the original problem? Is the solution unique?

13. Consider the problem in Question 12. Without adding extra variables other than a slack variable, show that the problem can be solved (as in (8.4.5)) by allowing the coefficients in the cost function to change between ± 1 as the iteration proceeds.

14. Given an optimal b.f.s. to a problem in standard form, consider perturbing the problem so that

    (i) $\mathbf{b}(\lambda) = \mathbf{b} + \lambda\mathbf{r}$
    (ii) $\mathbf{c}_N(\lambda) = \mathbf{c}_N + \lambda\mathbf{s}$
    (iii) $\mathbf{c}_B(\lambda) = \mathbf{c}_B + \lambda\mathbf{t}$

    where in each case λ is increased from zero. At what value of λ, if any (in each case), will the present solution no longer be an optimal b.f.s.? Consider the solution of Question 6 and the effect of (i) a reduction in the availability of labour, (ii) an increase in the profit of item a, and (iii) a decrease in the profit of item c. At what stage (in each case) do these changes cause a change in the type of item to be manufactured in the optimal schedule?
15. Use the active set method to solve the LP problem

$$\begin{aligned} &\text{maximize} \quad x_1 + x_2 \\ &\text{subject to } 3x_1 + x_2 \geqslant 3 \\ &\qquad\qquad x_1 + 4x_2 \geqslant 4, \qquad \mathbf{x} \geqslant \mathbf{0}. \end{aligned}$$

    Illustrate the result by sketching the set of feasible solutions.
16. Solve the LP problem

$$\begin{aligned} &\text{minimize} \quad -x_1 + x_2 \\ &\text{subject to } -x_1 + 2x_2 \leqslant 2 \\ &\qquad\qquad x_1 + x_2 \leqslant 4 \\ &\qquad\qquad x_1 \qquad \leqslant 3, \qquad \mathbf{x} \geqslant \mathbf{0} \end{aligned}$$

    both graphically and by the active set method. Correlate the stages of the active set method with the extreme points of the feasible region.
17. Consider the LP problem (8.3.1) with E empty, and compare its solution by the active set method as against the simplex method. By introducing slack variables $\mathbf{z}$, write the constraints in the simplex method as $[-\mathbf{A}^{\mathrm{T}} : \mathbf{I}]\begin{pmatrix}\mathbf{x}\\ \mathbf{z}\end{pmatrix} = -\mathbf{b}$ and $\mathbf{z} \geqslant \mathbf{0}$. Use a modification of the simplex method in which the variables $\mathbf{x}$ are allowed to be unrestricted (for example as in Question 8 with $u_i = -\ell_i = K$ for sufficiently large K). Choose the nonbasic variables as the slack variables z_i, $i \in \mathscr{A}$, where $\mathscr{A}$ is the active set in the active set method. Partition $\mathbf{A} = [\mathbf{A}_1 : \mathbf{A}_2]$, $\mathbf{b}^{\mathrm{T}} = (\mathbf{b}_1^{\mathrm{T}}, \mathbf{b}_2^{\mathrm{T}})$, and $\mathbf{z}^{\mathrm{T}} = (\mathbf{z}_1^{\mathrm{T}}, \mathbf{z}_2^{\mathrm{T}})$, corresponding to active and inactive constraints at $\mathbf{x}^{(k)}$ in the latter method. The quantities in the active set method are described in Section 8.3. In the simplex method, show that $\hat{\mathbf{b}}^{\mathrm{T}} = (\mathbf{x}^{(k)\mathrm{T}}, \mathbf{z}_2^{(k)\mathrm{T}})$, $\boldsymbol{\pi}^{\mathrm{T}} = (-\mathbf{c}^{\mathrm{T}}\mathbf{A}_1^{-\mathrm{T}}, \mathbf{0}^{\mathrm{T}})$, and $\hat{\mathbf{c}}_N = \mathbf{A}_1^{-1}\mathbf{c} = \boldsymbol{\lambda}$ in (8.3.3) and hence conclude that the same index q is chosen. Hence show that the vector $\mathbf{d}$ is given by $\mathbf{d}^{\mathrm{T}} = (\mathbf{s}_q^{\mathrm{T}}, \mathbf{s}_q^{\mathrm{T}}\mathbf{A}_2)$ (see (8.6.6)). Since there are no restrictions on the $\mathbf{x}$ variables, show that test (8.2.11) becomes

$$\min_{\substack{i:i \notin \mathscr{A} \\ \mathbf{a}_i^{\mathrm{T}}\mathbf{s}_q < 0}} \frac{z_i^{(k)}}{-\mathbf{a}_i^{\mathrm{T}}\mathbf{s}_q}$$

which is (8.3.6). Hence conclude that the same index p is chosen, and therefore that the sequence of values $\mathbf{x}^{(k)}$ and $\mathbf{z}^{(k)}$ in both methods is identical.

18. Use the method of artificial variables to find a b.f.s. for the LP problem in Question 1.
19. Find a basic feasible solution for the LP problem

$$\begin{aligned} &\text{minimize } -x_1 + x_2 + x_3 - 2x_4 \\ &\text{subject to } 4x_1 + x_2 - 2x_3 + 3x_4 \leqslant 8 \\ &\qquad x_1 - x_3 + x_4 = -2, \qquad \mathbf{x} \geqslant \mathbf{0} \end{aligned}$$

by the method of artificial variables, and hence solve the problem. Is the inequality constraint active at the solution?

20. Convert the LP problem

$$\begin{aligned} &\text{minimize } \quad x_1 + x_2 + 2x_3 \\ &\text{subject to } \quad x_1 + x_2 + x_3 \leqslant 9 \\ &\qquad 2x_1 - 3x_2 + 3x_3 = 1 \\ &\qquad -3x_1 + 6x_2 - 4x_3 = 3, \qquad \mathbf{x} \geqslant \mathbf{0} \end{aligned}$$

to standard form, and solve by a combined phase I–phase II simplex method, using the tableau form.

21. Convert the LP problem

$$\begin{aligned} &\text{minimize } -x_1 - x_2 \\ &\text{subject to } \quad x_1 + 2x_3 \leqslant 1 \\ &\qquad x_2 - x_3 \leqslant 1 \\ &\qquad x_1 + x_2 + x_3 = 2, \qquad \mathbf{x} \geqslant \mathbf{0} \end{aligned}$$

to standard form and find an initial b.f.s. by the method of artificial variables. Show that an artificial variable remains in the resulting basis at zero value. Why is this so? Show that the solution to the LP problem can nonetheless be obtained.

22. Convert the LP problem

$$\begin{aligned} &\text{minimize } 2x_1 + 4x_2 + 3x_3 \\ &\text{subject to } \quad x_1 - x_2 - x_3 \leqslant -2 \\ &\qquad 2x_1 + x_2 \geqslant 1, \qquad \mathbf{x} \geqslant \mathbf{0} \end{aligned}$$

to standard form. Solve using the simplex method
(i) starting with a basic feasible solution for which $x_1 = \frac{1}{2}$, $x_2 = 0$, $x_3 = 2\frac{1}{2}$,
(ii) using (8.4.5) with a basis of slack variables.

23. Compute $\mathbf{LL}^{\mathrm{T}}$ factors of the tridiagonal matrix $\mathbf{A}$ in which $A_{11} = 1$ and $A_{ii} = 2$ and $A_{i,i-1} = A_{i-1,i} = -1 \ \forall\ i > 1$. Observe that no sparsity is lost in the factors, but that the inverse matrix $\mathbf{L}^{-1}$ is always a full lower triangular matrix with a simple general form.
24. Consider the Gauss–Jordan factors (8.5.4) of a matrix $\mathbf{A}$. Express each $\mathbf{M}_i$ as $\mathbf{M}_i = \bar{\mathbf{U}}_i \bar{\mathbf{L}}_i$ where

$$\bar{\mathbf{L}}_i = \begin{bmatrix} 1 & & & & & \\ & \ddots & & & & \\ & & 1 & & & \\ & & -d_{i+1}/d_i & \ddots & & \\ & & \vdots & & \ddots & \\ & & -d_n/d_i & & & 1 \end{bmatrix}, \qquad \bar{\mathbf{U}}_i = \begin{bmatrix} 1 & & -d_1/d_i & & & \\ & \ddots & \vdots & & & \\ & & -d_{i-1}/d_i & & & \\ & & 1 & & & \\ & & & \ddots & & \\ & & & & & 1 \end{bmatrix}$$

(column i indicated in each matrix)

and show that $\bar{\mathbf{L}}_i\bar{\mathbf{U}}_j = \bar{\mathbf{U}}_j\bar{\mathbf{L}}_i$ for $j < i$. Hence rearrange (8.5.4) to give

$$\bar{\mathbf{U}}_n\bar{\mathbf{U}}_{n-1}\cdots\bar{\mathbf{U}}_1\bar{\mathbf{L}}_{n-1}\cdots\bar{\mathbf{L}}_2\bar{\mathbf{L}}_1\mathbf{A} = \mathbf{I}$$

or

$$\mathbf{A} = (\bar{\mathbf{L}}_1^{-1}\cdots\bar{\mathbf{L}}_{n-1}^{-1})(\bar{\mathbf{U}}_1^{-1}\cdots\bar{\mathbf{U}}_n^{-1}).$$

By comparing this with $\mathbf{A} = \mathbf{LU}$ where $\mathbf{L}$ and $\mathbf{U}$ are given in (8.5.5), show that the spike in $\bar{\mathbf{L}}_i$ is the ith column of $\mathbf{L}$, but that the same is not true for $\mathbf{U}$, since the spikes in $\mathbf{U}_i$ in (8.5.7) occur horizontally. However observe from the last equation that

$$\mathbf{U}^{-1} = \bar{\mathbf{U}}_n\bar{\mathbf{U}}_{n-1}\cdots\bar{\mathbf{U}}_1$$

in which the spikes in the $\bar{\mathbf{U}}_i$ are the columns of $\mathbf{U}^{-1}$. This justifies the assertion about the inefficiency of the Gauss–Jordan product form in Section 8.5.

25. Consider the LP problem due to Beale:

$$\begin{aligned} \text{minimize } & -\tfrac{3}{4}x_1 + 20x_2 - \tfrac{1}{2}x_3 + 6x_4 \\ \text{subject to } & \tfrac{1}{4}x_1 - 8x_2 - x_3 + 9x_4 \leqslant 0 \\ & \tfrac{1}{2}x_1 - 12x_2 - \tfrac{1}{2}x_3 + 3x_4 \leqslant 0 \\ & x_3 \leqslant 1, \qquad \mathbf{x} \geqslant \mathbf{0}. \end{aligned}$$

Add slack variables x_5, x_6, x_7 and show that the initial choice $B = \{5, 6, 7\}$ is feasible but non-optimal and degenerate. Solve the problem by the tableau method, resolving ties in (8.2.11) by choosing the first permitted basic variable in the tableau. Show that after six iterations the original tableau is restored so that cycling is established. (Hadley, 1962, gives the detailed tableaux for this example.)

26. For the example of Question 25, show that initial column orderings in the tableau $\{5, 6, 7, 1, 2, 3, 4\}$ and $\{1, 2, 3, 4, 5, 6, 7\}$ both have $\hat{\mathbf{b}}^{(1)}(\epsilon) \succ \mathbf{0}$. In the first case show that the tie break on iteration 1 is resolved by $p = 6$, which is different from the choice $p = 5$ made in the cycling iteration. Hence show that degeneracy is removed on the next iteration which solves the problem. In the second case show that the cycling iteration is followed for the first two iterations, but that the tie on iteration 3 is broken by $p = 2$ rather than by $p = 1$. Again show that degeneracy is removed on the next iteration which solves the problem. This example illustrates that different iteration sequences

which resolve degeneracy can be obtained by different column orderings in the perturbation method.

27. Set up subproblem (8.6.8) as a means of resolving degeneracy in the example of Question 25. Show that a solution is given by $\lambda_0 = \frac{8}{5}$, $\lambda_1 = 0$, and $\lambda_2 = \frac{2}{5}$. Show that this corresponds to a solution in which $s_2 = s_4 = 0$ and find s_1 and s_3.

Chapter 9

The Theory of Constrained Optimization

9.1 Lagrange Multipliers

There have been many contributions to the theory of constrained optimization. In this chapter a number of the most important results are developed; the presentation aims towards practicality and avoids undue generality. Perhaps the most important concept which needs to be understood is the way in which so-called *Lagrange multipliers* are introduced and this is the aim of this section. In order to make the underlying structure clear, this is done in a semi-rigorous way, with a fully rigorous treatment following in the next section. In Volume 1 it can be appreciated that the concept of a stationary point (for which $\mathbf{g}(\mathbf{x}) = \nabla f(\mathbf{x}) = \mathbf{0}$) is fundamental to the subject of unconstrained optimization, and is a necessary condition for a local minimizer. Lagrange multipliers arise when similar necessary conditions are sought for the solution $\mathbf{x}^*$ of the constrained minimization problem (7.1.1).

For unconstrained minimization in Volume 1, the necessary conditions illustrate the requirement for zero slope and positive curvature in any direction at $\mathbf{x}^*$, that is to say there is no descent direction at $\mathbf{x}^*$. In constrained optimization there is the additional complication of a feasible region. Hence a local minimizer must be a feasible point, and other necessary conditions illustrate the need for there to be *no feasible descent directions* at $\mathbf{x}^*$. However there are some difficulties on account of the fact that the boundary of the feasible region may be curved. In the first instance the simplest case of only equality constraints is presented (that is $I = \emptyset$ (the empty set)).

Suppose that a feasible incremental step $\boldsymbol{\delta}$ is taken from the minimizing point $\mathbf{x}^*$. By a Taylor series

$$c_i(\mathbf{x}^* + \boldsymbol{\delta}) = c_i^* + \boldsymbol{\delta}^{\mathrm{T}} \mathbf{a}_i^* + o(\|\boldsymbol{\delta}\|)$$

where $\mathbf{a}_i = \nabla c_i$ and $o(\cdot)$ indicates terms which can be ignored relative to $\boldsymbol{\delta}$ in the limit. By feasibility $c_i(\mathbf{x}^* + \boldsymbol{\delta}) = c_i^* = 0$, so that any feasible incremental step lies along a *feasible direction* $\mathbf{s}$ which satisfies

$$\mathbf{s}^{\mathrm{T}} \mathbf{a}_i^* = 0, \qquad \forall i \in E. \tag{9.1.1}$$

In a regular situation (for example if the vectors $\mathbf{a}_i^*$, $i \in E$, are independent), it is also possible to construct a feasible incremental step $\boldsymbol{\delta}$, given any such $\mathbf{s}$ (see Section

9.2). If in addition $f(\mathbf{x})$ has negative slope along $\mathbf{s}$, that is

$$\mathbf{s}^{\mathrm{T}}\mathbf{g}^* < 0, \tag{9.1.2}$$

then the direction $\mathbf{s}$ is a descent direction and the feasible incremental step along $\mathbf{s}$ reduces $f(\mathbf{x})$. This cannot happen, however, because $\mathbf{x}^*$ is a local minimizer. Therefore there can be no direction $\mathbf{s}$ which satisfies both (9.1.1) and (9.1.2). Now this statement is clearly true *if* $\mathbf{g}^*$ is a linear combination of the vectors $\mathbf{a}_i^*$, $i \in E$, that is if

$$\mathbf{g}^* = \sum_{i \in E} \mathbf{a}_i^* \lambda_i^* = \mathbf{A}^* \boldsymbol{\lambda}^*, \tag{9.1.3}$$

and in fact *only if* this condition holds, as is shown below. Therefore (9.1.3) is a *necessary condition for a local minimizer*. The multipliers λ_i^* in this linear combination are referred to as *Lagrange multipliers*, and the superscript * indicates that they are associated with the point $\mathbf{x}^*$. $\mathbf{A}^*$ denotes the matrix with columns $\mathbf{a}_i^*$, $i \in E$. Notice that there is a multiplier associated with each constraint function. In fact if $\mathbf{A}^*$ has full rank, then from (9.1.3) $\boldsymbol{\lambda}^*$ is defined *uniquely* by the expression

$$\boldsymbol{\lambda}^* = \mathbf{A}^{*+}\mathbf{g}^*$$

where $\mathbf{A}^+ = (\mathbf{A}^{\mathrm{T}}\mathbf{A})^{-1}\mathbf{A}^{\mathrm{T}}$ denotes the generalized inverse of $\mathbf{A}$ (see Question 9.15). Of course when there are no constraints present, then (9.1.3) reduces to the usual stationary point condition that $\mathbf{g}^* = \mathbf{0}$.

The formal proof that (9.1.3) is necessary is by contradiction. If (9.1.3) does not hold, the $\mathbf{g}^*$ can be expressed as

$$\mathbf{g}^* = \mathbf{A}^*\boldsymbol{\lambda} + \boldsymbol{\mu} \tag{9.1.4}$$

where $\boldsymbol{\mu} \neq \mathbf{0}$ is the component of $\mathbf{g}^*$ orthogonal to the vectors $\mathbf{a}_i^*$, so that $\mathbf{A}^{*\mathrm{T}}\boldsymbol{\mu} = \mathbf{0}$. Then $\mathbf{s} = -\boldsymbol{\mu}$ satisfies both (9.1.1) and (9.1.2). Hence by the regularity assumption there exists a feasible incremental step $\boldsymbol{\delta}$ along $\mathbf{s}$ which reduces $f(\mathbf{x})$, and this contradicts the fact that $\mathbf{x}^*$ is a local minimizer. Thus (9.1.3) is necessary. The conditions are illustrated in Figure 9.1.1. At $\mathbf{x}'$ which is not a local minimizer, $\mathbf{g}' \neq \mathbf{a}'\lambda$ and so there exists a non-zero vector $\boldsymbol{\mu}$ which is orthogonal to $\mathbf{a}'$, and an incremental step $\boldsymbol{\delta}$ along the feasible descent direction $-\boldsymbol{\mu}$ reduces $f(\mathbf{x})$. At $\mathbf{x}^*$, $\mathbf{g}^* = \mathbf{a}^*\lambda^*$ and no feasible descent direction exists. A numerical example is provided by the problem: minimize $f(\mathbf{x}) \triangleq x_1 + x_2$ subject to $c(\mathbf{x}) \triangleq x_1^2 - x_2 = 0$. In this case $\mathbf{x}^* = (-\frac{1}{2}, \frac{1}{4})^{\mathrm{T}}$ so that $\mathbf{g}^* = (1, 1)^{\mathrm{T}}$ and $\mathbf{a}^* = (-1, -1)^{\mathrm{T}}$, and (9.1.3) is thus satisfied with $\lambda^* = -1$. The regularity assumption clearly holds because $\mathbf{a}^*$ is independent; more discussion about the regularity condition is given in the next section. The use of incremental steps in the above can be expressed more rigorously by introducing directional sequences, as in the next section, or by means of differentiable arcs (Kuhn and Tucker, 1951). However the aim of this section is to avoid these technicalities as much as possible.

These conditions give rise to the classical *method of Lagrange multipliers* for solving equality constraint problems. The method is to find vectors $\mathbf{x}^*$ and $\boldsymbol{\lambda}^*$

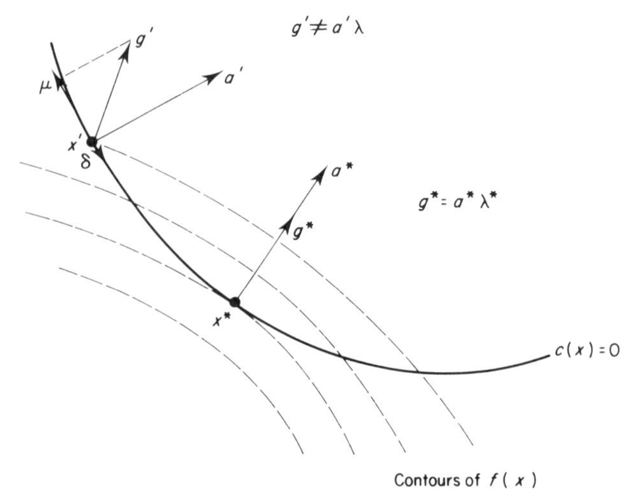

Figure 9.1.1 Existence of Lagrange multipliers

which solve the equations

$$\begin{aligned} \mathbf{g}(\mathbf{x}) &= \sum_{i \in E} \mathbf{a}_i(\mathbf{x})\lambda_i \\ c_i(\mathbf{x}) &= 0, \qquad i \in E \end{aligned} \tag{9.1.5}$$

which arise from (9.1.3) and feasibility. If there are m equality constraints then there are $n + m$ equations and $n + m$ unknowns $\mathbf{x}$ and $\boldsymbol{\lambda}$, so the system is well determined. However the system is non-linear (in $\mathbf{x}$) so in general may not be easy to solve (see Volume 1), although this may be possible in simple cases. An additional objection is that no second order information is taken into account, so (9.1.5) is also satisfied at a constrained saddle point or maximizer. The example of the previous paragraph can be used to illustrate the method. There are then three equations in (9.1.5), that is

$$\begin{pmatrix} 1 \\ 1 \end{pmatrix} = \begin{pmatrix} 2x_1 \\ -1 \end{pmatrix} \lambda$$

$$x_1^2 - x_2 = 0.$$

These can be solved in turn for the three variables $\lambda^* = -1, x_1^* = -\frac{1}{2}$, and $x_2^* = \frac{1}{4}$. It is instructive to see how the method differs from that of direct elimination. In this case $x_2 = x_1^2$ is used to eliminate x_2 from $f(\mathbf{x})$ leaving $f(x_1) = x_1 + x_1^2$ which is minimized by $x_1^* = -\frac{1}{2}$. Then back substitution gives $x_2^* = \frac{1}{4}$.

It is often convenient to restate these results by introducing the *Lagrangian function*

$$\mathscr{L}(\mathbf{x}, \boldsymbol{\lambda}) = f(\mathbf{x}) - \sum_i \lambda_i c_i(\mathbf{x}). \tag{9.1.6}$$

Then (9.1.5) becomes the very simple expression

$$\boldsymbol{\nabla}\mathscr{L}(\mathbf{x}^*, \boldsymbol{\lambda}^*) = \mathbf{0} \tag{9.1.7}$$

where

$$\boldsymbol{\nabla} = \begin{pmatrix} \nabla_x \\ \nabla_\lambda \end{pmatrix} \tag{9.1.8}$$

is a first derivative operator for the $n + m$ variable space. Hence a necessary condition for a local minimizer is that $\mathbf{x}^*$, $\boldsymbol{\lambda}^*$ is a stationary point of the Lagrangian function.

An alternative way of deriving these results starts by trying to find a stationary point of $f(\mathbf{x})$ subject to $\mathbf{c}(\mathbf{x}) = \mathbf{0}$, that is a feasible point $\mathbf{x}^*$ at which $f(\mathbf{x}^* + \boldsymbol{\delta}) = f(\mathbf{x}^*) + o(\|\boldsymbol{\delta}\|)$ for all feasible changes $\boldsymbol{\delta}$. The method of Lagrange multipliers finds a stationary point of $\mathscr{L}(\mathbf{x}, \boldsymbol{\lambda})$ by solving (9.1.7) or equivalently (9.1.5). Since $\nabla_\lambda \mathscr{L} = \mathbf{0}$ it follows that $\mathbf{x}^*$ is feasible. Since $\nabla_x \mathscr{L}(\mathbf{x}, \boldsymbol{\lambda}^*) = \mathbf{0}$ it follows that $\mathscr{L}(\mathbf{x}, \boldsymbol{\lambda}^*)$ is stationary at $\mathbf{x}^*$ for all changes $\boldsymbol{\delta}$, and hence for all feasible changes. But if $\boldsymbol{\delta}$ is a feasible change, $\mathscr{L}(\mathbf{x}^* + \boldsymbol{\delta}, \boldsymbol{\lambda}^*) = f(\mathbf{x}^* + \boldsymbol{\delta})$ and so $f(\mathbf{x})$ is stationary at $\mathbf{x}^*$ for all feasible changes $\boldsymbol{\delta}$. Thus if (9.1.7) can be solved, this solution $\mathbf{x}^*$ is a constrained stationary point of $f(\mathbf{x})$. The opposite may not always be true; it is possible (for example Questions 7.4 and 9.14) that $\mathbf{x}^*$ can be a minimizer but not satisfy (9.1.7). Also the examples with $f(\mathbf{x}) \triangleq \pm(x - 1)^3 + y$ and the same constraint show that $\mathbf{x}^* = (0, 1)^{\mathrm{T}}$ can be a constrained stationary point but not satisfy (9.1.7).

To get another insight into the meaning of Lagrange multipliers, consider what happens if the right hand sides of the constraints are perturbed, so that

$$c_i(\mathbf{x}) = \epsilon_i, \qquad i \in E. \tag{9.1.9}$$

Let $\mathbf{x}(\boldsymbol{\epsilon}), \boldsymbol{\lambda}(\boldsymbol{\epsilon})$ denote how the solution and multipliers change as $\boldsymbol{\epsilon}$ changes. The Lagrangian for this problem is

$$\mathscr{L}(\mathbf{x}, \boldsymbol{\lambda}, \boldsymbol{\epsilon}) = f(\mathbf{x}) - \sum_{i \in E} \lambda_i (c_i(\mathbf{x}) - \epsilon_i).$$

From (9.1.9), $f(\mathbf{x}(\boldsymbol{\epsilon})) = \mathscr{L}(\mathbf{x}(\boldsymbol{\epsilon}), \boldsymbol{\lambda}(\boldsymbol{\epsilon}), \boldsymbol{\epsilon})$, so using the chain rule and then (9.1.7) it follows that

$$\frac{\mathrm{d}f}{\mathrm{d}\epsilon_i} = \frac{\mathrm{d}\mathscr{L}}{\mathrm{d}\epsilon_i} = \frac{\partial \mathbf{x}^{\mathrm{T}}}{\partial \epsilon_i} \nabla_x \mathscr{L} + \frac{\partial \boldsymbol{\lambda}^{\mathrm{T}}}{\partial \epsilon_i} \nabla_\lambda \mathscr{L} + \frac{\partial \mathscr{L}}{\partial \epsilon_i}$$

$$= \lambda_i. \tag{9.1.10}$$

Thus the Lagrange multiplier of any constraint measures the rate of change in the objective function, consequent upon changes in that constraint function. This information can be valuable in that it indicates how sensitive the objective function is to changes in the different constraints: see Question 9.4 for example.

The additional complication of having inequality constraints present is now discussed. It is important to realize that only the *active constraints* $\mathscr{A}^*$ (see (7.1.2)) at $\mathbf{x}^*$ can influence matters. Denote the active inequality constraints at $\mathbf{x}^*$ by I^*

($= \mathscr{A}^* \cap I$). Since $c_i^* = 0$ and $c_i(\mathbf{x}^* + \boldsymbol{\delta}) \geqslant 0$ for $i \in I^*$, then any feasible incremental step $\boldsymbol{\delta}$ lies along a feasible direction s which satisfies

$$\mathbf{s}^{\mathrm{T}}\mathbf{a}_i^* \geqslant 0, \qquad \forall i \in I^* \tag{9.1.11}$$

in addition to (9.1.1). In this case it is clear that there is no direction s which satisfies (9.1.1), (9.1.11), and (9.1.2) together, *if* both

$$\mathbf{g}^* = \sum_{i \in \mathscr{A}^*} \mathbf{a}_i^* \lambda_i^* \tag{9.1.12}$$

and

$$\lambda_i^* \geqslant 0, \qquad i \in I^* \tag{9.1.13}$$

hold, and in fact *only if* these conditions hold, as is shown below. Hence these are therefore necessary conditions for a local minimizer. The only extra condition beyond (9.1.3) in this case, therefore, is that the multipliers of active inequality constraints must be non-negative. These conditions can be established by contradiction when the regularity assumption is made that the set of normal vectors $\mathbf{a}_i^*$, $i \in \mathscr{A}^*$, is independent. Equation (9.1.12) holds as for (9.13). Let (9.1.13) not hold so that there is some multiplier $\lambda_p^* < 0$. It is always possible to find a direction s for which $\mathbf{s}^{\mathrm{T}}\mathbf{a}_i^* = 0$, $i \in \mathscr{A}^*$, $i \neq p$, and $\mathbf{s}^{\mathrm{T}}\mathbf{a}_p^* = 1$ (for instance $\mathbf{s} = \mathbf{A}^{*+\mathrm{T}}\mathbf{e}_p$ where $\mathbf{A}^+ = (\mathbf{A}^{\mathrm{T}}\mathbf{A})^{-1}\mathbf{A}^{\mathrm{T}}$ denotes the generalized inverse of $\mathbf{A}$). Then s is feasible in (9.1.1) and (9.1.11), and from (9.1.12) it follows that

$$\mathbf{s}^{\mathrm{T}}\mathbf{g}^* = \mathbf{s}^{\mathrm{T}}\mathbf{a}_p^*\lambda_p^* = \lambda_p^* < 0. \tag{9.1.14}$$

Thus s is also downhill, and by the regularity assumption there exists a feasible incremental step $\boldsymbol{\delta}$ along s which reduces $f(\mathbf{x})$; this contradicts the fact that $\mathbf{x}^*$ is a local minimizer. Thus (9.1.13) is necessary. Note that this proof uses the independence of the vectors $\mathbf{a}_i^*$, $i \in \mathscr{A}^*$, in constructing the generalized inverse. A more general proof using Lemma 9.2.4 (Farkas' lemma) and its corollary does not make this assumption.

The need for condition (9.1.13) can also be deduced simply from (9.1.9) and (9.1.10). Let an active inequality constraint be perturbed from $c_i(\mathbf{x}) = 0$ to $c_i(\mathbf{x}) = \epsilon_i > 0$, $i \in I^*$. This induces a feasible change in $\mathbf{x}(\epsilon)$ so it is necessary that $f(\mathbf{x}(\epsilon))$ does not decrease. This implies that $\mathrm{d}f^*/\mathrm{d}\epsilon_i \geqslant 0$ at the solution and hence $\lambda_i^* \geqslant 0$. Thus the necessity of (9.1.13) has an obvious interpretation in these terms.

As an example of these conditions, consider the problem

$$\begin{aligned} &\text{minimize} \quad f(\mathbf{x}) \triangleq -x_1 - x_2 \\ &\text{subject to } c_1(\mathbf{x}) \triangleq x_2 - x_1^2 \geqslant 0 \\ &\qquad\qquad\; c_2(\mathbf{x}) \triangleq 1 - x_1^2 - x_2^2 \geqslant 0 \end{aligned} \tag{9.1.15}$$

which is illustrated in Figure 7.1.3. The solution is $\mathbf{x}^* = (1/\sqrt{2}, 1/\sqrt{2})^{\mathrm{T}}$ so c_1 is not active and hence $\mathscr{A}^* = \{2\}$. Then $\mathbf{g}^* = (-1, -1)^{\mathrm{T}}$ and $\mathbf{a}_2^* = (-\sqrt{2}, -\sqrt{2})^{\mathrm{T}}$ so (9.1.12) and (9.1.13) are satisfied with $\lambda_2^* = 1/\sqrt{2} \geqslant 0$. It is important to remember that a general inequality constraint must be correctly rearranged into the form $c_i(\mathbf{x}) \geqslant 0$ before the condition $\lambda_i \geqslant 0$ applies.

In fact the construction of a descent direction when $\lambda_p^* < 0$ above indicates another important property of Lagrange multipliers for inequality constraints. The conditions $\mathbf{s}^T\mathbf{a}_i^* = 0$, $i \neq p$, and $\mathbf{s}^T\mathbf{a}_p^* = 1$ indicate that the resulting feasible incremental step satisfies $c_i(\mathbf{x}^* + \boldsymbol{\delta}) = 0$ for $i \neq p$, and $c_p(\mathbf{x}^* + \boldsymbol{\delta}) > 0$. Thus it indicates that $f(\mathbf{x})$ can be reduced *by moving away from the boundary of constraint* p. This result also follows from (9.1.10) and is of great importance in the various active set methods (see Section 7.2) for handling inequality constraints. If conditions (9.1.13) are not satisfied then a constraint index p with $\lambda_p^* < 0$ can be removed from the active set. This result is also illustrated by the problem in the previous paragraph. Consider the feasible point $\mathbf{x}' \simeq (0.786, 0.618)$ to three decimal places, at which both constraints are active. Since $\mathbf{g}' = (-1, -1)^T$, $\mathbf{a}_1' = (-1.572, 1)^T$, and $\mathbf{a}_2' = (-1.572, -1.236)^T$, it follows that (9.1.12) is satisfied with $\boldsymbol{\lambda}' = (-0.096, 0.732)^T$. However (9.1.13) is not satisfied, so $\mathbf{x}'$ is not a local minimizer. Since $\lambda_1' < 0$ the objective function can be reduced by moving away from the boundary of constraint 1, along a direction for which $\mathbf{s}^T\mathbf{a}_1^* = 1$ and $\mathbf{s}^T\mathbf{a}_2^* = 0$. This is the direction $\mathbf{s} = (-0.352, 0.447)^T$ and is in fact the tangent to the circle at $\mathbf{x}'$. Moving round the arc of the circle in this direction leads to the solution point $\mathbf{x}^*$ at which only constraint 2 is active.

A further restatement of (9.1.12) and (9.1.13) is possible in terms of all the constraints rather than just the active ones. It is consistent to regard any inactive constraint as having a zero Lagrange multiplier, in which case (9.1.12), (9.1.13), and the feasibility conditions can be combined in the following theorem.

Theorem 9.1.1 (First order necessary conditions)

If $\mathbf{x}^*$ *is a local minimizer of problem (7.1.1) and if a regularity assumption (9.2.4) holds at* $\mathbf{x}^*$, *then there exist Lagrange multipliers* $\boldsymbol{\lambda}^*$ *such that* $\mathbf{x}^*$, $\boldsymbol{\lambda}^*$ *satisfy the following system:*

$$\begin{aligned} \nabla_x \mathscr{L}(\mathbf{x}, \boldsymbol{\lambda}) &= \mathbf{0} & & \\ c_i(\mathbf{x}) &= 0, & & i \in E \\ c_i(\mathbf{x}) &\geqslant 0, & & i \in I \\ \lambda_i &\geqslant 0, & & i \in I \\ \lambda_i c_i(\mathbf{x}) &= 0 & & \forall i. \end{aligned} \tag{9.1.16}$$

These are often described as Kuhn–Tucker (KT) conditions (Kuhn and Tucker, 1951) and a point $\mathbf{x}^*$ which satisfies the conditions is sometimes referred to as a *KT point*. The regularity assumption (9.2.4) is implied by the vectors $\mathbf{a}_i^*$, $i \in \mathscr{A}^*$, being independent and is discussed in detail in the next section where a more rigorous proof is given. The final condition $\lambda_i^* c_i^* = 0$ is referred to as the *complementarity condition* and states that both λ_i^* and c_i^* cannot be non-zero, or equivalently that inactive constraints have a zero multiplier. If there is no i such that $\lambda_i^* = c_i^* = 0$ then *strict complementarity* is said to hold. The case $\lambda_i^* = c_i^* = 0$ is an intermediate state between a constraint being strongly active and being inactive, as indicated in Figure 9.1.2.

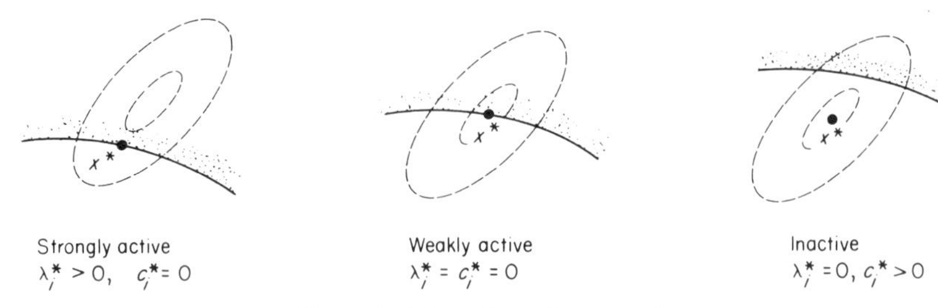

Figure 9.1.2 Complementarity

So far only first order (that is first derivative) conditions have been considered. It is also possible to state second order conditions which give information about the curvature of the objective and constraint functions at a local minimizer. This subject is discussed in Section 9.3. It is also possible to make even stronger statements when the original problem is a *convex programming problem*, and the more simple results of convexity and its application to optimization theory are developed in Section 9.4. For certain convex programming problems it is possible to state useful alternative (dual) problems from which the solution to the original (primal) problem can be obtained. These problems involve the Lagrange multipliers as dual variables, much in the way that they arise in the method of Lagrange multipliers. The subject of duality is discussed further in Section 9.5. In fact the literature on convexity and duality is very extensive and often very theoretical, so as to become a branch of pure mathematics. In this volume I have attempted to describe those aspects of these subjects which are of most relevance to practical algorithms.

9.2 First Order Conditions

In this section the results of the previous section are considered in more technical detail. First of all it is important to have a more rigorous notion of what is meant by a feasible incremental step. Consider any feasible point $\mathbf{x}'$ and any infinite sequence of feasible points $\{\mathbf{x}^{(k)}\} \to \mathbf{x}'$ where $\mathbf{x}^{(k)} \neq \mathbf{x}'$ for all k. Then it is possible to write

$$\mathbf{x}^{(k)} - \mathbf{x}' = \delta^{(k)}\mathbf{s}^{(k)} \qquad \forall k \tag{9.2.1}$$

where $\delta^{(k)} > 0$ is a scalar and $\mathbf{s}^{(k)}$ is a vector of any fixed length $\sigma > 0$ ($\|\mathbf{s}^{(k)}\|_2 = \sigma$). It follows that $\delta^{(k)} \to 0$. A *directional sequence* can be defined as any such sequence for which $\mathbf{s}^{(k)} \to \mathbf{s}$. The limiting vector $\mathbf{s}$ is referred to as a *feasible direction*, and $\mathscr{F}(\mathbf{x}')$ or $\mathscr{F}'$ is used to denote the *set of feasible directions* at $\mathbf{x}'$. Taking the limit in (9.2.1) corresponds to the notion of making a feasible incremental step along $\mathbf{s}$. Clearly the length of $\mathbf{s}$ is arbitrary and it is possible to restrict the discussion to any fixed normalization, for example $\sigma = 1$. In some texts $\mathscr{F}'$ is defined so that it also includes the zero vector, and is then referred to as the *tangent cone* at $\mathbf{x}'$.

The set $\mathscr{F}'$ is not very amenable to manipulation, however, and it is convenient to consider a related set of feasible directions which are obtained if the constraints are linearized. The linearized constraint function is given by (7.1.3) so clearly the set of feasible directions for the linearized constraint set can be written as

$$F(\mathbf{x}') = F' = \{\mathbf{s} \mid \mathbf{s} \neq \mathbf{0},\ \ \mathbf{s}^{\mathrm{T}}\mathbf{a}_i' = 0,\ i \in E, \\ \mathbf{s}^{\mathrm{T}}\mathbf{a}_i' \geqslant 0,\ i \in I'\}. \tag{9.2.2}$$

where $I' = \mathscr{A}' \cap I$ denotes the set of active inequality constraints at $\mathbf{x}'$. (A vector $\mathbf{s} \in F'$ corresponds to a directional sequence along a half-line through $\mathbf{x}'$ in the direction $\mathbf{s}$, which is clearly feasible, and it is straightforward to contradict feasibility in the linearized set for any directional sequence for which $\mathbf{s} \notin F'$.) It is very convenient if the sets F' and $\mathscr{F}'$ are the same, so it is important to consider the extent to which this is true.

Lemma 9.2.1

$F' \supseteq \mathscr{F}'$

Proof

Let $\mathbf{s} \in \mathscr{F}'$: then $\exists$ a directional sequence $\mathbf{x}^{(k)} \to \mathbf{x}'$ such that $\mathbf{s}^{(k)} \to \mathbf{s}$. A Taylor series about $\mathbf{x}'$ using (9.2.1) gives

$$c_i(\mathbf{x}^{(k)}) = c_i' + \delta^{(k)}\mathbf{s}^{(k)\mathrm{T}}\mathbf{a}_i' + o(\delta^{(k)}).$$

Now $c_i(\mathbf{x}^{(k)}) = c_i' = 0$ for $i \in E$ and $c_i(\mathbf{x}^{(k)}) \geqslant c_i' = 0$ for $i \in I'$, so dividing by $\delta^{(k)} > 0$ it follows that

$$\mathbf{s}^{(k)\mathrm{T}}\mathbf{a}_i' + o(1) = 0, \qquad i \in E$$
$$\mathbf{s}^{(k)\mathrm{T}}\mathbf{a}_i' + o(1) \geqslant 0, \qquad i \in I'.$$

Taking limits as $k \to \infty$, $\mathbf{s}^{(k)} \to \mathbf{s}$, $o(1) \to 0$, then $\mathbf{s} \in F'$ from (9.2.2). □

Unfortunately a result going the other way ($\mathscr{F}' \supseteq F'$) is not in general true and it is this that is required in the proof of Theorem 9.1.1. To get round this difficulty, Kuhn and Tucker (1951) make an assumption that $F' = \mathscr{F}'$, which they refer

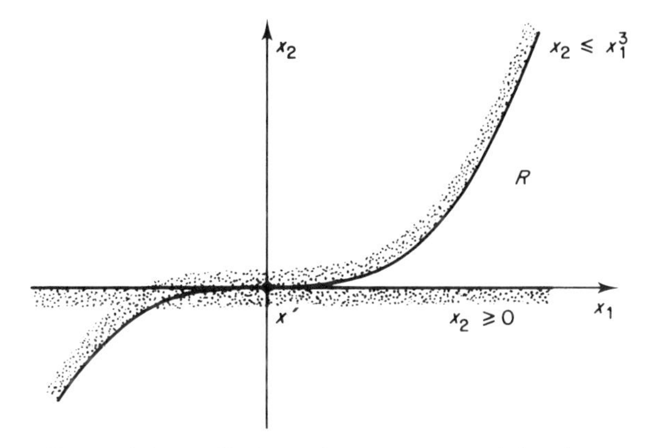

Figure 9.2.1 Failure of constraint qualification

to as a *constraint qualification* at $\mathbf{x}'$. That is to say, for any $\mathbf{s} \in F'$, the existence of a feasible directional sequence with feasible direction $\mathbf{s}$ is assumed. They also give an example in which this property does not hold which is essentially that illustrated in Figure 9.2.1. Clearly at $\mathbf{x}' = \mathbf{0}$, the direction $\mathbf{s} = (-1, 0)^T$ is in F' but there is no corresponding feasible directional sequence, and so $\mathbf{s} \notin \mathscr{F}'$. However it is important to realize that this failing case is an unlikely situation and that it is usually valid to assume that $F' = \mathscr{F}'$. Indeed this result can be guaranteed if certain linearity or independence conditions hold, as the following result shows.

Lemma 9.2.2

Sufficient conditions for $F' = \mathscr{F}'$ at a feasible point $\mathbf{x}'$ are either

(i) *the constraints $i \in \mathscr{A}'$ are all linear constraints or*
(ii) *the vectors $\mathbf{a}_i'$, $i \in \mathscr{A}'$, are linearly independent.*

Proof

Case (i) is clear by definition of F'. For case (ii) a feasible directional sequence with feasible direction $\mathbf{s}$ is constructed for any $\mathbf{s} \in F'$. Let $\mathbf{s} \in F'$, and consider the nonlinear system

$$\mathbf{r}(\mathbf{x}, \theta) = \mathbf{0} \tag{9.2.3}$$

defined by

$$r_i(\mathbf{x}, \theta) = c_i(\mathbf{x}) - \theta \mathbf{s}^T \mathbf{a}_i', \qquad i = 1, 2, \ldots, m$$
$$r_i(\mathbf{x}, \theta) = (\mathbf{x} - \mathbf{x}')^T \mathbf{b}_i - \theta \mathbf{s}^T \mathbf{b}_i, \qquad i = m+1, \ldots, n$$

where it is assumed that $\mathscr{A}' = \{1, 2, \ldots, m\}$. The system (9.2.3) is solved by $\mathbf{x}'$ when $\theta = 0$, and any solution $\mathbf{x}$ is also a feasible point in (7.1.1) when $\theta \geqslant 0$ is sufficiently small. Writing $\mathbf{A} = [\mathbf{a}_1, \ldots, \mathbf{a}_m]$ and $\mathbf{B} = [\mathbf{b}_{m+1}, \ldots, \mathbf{b}_n]$ then the Jacobian matrix $\mathbf{J}(\dot{\mathbf{x}}, \theta) = \nabla_x \mathbf{r}^T(\mathbf{x}, \theta) = [\mathbf{A} : \mathbf{B}]$. Since $\mathbf{A}'$ has full rank by case (ii), it is possible to choose $\mathbf{B}$ so that $\mathbf{J}'$ is non-singular. Hence by the implicit function theorem (Apostol (1957) for example) there exist open neighbourhoods Ω_x about $\mathbf{x}'$ and Ω_θ about $\theta = 0$ such that for any $\theta \in \Omega_\theta$, a unique solution $\mathbf{x}(\theta) \in \Omega_x$ exists to (9.2.3), and $\mathbf{x}(\theta)$ is a $\mathbb{C}^1$ function of θ. From (9.2.3) and using the chain rule

$$0 = \frac{\mathrm{d}r_i}{\mathrm{d}\theta} = \sum_j \frac{\partial r_i}{\partial x_j} \frac{\mathrm{d}x_j}{\mathrm{d}\theta} + \frac{\partial r_i}{\partial \theta}$$

so that

$$\mathbf{0} = \mathbf{J}^T \frac{\mathrm{d}\mathbf{x}}{\mathrm{d}\theta} - \mathbf{J}'^T \mathbf{s}.$$

Thus $\mathrm{d}\mathbf{x}/\mathrm{d}\theta = \mathbf{s}$ at $\theta = 0$. Hence if $\theta^{(k)} \downarrow 0$ is any sequence then $\mathbf{x}^{(k)} = \mathbf{x}(\theta^{(k)})$ is a feasible directional sequence with feasible direction $\mathbf{s}$. $\square$

In moving on to discuss necessary conditions at a local solution it is convenient to define the *set of descent directions*

$$\mathscr{D}(\mathbf{x}') = \mathscr{D}' = \{\mathbf{s} \mid \mathbf{s}^{\mathrm{T}}\mathbf{g}' < 0\}.$$

Then the most basic necessary condition is the following.

Lemma 9.2.3

If $\mathbf{x}^*$ is a local minimizer, then $\mathscr{F}^* \cap \mathscr{D}^* = \emptyset$ (no feasible descent directions).

Proof

Let $\mathbf{s} \in \mathscr{F}^*$ so there exists a feasible sequence $\mathbf{x}^{(k)} \to \mathbf{x}^*$ such that $\mathbf{s}^{(k)} \to \mathbf{s}$. By a Taylor series about $\mathbf{x}^*$,

$$f(\mathbf{x}^{(k)}) = f^* + \delta^{(k)}\mathbf{s}^{(k)^{\mathrm{T}}}\mathbf{g}^* + o(\delta^{(k)}).$$

Because $\mathbf{x}^*$ is a local minimizer, $f(\mathbf{x}^{(k)}) \geqslant f^*$ for all k sufficiently large, so dividing by $\delta^{(k)} > 0$,

$$\mathbf{s}^{(k)^{\mathrm{T}}}\mathbf{g}^* + o(1) \geqslant 0.$$

In the limit, $\mathbf{s}^{(k)} \to \mathbf{s}$, $o(1) \to 0$ so $\mathbf{s} \notin \mathscr{D}^*$ and hence $\mathscr{F}^* \cap \mathscr{D}^* = \emptyset$. □

Unfortunately it is not possible to proceed further without making a *regularity assumption*

$$F^* \cap \mathscr{D}^* = \mathscr{F}^* \cap \mathscr{D}^* \tag{9.2.4}$$

This assumption is clearly implied by the Kuhn–Tucker constraint qualification ($F^* = \mathscr{F}^*$) at $\mathbf{x}^*$, but (9.2.4) may hold when $F^* = \mathscr{F}^*$ does not, for example at $\mathbf{x}^* = \mathbf{0}$ in the problem: minimize x_2 subject to the constraints of Figure 9.2.1. Also the problem: minimize x_1 subject to the same constraints, illustrates the need for a regularity assumption. Here $\mathbf{s} = (-1, 0)^{\mathrm{T}} \in F^* \cap \mathscr{D}^*$ at $\mathbf{x}^* = \mathbf{0}$ so this set is not empty and in fact $\mathbf{x}^*$ is not a KT point, although it is a minimizer and $\mathscr{F}^* \cap \mathscr{D}^*$ is empty.

With assumption (9.2.4) the necessary condition from lemma 9.2.3 becomes $F^* \cap \mathscr{D}^* = \emptyset$ (no linearized feasible descent directions). It is now possible to relate this condition to the existence of multipliers in (9.1.12) and (9.1.13). In fact the construction given after (9.1.13) can be used to do this when the set $\mathbf{a}_i^*$, $i \in \mathscr{A}^*$, is independent. However the following lemma shows that the result is more general than this.

Lemma 9.2.4 (Farkas' lemma)

Given any vectors $\mathbf{a}_1, \mathbf{a}_2, \ldots, \mathbf{a}_m$ and $\mathbf{g}$ then the set

$$S = \{\mathbf{s} \mid \mathbf{s}^{\mathrm{T}}\mathbf{g} < 0, \tag{9.2.5}$$

$$\mathbf{s}^{\mathrm{T}}\mathbf{a}_i \geqslant 0,\ i = 1, 2, \ldots, m\} \tag{9.2.6}$$

is empty if and only if there exist multipliers $\lambda_i \geqslant 0$ *such that*

$$\mathbf{g} = \sum_{i=1}^{m} \mathbf{a}_i \lambda_i. \tag{9.2.7}$$

Remark

In the context of this section, S is the set of linearized feasible descent directions for a problem with no equality constraints. The extension to include equality constraints is made in the corollary.

Proof

The 'if' part is straightforward since (9.2.7) implies that $\mathbf{s}^{\mathrm{T}}\mathbf{g} = \sum \mathbf{s}^{\mathrm{T}}\mathbf{a}_i\lambda_i \geqslant 0$ by (9.2.6) and $\lambda_i \geqslant 0$. Thus (9.2.5) is not true and so S is empty. The converse result is established by showing that if (9.2.7) with $\lambda_i \geqslant 0$ does not hold, then there is a vector $\mathbf{s} \in S$. Geometrically, the result is easy to visualize. The set of vectors

$$C = \{\mathbf{v} \mid \mathbf{v} = \sum_{i=1}^{m} \mathbf{a}_i\lambda_i,\ \lambda_i \geqslant 0\}$$

is known as a *polyhedral cone* and is closed and convex (see Section 9.4). From Figure 9.2.2 it is clear that if $\mathbf{g} \notin C$ then there exists a hyperplane with normal vector $\mathbf{s}$ which *separates* C and $\mathbf{g}$, and for which $\mathbf{s}^{\mathrm{T}}\mathbf{a}_i \geqslant 0$, $i = 1, 2, \ldots, m$, and $\mathbf{s}^{\mathrm{T}}\mathbf{g} < 0$. Thus $\mathbf{s} \in S$ exists and the lemma is proved. For completeness, the general proof of the existence of a separating hyperplane is given below in Lemma 9.2.5. □

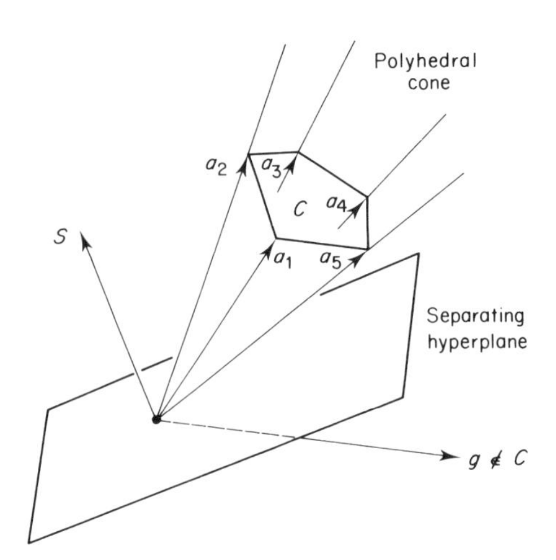

Figure 9.2.2 Existence of a separating hyperplane

Corollary

The set

$$S = \{\mathbf{s} \mid \mathbf{s}^T\mathbf{g}^* < 0$$
$$\mathbf{s}^T\mathbf{a}_i^* = 0,\ i \in E$$
$$\mathbf{s}^T\mathbf{a}_i^* \geqslant 0,\ i \in I^*\}$$

is empty if and only if there exist multipliers λ_i^* *such that (9.1.12) and (9.1.13) hold.*

Proof

Ignoring superscript *, then $\mathbf{s}^T\mathbf{a}_i = 0$, $i \in E$, can be written as $\mathbf{s}^T\mathbf{a}_i \geqslant 0$ and $-\mathbf{s}^T\mathbf{a}_i \geqslant 0$, $i \in E$. Then by Farkas' lemma, this is equivalent to the fact that there exist non-negative multipliers λ_i, $i \in I^*$, λ_i^+, λ_i^-, $i \in E$, such that

$$\mathbf{g} = \sum_{i \in I^*} \mathbf{a}_i\lambda_i + \sum_{i \in E} \mathbf{a}_i\lambda_i^+ + \sum_{i \in E} -\mathbf{a}_i\lambda_i^- .$$

But defining $\lambda_i = \lambda_i^+ - \lambda_i^-$, $i \in E$, gives (9.1.12) and (9.1.13) so the corollary is proved. □

Lemma 9.2.5 (Existence of a separating hyperplane)

There exists a hyperplane $\mathbf{s}^T\mathbf{x} = 0$ *which separates a closed convex cone C and a non-zero vector* $\mathbf{g} \notin C$.

Proof

By construction. Consider minimizing $\|\mathbf{g} - \mathbf{x}\|_2^2$ for all $\mathbf{x} \in C$, and let $\mathbf{x}_1 \in C$. Since the solution satisfies $\|\mathbf{g} - \mathbf{x}\|_2 \leqslant \|\mathbf{g} - \mathbf{x}_1\|_2$ it is bounded and so by continuity of $\|\cdot\|_2$ a minimizing point, $\hat{\mathbf{g}}$ say, exists. Because $\lambda\hat{\mathbf{g}} \in C$ for all $\lambda \geqslant 0$, and because $\|\lambda\hat{\mathbf{g}} - \mathbf{g}\|_2^2$ has a minimum at $\lambda = 1$, it follows by setting $d/d\lambda = 0$ that

$$\hat{\mathbf{g}}^T(\hat{\mathbf{g}} - \mathbf{g}) = 0. \tag{9.2.8}$$

Let $\mathbf{x} \in C$; then $\hat{\mathbf{g}} + \theta(\mathbf{x} - \hat{\mathbf{g}}) \in C$ for $\theta \in (0, 1)$ by convexity, and hence

$$\|\theta(\mathbf{x} - \hat{\mathbf{g}}) + \hat{\mathbf{g}} - \mathbf{g}\|_2^2 \geqslant \|\hat{\mathbf{g}} - \mathbf{g}\|_2^2.$$

Simplifying and taking the limit $\theta \downarrow 0$ it follows that

$$0 \leqslant (\mathbf{x} - \hat{\mathbf{g}})^T(\hat{\mathbf{g}} - \mathbf{g}) = \mathbf{x}^T(\hat{\mathbf{g}} - \mathbf{g})$$

from (9.2.8). Thus the vector $\mathbf{s} = \hat{\mathbf{g}} - \mathbf{g}$ is such that $\mathbf{s}^T\mathbf{x} \geqslant 0$ for all $\mathbf{x} \in C$. But $\mathbf{g}^T\mathbf{s} = \mathbf{s}^T\mathbf{s}$ from (9.2.8), and $\mathbf{s} \neq \mathbf{0}$ since $\mathbf{g} \notin C$. Thus $\mathbf{g}^T\mathbf{s} < 0$ and hence the hyperplane $\mathbf{s}^T\mathbf{x}$ separates C and $\mathbf{g}$. □

Geometrically the vector $\mathbf{s}$ is along the perpendicular from $\mathbf{g}$ to C. In Figure

9.2.2 $\hat{\mathbf{g}}$ would be a multiple of the vector $\mathbf{a}_5$ and the resulting hyperplane (not the one illustrated) would touch the cone along $\mathbf{a}_5$.

It is now possible to bring together the various aspects of this section in proving that the first order conditions (9.1.12) and (9.1.13) (or equivalently the conditions (9.1.16) of theorem 9.1.1) are necessary at a local minimizing point $\mathbf{x}^*$. At $\mathbf{x}^*$ there are no feasible descent directions ($\mathscr{F}^* \cap \mathscr{D}^* = \emptyset$) by lemma 9.2.3. A regularity assumption (9.2.4) is made that there are no linearized feasible descent directions. Then by the corollary to Farkas' lemma it follows that (9.1.12) and (9.1.13) hold. Thus these results have been established in a quite rigorous way.

9.3 Second Order Conditions

A natural progression from the previous section is to examine the effect of second order (curvature) terms in the neighbourhood of a local solution. It can readily be seen for unconstrained optimization (Volume 1) that the resulting sufficient condition that $\mathbf{G}^*$ is positive definite has significant implications for the design of satisfactory algorithms, and the same is true for constrained optimization. It is important to realize first of all that constraint curvature plays an important role, and that it is not possible to examine the curvature of $f(\mathbf{x})$ in isolation. For example realistic problems exist for which $\mathbf{G}^*$ is positive definite at a Kuhn–Tucker point $\mathbf{x}^*$, which is not, however, a local minimizer (see (9.3.6) below). As in Section 9.1, it is possible to present the essence of the situation in a fairly straightforward way, to be followed by a more general and more rigorous treatment later in the section. It is assumed that $f(\mathbf{x})$ and $c_i(\mathbf{x})$ for all i are $\mathbb{C}^2$ functions.

Suppose that there are only equality constraints present, and that a local solution $\mathbf{x}^*$ exists at which the vectors $\mathbf{a}_i^*$, $i \in E$, are independent, so that a unique vector of multipliers $\boldsymbol{\lambda}^*$ exists in (9.1.3). Under these conditions a feasible incremental step $\boldsymbol{\delta}$ can be taken along any feasible direction $\mathbf{s}$ at $\mathbf{x}^*$. By feasibility and (9.1.6) it follows that $f(\mathbf{x}^* + \boldsymbol{\delta}) = \mathscr{L}(\mathbf{x}^* + \boldsymbol{\delta}, \boldsymbol{\lambda})$. Also since $\mathscr{L}$ is stationary at $\mathbf{x}^*, \boldsymbol{\lambda}^*$ (equation (9.1.7)), a Taylor expansion of $\mathscr{L}(\mathbf{x}, \boldsymbol{\lambda}^*)$ about $\mathbf{x}^*$ enables the second order terms to be isolated. Hence

$$\begin{aligned} f(\mathbf{x}^* + \boldsymbol{\delta}) &= \mathscr{L}(\mathbf{x}^* + \boldsymbol{\delta}, \boldsymbol{\lambda}^*) \\ &= \mathscr{L}(\mathbf{x}^*, \boldsymbol{\lambda}^*) + \boldsymbol{\delta}^{\mathrm{T}} \nabla_x \mathscr{L}(\mathbf{x}^*, \boldsymbol{\lambda}^*) + \tfrac{1}{2}\boldsymbol{\delta}^{\mathrm{T}} \mathbf{W}^* \boldsymbol{\delta} + o(\boldsymbol{\delta}^{\mathrm{T}}\boldsymbol{\delta}) \\ &= f^* + \tfrac{1}{2}\boldsymbol{\delta}^{\mathrm{T}} \mathbf{W}^* \boldsymbol{\delta} + o(\boldsymbol{\delta}^{\mathrm{T}}\boldsymbol{\delta}) \end{aligned} \tag{9.3.1}$$

where $\mathbf{W}^* = \nabla_x^2 \mathscr{L}(\mathbf{x}^*, \boldsymbol{\lambda}^*) = \nabla^2 f(\mathbf{x}^*) - \Sigma_i \lambda_i^* \nabla^2 c_i(\mathbf{x}^*)$ denotes the Hessian matrix with respect to $\mathbf{x}$ of the Lagrangian function. It follows by the minimality of f^*, and taking the limit in (9.3.1), that

$$\mathbf{s}^{\mathrm{T}} \mathbf{W}^* \mathbf{s} \geqslant 0. \tag{9.3.2}$$

As in Section 9.1 a feasible direction satisfies

$$\mathbf{a}_i^{*\mathrm{T}} \mathbf{s} = 0, \qquad i \in E \tag{9.3.3}$$

which can be written in matrix notation as

$$\mathbf{A}^{*\mathrm{T}} \mathbf{s} = \mathbf{0}. \tag{9.3.4}$$

Thus a *second order necessary condition for a local minimizer* is that (9.3.2) must hold for any $\mathbf{s}$ which satisfies (9.3.4). That is to say, the Lagrangian function must have non-negative curvature for all feasible directions at $\mathbf{x}^*$. Of course when no constraints are present then (9.3.2) reduces to the usual condition that $\mathbf{G}^*$ is positive semi-definite.

As for unconstrained optimization in Volume 1, Section 2.1, it is also possible to state very similar conditions which are sufficient. A *sufficient condition for an isolated local minimizer* is that if (9.1.3) holds at any feasible point $\mathbf{x}^*$ and if

$$\mathbf{s}^{\mathrm{T}}\mathbf{W}^*\mathbf{s} > 0 \tag{9.3.5}$$

for all $\mathbf{s}$ ($\neq \mathbf{0}$) in (9.3.4), then $\mathbf{x}^*$ is an isolated local minimizer. The proof of this makes use of the fact that (9.3.5) implies the existence of a constant $a > 0$ for which $\mathbf{s}^{\mathrm{T}}\mathbf{W}^*\mathbf{s} \geqslant a\mathbf{s}^{\mathrm{T}}\mathbf{s}$ for all $\mathbf{s}$ ($\neq \mathbf{0}$) in (9.3.4). Then for any feasible step $\boldsymbol{\delta}$, (9.3.1) holds, and if $\boldsymbol{\delta}$ is sufficiently small it follows that $f(\mathbf{x}^* + \boldsymbol{\delta}) > f^*$. No regularity assumptions are made in this proof.

A simple but effective illustration of these conditions is given by Fiacco and McCormick (1968). Consider the problem

$$\begin{aligned} &\text{minimize } f(\mathbf{x}) \triangleq \tfrac{1}{2}((x_1 - 1)^2 + x_2^2) \\ &\text{subject to } c(\mathbf{x}) \triangleq -x_1 + \beta x_2^2 = 0 \end{aligned} \tag{9.3.6}$$

where β is fixed, and examine for what values of β is $\mathbf{x}^* = \mathbf{0}$ a local minimizer. The cases $\beta = \frac{1}{4}$ ($\mathbf{x}^*$ is a local minimizer) and $\beta = 1$ (not a minimizer) are illustrated in Figure 9.3.1. At $\mathbf{x}^*$, $\mathbf{g}^* = \mathbf{a}^* = (-1, 0)^{\mathrm{T}}$ so the first order conditions (9.1.3) are satisfied with $\lambda^* = 1$, and $\mathbf{x}^*$ is also feasible. The set of feasible directions in (9.3.4) is $\mathbf{s} = (0, s_2)^{\mathrm{T}}$ for all $s_2 \neq 0$. Now $\mathbf{W}^* = \begin{bmatrix} 1 & 0 \\ 0 & 1 - 2\beta \end{bmatrix}$ so $\mathbf{s}^{\mathrm{T}}\mathbf{W}^*\mathbf{s} = s_2^2(1 - 2\beta)$. Thus

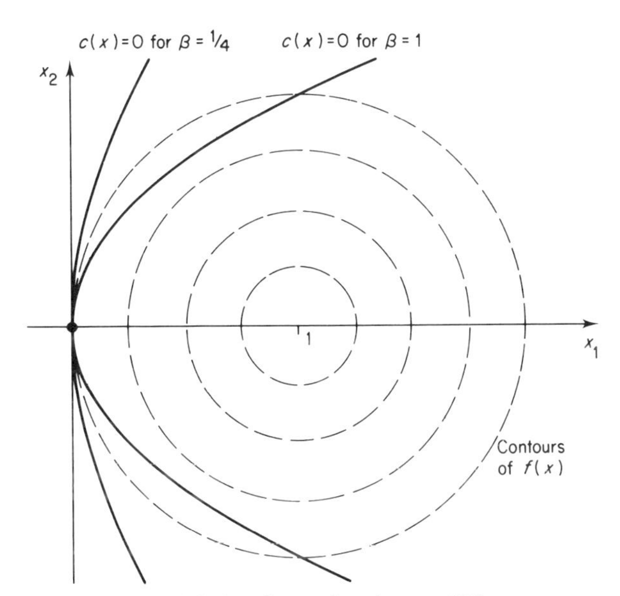

Figure 9.3.1 Second order conditions

the second order necessary conditions are violated when $\beta > \frac{1}{2}$, in which case it can be concluded that $\mathbf{x}^*$ is not a local minimizer. When $\beta < \frac{1}{2}$ then the sufficient conditions (9.3.5) and (9.3.4) are satisfied and it follows that $\mathbf{x}^*$ is a local minimizer. Only when $\beta = \frac{1}{2}$ is the result not determined by the second order conditions. This corresponds to zero curvature of $\mathscr{L}$ existing along a feasible direction, so that higher order terms become significant.

An important generalization of these conditions is to allow inequality constraints to be present. Now second order conditions are only operative along feasible stationary directions ($\mathbf{s}^{\mathrm{T}}\mathbf{g}^* = 0$) and not along ascent directions. If an inequality constraint $c_i(\mathbf{x}) \geqslant 0$ is present, and if its multiplier is $\lambda_i^* > 0$, then feasible directions for which $\mathbf{s}^{\mathrm{T}}\mathbf{a}_i^* > 0$ are ascent directions (see the discussion regarding equation (9.1.14)). Thus usually the stationary directions will satisfy

$$\mathbf{s}^{\mathrm{T}}\mathbf{a}_i^* = 0, \qquad i \in \mathscr{A}^* \tag{9.3.7}$$

and second order necessary conditions are that (9.3.2) holds for all $\mathbf{s}$ in (9.3.7). Another way of looking at this is that if $\mathbf{x}^*$ solves (7.1.1) locally, then it must solve

$$\begin{aligned} &\text{minimize } f(\mathbf{x}) \\ &\text{subject to } c_i(\mathbf{x}) = 0, \qquad i \in \mathscr{A}^* \end{aligned} \tag{9.3.8}$$

locally, and these conditions follow from (9.3.2) and (9.3.3). For sufficient conditions, if $\mathbf{x}^*$ is feasible, if (9.1.12) and (9.1.13) hold, and if $\lambda_i^* > 0 \ \forall \ i \in I^*$ (KT conditions with strict complementarity), then (9.3.5) for all $\mathbf{s} \neq \mathbf{0}$ in (9.3.7) is a sufficient condition for an isolated local minimizer. Alternatively positive curvature can be assumed on a larger subspace in which the conditions corresponding to $\lambda_i^* = 0$ are excluded. These results are justified below. An illustration of these conditions is also given by problem (9.3.6) if the constraint is changed to read $c(\mathbf{x}) \geqslant 0$. A feasible direction $\mathbf{s}$ can then have any $s_1 \leqslant 0$. However because $\lambda^* = 1 > 0$, these directions are uphill unless $s_1 = 0$ and hence stationary feasible directions are given by $\mathbf{s} = (0, s_2)^{\mathrm{T}}$ as in the equality constraint case. Thus the same conclusions about β can be deduced.

It is not difficult to make the further generalization which includes the possibility that there exists a $\lambda_i^* = 0$, $i \in \mathscr{A}^*$, in which case a stationary direction exists for which $\mathbf{s}^{\mathrm{T}}\mathbf{a}_i^* > 0$. It is also possible to allow for non-unique λ_i^* in the necessary conditions. A rigorous derivation of the second order conditions is now set out which includes these features. Given any fixed vector $\boldsymbol{\lambda}^*$, it is possible to define the set of *strictly* (or *strongly*) *active constraints*

$$\mathscr{A}_+^* = \{i \mid i \in E \text{ or } \lambda_i^* > 0\} \tag{9.3.9}$$

which is obtained by deleting indices for which $\lambda_i^* = 0$, $i \in I^*$ from $\mathscr{A}^*$. Consider all feasible directional sequences $\mathbf{x}^{(k)} \to \mathbf{x}^*$ for which

$$c_i(\mathbf{x}^{(k)}) = 0 \qquad \forall\, i \in \mathscr{A}_+^* \tag{9.3.10}$$

also holds. Define $\mathscr{G}^*$ as the resulting set of feasible directions. As in Section 9.2, consider also the set of feasible directions

$$G^* = \{\mathbf{s} \mid \mathbf{s} \neq \mathbf{0},\ \ \mathbf{a}_i^{*\mathrm{T}}\mathbf{s} = 0,\ \ i \in \mathscr{A}_+^*$$
$$\mathbf{a}_i^{*\mathrm{T}}\mathbf{s} \geqslant 0,\ \ i \in \mathscr{A}^* - \mathscr{A}_+^*\} \tag{9.3.11}$$

in which the constraints which determine $\mathscr{G}^*$ (including (9.3.10)) are linearized. By an identical argument to lemma 9.2.1 it follows that $G^* \supseteq \mathscr{G}^*$. However to state the second order necessary conditions a result going the other way is required, so another regularity assumption is made, namely that

$$G^* = \mathscr{G}^*. \tag{9.3.12}$$

Again this is a reasonable assumption to make as it is also implied by the mild conditions of lemma 9.2.2 in a similar way.

It is now possible to state the main results of this section in their full generality.

Theorem 9.3.1 (Second order necessary conditions)

If $\mathbf{x}^*$ *is a local solution to (7.1.1), and if (9.2.4) holds, then there exist multipliers* $\boldsymbol{\lambda}^*$ *such that theorem 9.1.1 is valid. For any such* $\boldsymbol{\lambda}^*$, *if (9.3.12) holds, it follows that*

$$\mathbf{s}^{\mathrm{T}}\mathbf{W}^*\mathbf{s} \geqslant 0 \qquad \forall\, \mathbf{s} \in G^*. \tag{9.3.13}$$

Proof

Let $\mathbf{s} \in G^*$. Then by (9.3.12), $\mathbf{s} \in \mathscr{G}^*$, and $\exists$ a feasible directional sequence with $\mathbf{s}^{(k)} \to \mathbf{s}$, for which (9.3.10) holds. Since either $c_i^{(k)} = 0$ for $i \in \mathscr{A}_+^*$, or $\lambda_i^* = 0$ otherwise, it follows that $f^{(k)} = \mathscr{L}(\mathbf{x}^{(k)}, \boldsymbol{\lambda}^*)$. Using (9.2.1) and (9.1.16), a Taylor series for $\mathscr{L}(\mathbf{x}, \boldsymbol{\lambda}^*)$ about $\mathbf{x}^*$ gives

$$\begin{aligned}\mathscr{L}(\mathbf{x}^{(k)}, \boldsymbol{\lambda}^*) &= \mathscr{L}(\mathbf{x}^*, \boldsymbol{\lambda}^*) + \delta^{(k)}\mathbf{s}^{(k)\mathrm{T}}\nabla_x \mathscr{L}(\mathbf{x}^*, \boldsymbol{\lambda}^*) \\ &\qquad + \tfrac{1}{2}\delta^{(k)2}\mathbf{s}^{(k)\mathrm{T}}\mathbf{W}^*\mathbf{s}^{(k)} + o(\delta^{(k)2}) \\ &= f^* + \tfrac{1}{2}\delta^{(k)2}\mathbf{s}^{(k)\mathrm{T}}\mathbf{W}^*\mathbf{s}^{(k)} + o(\delta^{(k)2}).\end{aligned} \tag{9.3.14}$$

Since $\mathbf{x}^*$ is a local minimizer, it follows for all k sufficiently large that $f^{(k)} \geqslant f^*$, and hence that

$$\mathbf{s}^{(k)\mathrm{T}}\mathbf{W}^*\mathbf{s}^{(k)} + o(1) \geqslant 0.$$

Then (9.3.13) follows in the limit. □

Theorem 9.3.2 (Second order sufficient conditions)

If at $\mathbf{x}^*$ *there exists multipliers* $\boldsymbol{\lambda}^*$ *such that conditions (9.1.16) hold, and if*

$$\mathbf{s}^{\mathrm{T}}\mathbf{W}^*\mathbf{s} > 0 \qquad \forall\, \mathbf{s} \in G^*, \tag{9.3.15}$$

then $\mathbf{x}^*$ *is an isolated local solution to (7.1.1).*

Proof

Assume $\mathbf{x}^*$ is not an isolated local minimizer, so that $\exists$ a feasible sequence $\mathbf{x}^{(k)} \to \mathbf{x}^*$ such that $f^{(k)} \leqslant f^*$. Fixing $\| \mathbf{s}^{(k)} \| = 1$, say, in (9.2.1) then this bound implies that $\exists$ a subsequence such that $\mathbf{s}^{(k)}$ converges to $\mathbf{s}$, say. By lemma 9.2.1, $\mathbf{s} \in F^*$, and by a similar argument to that in lemma 9.2.3, $\mathbf{s}^{\mathrm{T}}\mathbf{g}^* \leqslant 0$. Two cases occur, both of which imply a contradiction:

(i) $\mathbf{s} \notin G^*$; then $\exists\, i : \lambda_i^* > 0$ and $\mathbf{a}_i^{*\mathrm{T}}\mathbf{s} > 0$ whence $0 \geqslant \mathbf{s}^{\mathrm{T}}\mathbf{g}^* = \Sigma\, \mathbf{s}^{\mathrm{T}}\mathbf{a}_i^*\lambda_i^* > 0$.
(ii) $\mathbf{s} \in G^*$; by feasibility of $\mathbf{x}^{(k)}$, $\mathscr{L}(\mathbf{x}^{(k)}, \boldsymbol{\lambda}^*) < f^{(k)}$, so from (9.3.14) it follows that $0 \geqslant f^{(k)} - f^* = \frac{1}{2}\delta^{(k)^2}\mathbf{s}^{(k)^{\mathrm{T}}}\mathbf{W}^*\mathbf{s}^{(k)} + o(\delta^{(k)^2})$, and dividing by $\delta^{(k)^2}$ and taking the limit contradicts (9.3.15). □

Notice that a sufficient condition for (9.3.15) is that $\mathbf{s}^{\mathrm{T}}\mathbf{W}^*\mathbf{s} > 0\ \forall\, \mathbf{s} \neq \mathbf{0}$ such that $\mathbf{s}^{\mathrm{T}}\mathbf{a}_i^* = 0$, $i \in \mathscr{A}_+^*$, which is more convenient to verify in practice.

This treatment of second order conditions owes a lot to the presentation given by Fiacco and McCormick (1968). However it is worth pointing out that the statement of the second order necessary conditions given here is an improvement. Fiacco and McCormick define feasible directions from arcs which satisfy the conditions $c_i(\mathbf{x}) = 0$, $i \in \mathscr{A}^*$, rather than using $\mathscr{A}_+^*$ as in (9.3.10). Although this has the advantage of not involving $\boldsymbol{\lambda}^*$ and hence f, it neglects the stronger implications which can be made when $\lambda_i^* = 0$ in regular situations. Furthermore in a degenerate case it is extremely unlikely that any feasible arc will exist at all, so that the regularity assumption which they make (the second order constraint qualification) cannot usually hold. Both these objections are overcome when $\mathscr{A}_+^*$ is used. The situation is shown by the example

$$\begin{aligned} \text{minimize} \quad & f(\mathbf{x}) \triangleq x_3 - \tfrac{1}{2}x_1^2 \\ \text{subject to} \quad & c_1(\mathbf{x}) \triangleq x_3 + x_2 + x_1^2 \geqslant 0 \\ & c_2(\mathbf{x}) \triangleq x_3 - x_2 + x_1^2 \geqslant 0 \\ & c_3(\mathbf{x}) \triangleq x_3 \geqslant 0 \end{aligned} \tag{9.3.16}$$

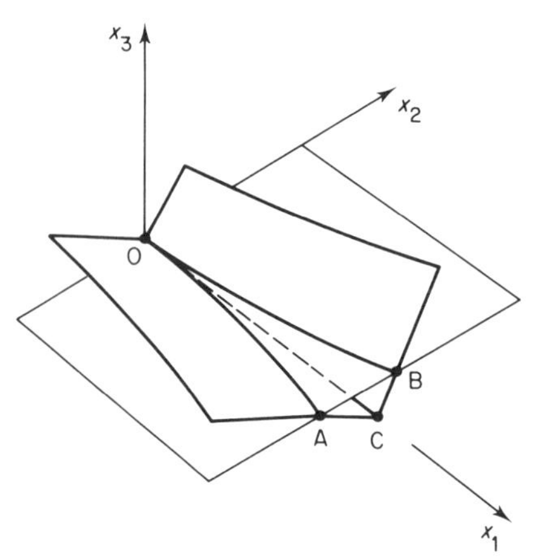

Figure 9.3.2 Regularity assumptions for second order necessary conditions

illustrated in Figure 9.3.2. Clearly $\mathbf{x}^* = \mathbf{0}$ is not a minimizing point. Yet $F^* = \mathscr{F}^*$ and $\mathbf{g}^* = (0, 0, 1)^T$, $\mathbf{a}_1^* = (0, 1, 1)^T$, $\mathbf{a}_2^* = (0, -1, 1)^T$, $\mathbf{a}_3^* = (0, 0, 1)^T$. All constraints are active and the first order conditions at $\mathbf{x}^*$ are satisfied non-uniquely by any vector $\boldsymbol{\lambda}^*$ which is a convex combination of $(0, \frac{1}{2}, \frac{1}{2})^T$ and $(0, 0, 1)^T$. Since, however, there is no point other than $\mathbf{x}^*$ which satisfies $c_i(\mathbf{x}) = 0$ for all $i \in \mathscr{A}^*$, Fiacco and McCormick's second order constraint qualification does not hold, and no implication can be made. However consider theorem 9.3.1 using the extreme multiplier vectors. In both cases the set G^* comprises any vector $\mathbf{s}$ with $s_1 \neq 0$ and $s_2 = s_3 = 0$. For $\boldsymbol{\lambda}^* = (0, \frac{1}{2}, \frac{1}{2})^T$ there is an arc (OC in Figure 9.3.2) which satisfies $c_i(\mathbf{x}) = 0$ for all $i \in \mathscr{A}_+^*$, but which is not feasible in $c_i(\mathbf{x}) \geqslant 0$, $i \in \mathscr{A}^* - \mathscr{A}_+^*$. Hence (9.3.12) does not hold for this arc and so no implication can be made. However for $\boldsymbol{\lambda}^* = (0, 0, 1)^T$, either of the arcs OA or OB in Figure 9.3.2, or any intermediate arc with $x_3 = 0$, provides a suitable arc when $s_1 > 0$ and an opposite arc can be used when $s_1 < 0$. Then since $W_{11}^* = -1$, (9.3.13) does not hold, so it can be concluded correctly that $\mathbf{x}^*$ is not a local solution to problem (9.3.16).

9.4 Convexity

The subject of convexity is often treated quite extensively in texts on optimization. My experience, however, is that much of this theory contributes little to the development and use of optimization algorithms. Applications of convexity are expressed in terms of a so-called convex programming problem. Into this category come linear programming, certain quadratic programming problems, and some more general problems, more often with linear constraints. Unfortunately many real life problems do not fit into this category, and this is especially so when the constraint functions are non-linear. On the other hand, it is possible to give quite strong (and simple) results for a convex programming problem about the global nature of solutions and the sufficiency of first order conditions. Therefore a fairly simple treatment of convexity is given in this section, aimed mainly at establishing these results for smooth problems. Some extensions of convexity theory which are helpful for handling non-smooth problems are given in Section 14.2.

First of all, a *convex set* K in $\mathbb{R}^n$ is defined by the property that for all $\mathbf{x}_0, \mathbf{x}_1 \in K$, it follows that $\mathbf{x}_\theta \in K$ where

$$\mathbf{x}_\theta = (1 - \theta)\mathbf{x}_0 + \theta\mathbf{x}_1 \qquad \forall \theta \in [0, 1]. \tag{9.4.1}$$

It follows from this that K can have no re-entrant corners (see Figure 9.4.1). A more general definition of a convex set which readily follows is that for all $\mathbf{x}_0, \mathbf{x}_1, \ldots, \mathbf{x}_m \in K$ it follows that $\mathbf{x}_\theta \in K$ where

$$\mathbf{x}_\theta = \sum_{i=0}^{m} \theta_i \mathbf{x}_i, \qquad \sum_{i=0}^{m} \theta_i = 1, \qquad \theta_i \geqslant 0. \tag{9.4.2}$$

The vector $\mathbf{x}_\theta$ in (9.4.1) or (9.4.2) is referred to as a *convex combination* of the points $\mathbf{x}_0, \mathbf{x}_1$, etc. If $\mathbf{x}_0, \mathbf{x}_1, \ldots, \mathbf{x}_m$ is a given set of points, then the set of all vectors $\mathbf{x}_\theta$ defined by (9.4.2) is a convex set referred to as the *convex hull* of the set of points. Examples of convex sets are many and include the empty set, a point,

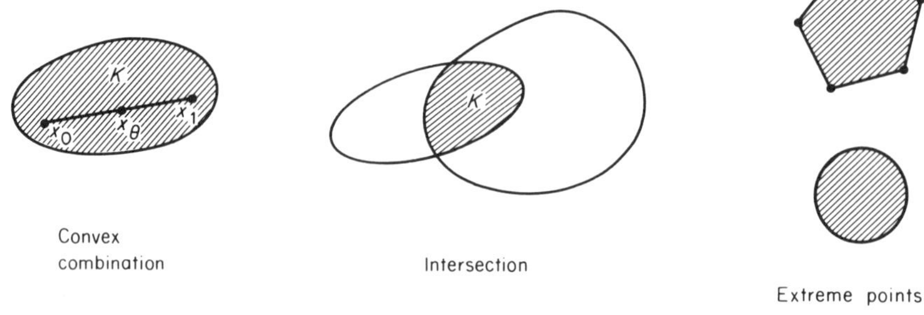

Figure 9.4.1 Convex sets

the whole of $\mathbb{R}^n$, a line or line segment, a hyperplane (or linear equation) $\mathbf{a}^T\mathbf{x} = b$, the half-space (or linear inequality) $\mathbf{a}^T\mathbf{x} \geqslant b$, the ball $\| \mathbf{x} - \mathbf{x}' \|_2 \leqslant h$, a cone, and many others. A simple result is the following.

Lemma 9.4.1

If K_i, $i = 1, 2, \ldots, m$, are convex sets then the intersection $K = \cap_i^m K_i$ is also a convex set.

Proof

Because $\mathbf{x}_0, \mathbf{x}_1 \in K$ implies $\mathbf{x}_0, \mathbf{x}_1 \in K_i$, $i = 1, 2, \ldots, m$. □

This result is also illustrated in Figure 9.4.1. It shows that the feasible region in a linear or quadratic programming problem is a convex set because it is the intersection of hyperplanes and half-spaces.

Another useful concept is that of an *extreme point* of a convex set K. An extreme point $\mathbf{x}$ is one which may not lie interior to any line segment contained in K, that is $\mathbf{x} = (1 - \theta)\mathbf{x}_0 + \theta\mathbf{x}_1$ for $\mathbf{x}_0, \mathbf{x}_1 \in K$, $\theta \in (0, 1)$ implies that $\mathbf{x} = \mathbf{x}_0 = \mathbf{x}_1$. The vertices of a regular polygon or any point on the circumference of a circle are examples of extreme points (Figure 9.4.1). Another example is the basic feasible solution $\mathbf{x}$ of the convex set K defined by the feasible region R of a linear programming problem in standard form (8.1.1). Details of the relationship between an extreme point and a b.f.s. are sketched out in Questions 9.20, 9.21, and 9.22.

The other fundamental idea is that of a *convex function*. The discussion is limited to continuous functions defined on a convex set K, to eliminate trivial cases. Then a convex function $f(\mathbf{x})$ is defined by the condition that for any $\mathbf{x}_0, \mathbf{x}_1 \in K$ it follows that

$$f_\theta \leqslant (1 - \theta)f_0 + \theta f_1 \qquad \forall \theta \in [0, 1] \tag{9.4.3}$$

where f_θ refers to $f(\mathbf{x}_\theta)$, etc., and where $\mathbf{x}_\theta$ is given by (9.4.1). The right hand side of (9.4.3) is the chord joining $(\mathbf{x}_0, f_0)$ to $(\mathbf{x}_1, f_1)$ on the graph of $f(\mathbf{x})$ (see Figure 9.4.2), and the inequality expresses the fact that the graph of a convex function

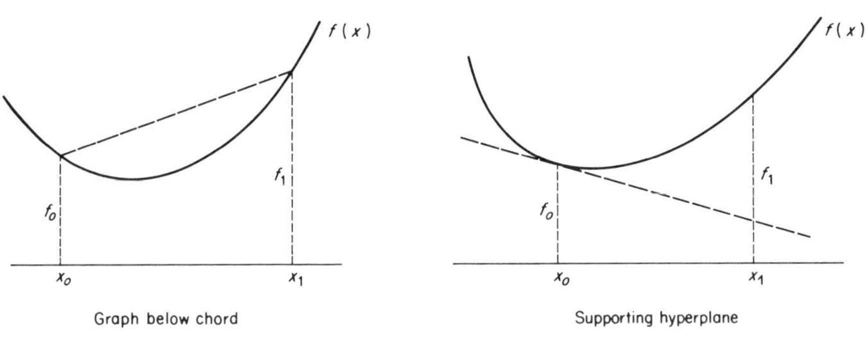

Figure 9.4.2 Convex functions

always lies below (or along) the chord. If K is an open set and $f(\mathbf{x})$ is differentiable ($\mathbb{C}^1$) on K, an equivalent definition of convexity is that for all $\mathbf{x}_0, \mathbf{x}_1 \in K$ it follows that

$$f_1 \geqslant f_0 + (\mathbf{x}_1 - \mathbf{x}_0)^T \nabla f_0. \tag{9.4.4}$$

This definition shows that the graph of $f(\mathbf{x})$ must lie above (or along) the linearization of $f(\mathbf{x})$ about $\mathbf{x}_0$, and hence that this linearization acts as a *supporting hyperplane* for the convex function. The equivalence of (9.4.3) and (9.4.4) is readily demonstrated. It follows from (9.4.3) that

$$\frac{f_\theta - f_0}{\theta} \leqslant f_1 - f_0 \, .$$

But regarding $\mathbf{x}_\theta$ as a point on the line $\mathbf{x}_\theta = \mathbf{x}_0 + \theta(\mathbf{x}_1 - \mathbf{x}_0)$, and taking the limit $\theta \downarrow 0$, then (9.4.4) follows. Conversely if (9.4.4) holds then expanding about $\mathbf{x}_\theta$,

$$f_1 \geqslant f_\theta + (\mathbf{x}_1 - \mathbf{x}_\theta)^T \nabla f_\theta$$
$$f_0 \geqslant f_\theta + (\mathbf{x}_0 - \mathbf{x}_\theta)^T \nabla f_\theta$$

so that

$$(1-\theta) f_0 + \theta f_1 \geqslant f_\theta + ((1-\theta)(\mathbf{x}_0 - \mathbf{x}_\theta) + \theta(\mathbf{x}_1 - \mathbf{x}_\theta))^T \nabla f_\theta = f_\theta$$

which is (9.4.3).

Another result which follows from (9.4.4) is that

$$(\mathbf{x}_1 - \mathbf{x}_0)^T \nabla f_1 \geqslant f_1 - f_0 \geqslant (\mathbf{x}_1 - \mathbf{x}_0)^T \nabla f_0. \tag{9.4.5}$$

This illustrates the fact that the slope of a convex function is non-decreasing along any line. In fact this result (for the directional derivative) can also be proved to hold for a non-differentiable convex function. Finally for twice differentiable ($\mathbb{C}^2$) convex functions and K open, another equivalent definition of a convex function is that

$$\nabla^2 f_0 \text{ is positive semi-definite} \qquad \forall \mathbf{x}_0 \in K. \tag{9.4.6}$$

Thus convex functions are typified by having non-negative curvature. To establish this result, let $\mathbf{s} \neq \mathbf{0}$ and let $\mathbf{x}_1 = \mathbf{x}_0 + \alpha\mathbf{s}$. Then a Taylor series for ∇f_1 gives

$$\nabla f_1 = \nabla f_0 + \alpha \nabla^2 f_0 \mathbf{s} + o(\alpha). \tag{9.4.7}$$

Substituting in (9.4.5) and taking the limit $\alpha \to 0$ gives $\mathbf{s}^{\mathrm{T}} \nabla^2 f_0 \mathbf{s} \geqslant 0$, which is (9.4.6). Conversely, a Taylor series for f_1 (with $\theta \in [0, 1]$) and (9.4.6) yields

$$\begin{aligned} f_1 &= f_0 + (\mathbf{x}_1 - \mathbf{x}_0)^{\mathrm{T}} \nabla f_0 + \tfrac{1}{2}\alpha^2 \mathbf{s}^{\mathrm{T}} \nabla^2 f_\theta \mathbf{s} \\ &\geqslant f_0 + (\mathbf{x}_1 - \mathbf{x}_0)^{\mathrm{T}} \nabla f_0, \end{aligned}$$

which is (9.4.4).

Other definitions which are closely related to that of a convex function are the following. A *strictly convex function* is defined whenever the inequality in (9.4.3) is strict for all distinct $\mathbf{x}_0, \mathbf{x}_1$ and $\theta \in (0, 1)$. For $\mathbb{C}^1$ functions (9.4.4) again provides an equivalent definition when the inequality is strict and $\mathbf{x}_0 \neq \mathbf{x}_1$. However for $\mathbb{C}^2$ functions, (9.4.6) is not equivalent, since although $\nabla^2 f_0$ positive definite $\forall\, \mathbf{x}_0 \in K$ implies that $f(\mathbf{x})$ is strictly convex, the converse result does not hold (for example x^4 is strictly convex but has zero Hessian at $x = 0$). A *concave function* $f(\mathbf{x})$ is defined as one for which $-f(\mathbf{x})$ is convex, and so is associated with non-increasing slope or non-positive curvature. Likewise a *strictly concave function* $f(\mathbf{x})$ has $-f(\mathbf{x})$ strictly convex.

Examples of convex functions include the linear function, which is both convex and concave. A quadratic function is convex when the Hessian is positive semi-definite and strictly convex when the Hessian is positive definite. Another convex function is $\|\mathbf{x}\|$ (for any norm). However $\|\mathbf{c}(\mathbf{x})\|$, where $\mathbf{c}(\mathbf{x})$ maps $\mathbb{R}^n$ on $\mathbb{R}^m$, is not generally convex, except when $\mathbf{c}(\mathbf{x})$ is a linear function. A transformation which preserves convexity is expressed in the next result.

Lemma 9.4.2

If $f_i(\mathbf{x})$, $i = 1, 2, \ldots, m$, are convex functions on a convex set K, and if $\lambda_i \geqslant 0$, then $\Sigma_i \lambda_i f_i(\mathbf{x})$ is a convex function on K.

Proof

Take $\mathbf{x}_\theta$ as in (9.4.1) and use the definition of a convex function. □

The problem of minimizing a convex function on a convex set K is said to be a *convex programming problem*. Such a problem arises when (7.1.1) can be expressed as

$$\begin{aligned} &\text{minimize } f(\mathbf{x}) \\ &\text{subject to } \mathbf{x} \in K \triangleq \{\mathbf{x} \mid c_i(\mathbf{x}) \geqslant 0,\ i = 1, 2, \ldots, m\} \end{aligned} \tag{9.4.8}$$

where $f(\mathbf{x})$ is a convex function on K, and the functions $c_i(\mathbf{x})$, $i = 1, 2, \ldots, m$, are concave on $\mathbb{R}^n$. That the feasible region in (9.4.8) is a convex set is a consequence of the following lemma and of lemma 9.4.1.

Lemma 9.4.3

If $c(\mathbf{x})$ is a concave function then the set

$$S(k) = \{\mathbf{x} \mid c(\mathbf{x}) \geqslant k\}$$

is convex.

Proof

For $\mathbf{x}_0, \mathbf{x}_1 \in S(k)$, and if $\mathbf{x}_\theta$ is given by (9.4.1), it follows by concavity that

$$c_\theta \geqslant (1-\theta)c_0 + \theta c_1 \geqslant (1-\theta)k + \theta k$$

by definition of $S(k)$. Thus $\mathbf{x}_\theta \in S(k)$ which is therefore convex. □

Notice that the system (9.4.8) does not allow general equality constraints, although it is possible to include any linear equality $c(\mathbf{x}) = 0$ as the intersection of $c(\mathbf{x}) \geqslant 0$ and $-c(\mathbf{x}) \geqslant 0$. An example of a convex programming problem is therefore the linear programming problem (linear objective function, linear equality and inequality constraints). Another example is the quadratic programming problem (quadratic objective function, linear constraints) when the Hessian of the quadratic function is positive semi-definite. However it should be noticed that quadratic programming problems can (and do) exist which have well-behaved local (or even unique global) solutions, yet for which the Hessian is indefinite. It is then erroneous to assume that some of the consequences of convexity (see next section) will apply in this case.

One main attraction of convexity is that it provides an overall assumption whereby the existence of local but not global solutions can be excluded, as the next theorem shows. An additional assumption, given in the corollary, enables uniqueness of global solutions to be established.

Theorem 9.4.1

Every local solution $\mathbf{x}^*$ to a convex programming problem (9.4.8) is a global solution, and the set of global solutions S is convex.

Proof

Let $\mathbf{x}^*$ be a local but not global solution. Then $\exists \mathbf{x}_1 \in K$ such that $f_1 < f^*$. For $\theta \in [0, 1]$, consider $\mathbf{x}_\theta = (1-\theta)\mathbf{x}^* + \theta\mathbf{x}_1 \in K$ by convexity of K. By convexity of f, $f_\theta \leqslant (1-\theta)f^* + \theta f_1 = f^* + \theta(f_1 - f^*) < f^*$. In the limit $\theta \to 0$ the local solution property is contradicted. Thus local solutions are global. Now let $\mathbf{x}_0, \mathbf{x}_1 \in S$ and define $\mathbf{x}_\theta$ by (9.4.1). By the global solution property, $f_\theta \geqslant f_0 = f_1$. By convexity $f_\theta \leqslant (1-\theta)f_0 + \theta f_1 = f_0 = f_1$. Thus $f_\theta = f_0 = f_1$ and so $\mathbf{x}_\theta \in S$, so S is convex. □

Corollary

If also $f(\mathbf{x})$ is strictly convex on K then any global solution is unique.

Proof

Let $\mathbf{x}_0 \neq \mathbf{x}_1 \in S$ and $\theta \in (0, 1)$. As above, but using strict convexity, both $f_\theta \geqslant f_0 = f_1$ and $f_\theta < f_0 = f_1$, which is a contradiction. □

A second attraction of a convexity assumption is that it provides a framework within which the first order (Kuhn–Tucker) conditions are sufficient for a local solution, as the next theorem shows. In common with other sufficient conditions (theorem 9.3.2), no regularity assumption is required.

Theorem 9.4.2

In the convex programming problem (9.4.8), if $f(\mathbf{x})$ and $c_i(\mathbf{x})$, $i = 1, 2, \ldots, m$, are $\mathbb{C}^1$ functions on K and if conditions (9.1.16) hold at $\mathbf{x}^*$, then $\mathbf{x}^*$ is a global solution to (9.4.8).

Proof

Let $\mathbf{x}' \in K$, $\mathbf{x}' \neq \mathbf{x}^*$. Then since $\lambda_i^* \geqslant 0$ and $c_i' \geqslant 0$,

$$
\begin{aligned}
f' &\geqslant f' - \sum_{i=1}^{m} \lambda_i^* c_i' \\
&\geqslant f^* + (\mathbf{x}' - \mathbf{x}^*)^{\mathrm{T}} \mathbf{g}^* - \Sigma \lambda_i^* (c_i^* + (\mathbf{x}' - \mathbf{x}^*)^{\mathrm{T}} \mathbf{a}_i^*),
\end{aligned}
$$

using (9.4.4), since f is convex and the c_i are concave. Then from (9.1.16), $\lambda_i^* c_i^* = 0$ and $\mathbf{g}^* = \Sigma \mathbf{a}_i^* \lambda_i^*$ show that $f' \geqslant f^*$ and hence $\mathbf{x}^*$ is a global solution. □

To summarize, for $\mathbf{x}^*$ to solve the convex programming problem (9.4.8), conditions (9.1.16) and a regularity assumption (9.2.4) are necessary, whilst (9.1.16) alone are sufficient. Of course (9.2.4) is implied by the constraint qualification $F^* = \mathscr{F}^*$ which in turn is implied by the assumptions of lemma 9.2.2, as before. It should be emphasized that it is not possible to dispense with the regularity assumption (9.2.4). An example of a convex feasible region in which $F^* \neq \mathscr{F}^*$ (in both $\mathbb{R}^2$ and $\mathbb{R}^3$) is given by the inequalities $x_2 \geqslant x_1^2$ and $x_2 \leqslant 0$ at $\mathbf{x}^* = \mathbf{0}$. An illustration of a (regular) convex programming problem is provided by problem (9.1.15). The objective function is linear and hence convex. The Hessian matrices $\nabla^2 c_1 = \begin{bmatrix} -2 & 0 \\ 0 & 0 \end{bmatrix}$ and $\nabla^2 c_2 = \begin{bmatrix} -2 & 0 \\ 0 & -2 \end{bmatrix}$ are negative semi-definite so the constraint functions are concave. It can be seen from Figure 7.1.3 that the feasible region is a convex set. Since first order conditions hold at $\mathbf{x}^* = (1/\sqrt{2}, 1/\sqrt{2})^{\mathrm{T}}$ it follows from theorem 9.4.2 that $\mathbf{x}^*$ is a global solution to the problem.

It can be seen from the above theorems that a convexity assumption is in the nature of a curvature or second order assumption, in that the directional derivative is non-decreasing along any line. Therefore, although convexity gives useful results for certain special types of problem, it is an assumption which does not often hold in the general case. A weakening of the assumptions is to require convexity of $f(\mathbf{x})$ and $-c_i(\mathbf{x})$, $i = 1, 2, \ldots, m$, on a ball about $\mathbf{x}^*$. In this case theorem 9.4.2 can be interpreted as stating that local convexity and (9.1.16) is sufficient for a local solution. Even in this form, however, the assumption is not valid for many problems. The requirement is essentially that the matrices $\nabla^2 f^*$ and $-\nabla^2 c_i^*$, $i = 1, 2, \ldots, m$, are positive semi-definite. The second order conditions of theorem 9.3.2 involve the much weaker assumption (9.3.14) that $\mathbf{W}^* = \nabla^2 f^* - \Sigma \lambda_i^* \nabla^2 c_i^*$ is positive definite only on a restricted subspace, and it is only rarely that this condition does not hold at a local solution. Thus convexity does not provide a valid model situation for the general non-linear programming problem from which to derive an algorithm, whereas the second order conditions can do this (see Section 12.3).

9.5 Duality

The concept of duality occurs widely in the mathematical programming literature. The aim is to provide an alternative formulation of a mathematical programming problem which is more convenient computationally or has some theoretical significance. The original problem is referred to as the *primal* and the transformed problem as the *dual*. Usually the variables in the dual (or some of them) can be interpreted as Lagrange multipliers and take the value $\boldsymbol{\lambda}^*$ at the dual solution, where $\boldsymbol{\lambda}^*$ is a multiplier vector associated with a primal solution $\mathbf{x}^*$. In this sense the method of Lagrange multipliers (Section 9.1) might be thought of as a dual method. Usually, however, there is also present an objective function (often related to the Lagrangian function (9.1.6)) which has to be optimized. Duality theory of this kind is associated with a convex programming problem as the primal, and it is important to realize that if the primal is not convex then the dual problem may well not have a solution from which the primal solution can be derived (see Question 9.23). Thus it is not valid to apply the duality transformation as a general purpose solution technique.

In this book the emphasis is on duality transformations which are convenient computationally, and it seems that these can largely be deduced as a consequence of one particular form known as the *Wolfe dual* (Wolfe, 1961). This is a very simple result, closely related to the first order conditions (9.1.16), but which replaces constraint conditions by an optimality requirement on the Lagrangian function.

Theorem 9.5.1

If $\mathbf{x}^*$ solves the convex programming primal problem (9.4.8), if f and c_i, $i = 1, 2, \ldots, m$, are $\mathbb{C}^1$ functions, and if the regularity assumption (9.2.4) holds, then

$\mathbf{x}^*$, $\boldsymbol{\lambda}^*$ *solves the dual problem*

$$\begin{aligned} &\underset{\mathbf{x},\boldsymbol{\lambda}}{\text{maximize}}\ \mathscr{L}(\mathbf{x},\boldsymbol{\lambda}) \\ &\text{subject to } \nabla_x \mathscr{L}(\mathbf{x},\boldsymbol{\lambda}) = \mathbf{0}, \qquad \boldsymbol{\lambda} \geqslant \mathbf{0}. \end{aligned} \tag{9.5.1}$$

Furthermore the minimum primal and maximum dual function values are equal, that is $f^* = \mathscr{L}(\mathbf{x}^*, \boldsymbol{\lambda}^*)$.

Proof

The conditions of the theorem are those of theorem 9.1.1 so it follows that multipliers $\boldsymbol{\lambda}^* \geqslant \mathbf{0}$ exist such that $\nabla_x \mathscr{L}(\mathbf{x}^*, \boldsymbol{\lambda}^*) = \mathbf{0}$ (dual feasibility), and $\lambda_i^* c_i^* = 0$, $i = 1, 2, \ldots, m$, from which $f^* = \mathscr{L}(\mathbf{x}^*, \boldsymbol{\lambda}^*)$ follows. Now let $\mathbf{x}$, $\boldsymbol{\lambda}$ be dual feasible. Then using $\boldsymbol{\lambda} \geqslant \mathbf{0}$, convexity of $\mathscr{L}$ (lemma 9.4.2), and $\nabla_x \mathscr{L} = \mathbf{0}$ in turn,

$$\begin{aligned} \mathscr{L}(\mathbf{x}^*, \boldsymbol{\lambda}^*) = f^* \geqslant f^* - \sum_i \lambda_i c_i^* &= \mathscr{L}(\mathbf{x}^*, \boldsymbol{\lambda}) \\ &\geqslant \mathscr{L}(\mathbf{x}, \boldsymbol{\lambda}) + (\mathbf{x}^* - \mathbf{x})^{\mathrm{T}} \nabla_x \mathscr{L}(\mathbf{x}, \boldsymbol{\lambda}) \\ &= \mathscr{L}(\mathbf{x}, \boldsymbol{\lambda}). \end{aligned}$$

Hence $\mathscr{L}(\mathbf{x}^*, \boldsymbol{\lambda}^*) \geqslant \mathscr{L}(\mathbf{x}, \boldsymbol{\lambda})$ and so $\mathbf{x}^*$, $\boldsymbol{\lambda}^*$ solves the dual. □

An apparent disadvantage of the Wolfe dual is that the symmetry in which the primal is the dual of the dual is not generally present, since the dual may not even be a convex programming problem. However this does not affect the computational value of the Wolfe dual, and in fact for some of the transformed problems which can be deduced directly from the Wolfe dual, this symmetry does hold. It is also important to consider what happens in the dual if the primal has no solution; this point is taken up at the end of this section.

The first example of the dual transformation is in linear programming (LP). The primal problem

$$\begin{aligned} &\underset{\mathbf{x}}{\text{minimize}}\ f_0 + \mathbf{c}^{\mathrm{T}}\mathbf{x} \\ &\text{subject to } \mathbf{A}^{\mathrm{T}}\mathbf{x} \geqslant \mathbf{b} \end{aligned} \tag{9.5.2}$$

is not in standard form. Section 7.2 describes how to obtain a standard form by including both slack variables and also non-negative variables $\mathbf{x}^+$ and $\mathbf{x}^-$ (since the bounds $\mathbf{x} \geqslant \mathbf{0}$ are not present). However if there are many more inequalities than variables ($m \gg n$) it is much more attractive to disregard this transformation and instead to use the Wolfe dual. This is valid because the linear functions in (9.5.2) imply both that (9.5.2) is a convex programming problem, and that the regularity assumption (9.2.4) is true. The dual problem (9.5.1) becomes

$$\begin{aligned} &\underset{\mathbf{x},\boldsymbol{\lambda}}{\text{maximize}}\ f_0 + \mathbf{c}^{\mathrm{T}}\mathbf{x} - \boldsymbol{\lambda}^{\mathrm{T}}(\mathbf{A}^{\mathrm{T}}\mathbf{x} - \mathbf{b}) \\ &\text{subject to } \mathbf{c} - \mathbf{A}\boldsymbol{\lambda} = \mathbf{0}, \qquad \boldsymbol{\lambda} \geqslant \mathbf{0}. \end{aligned}$$

On substituting for $\mathbf{c}$ in the objective function, the problem becomes independent of $\mathbf{x}$, giving

$$\begin{aligned} &\underset{\boldsymbol{\lambda}}{\text{maximize}}\ f_0 + \mathbf{b}^{\mathrm{T}}\boldsymbol{\lambda} \\ &\text{subject to } \mathbf{A}\boldsymbol{\lambda} = \mathbf{c}, \qquad \boldsymbol{\lambda} \geqslant \mathbf{0} \end{aligned} \tag{9.5.3}$$

which is an LP problem in standard form. Once this problem has been solved for $\boldsymbol{\lambda}^*$, the variables λ_i^*, $i \in B^*$, will have $\lambda_i^* > 0$ (ignoring the possibility of degeneracy) which implies that $c_i^* = 0$ from (9.1.16). Thus the solution of the square system of equations $\mathbf{a}_i^{\mathrm{T}}\mathbf{x} = b_i$, $i \in B^*$, determines the vector $\mathbf{x}^*$ which minimizes the primal.

Another example in linear programming arises by considering the primal problem

$$\begin{aligned} &\underset{\mathbf{x}}{\text{minimize}}\ f_0 + \mathbf{c}^{\mathrm{T}}\mathbf{x} \\ &\text{subject to } \mathbf{A}^{\mathrm{T}}\mathbf{x} \geqslant \mathbf{b}, \qquad \mathbf{x} \geqslant \mathbf{0}. \end{aligned} \tag{9.5.4}$$

Introducing multipliers $\boldsymbol{\lambda}$ and $\boldsymbol{\pi}$ respectively, then the Wolfe dual (9.5.1) becomes

$$\begin{aligned} &\underset{\mathbf{x},\,\boldsymbol{\lambda},\,\boldsymbol{\pi}}{\text{maximize}}\ f_0 + \mathbf{c}^{\mathrm{T}}\mathbf{x} - \boldsymbol{\lambda}^{\mathrm{T}}(\mathbf{A}^{\mathrm{T}}\mathbf{x} - \mathbf{b}) - \boldsymbol{\pi}^{\mathrm{T}}\mathbf{x} \\ &\text{subject to } \mathbf{c} - \mathbf{A}\boldsymbol{\lambda} - \boldsymbol{\pi} = \mathbf{0}, \qquad \boldsymbol{\lambda} \geqslant \mathbf{0},\ \boldsymbol{\pi} \geqslant \mathbf{0}. \end{aligned}$$

Substituting for $\mathbf{c}$ in the objective eliminates both the $\mathbf{x}$ and $\boldsymbol{\pi}$ variables to give the problem

$$\begin{aligned} &\underset{\boldsymbol{\lambda}}{\text{maximize}}\ f_0 + \mathbf{b}^{\mathrm{T}}\boldsymbol{\lambda} \\ &\text{subject to } \mathbf{A}\boldsymbol{\lambda} \leqslant \mathbf{c}, \qquad \boldsymbol{\lambda} \geqslant 0. \end{aligned} \tag{9.5.5}$$

Since this problem is like (9.5.4) in having both inequality constraints and bounds it is often referred to as the *symmetric dual*. Its use is advantageous when $\mathbf{A}$ has many fewer rows than columns. Then (after adding slack variables) the standard form arising from (9.5.5) is much smaller than that arising from (9.5.4).

An extension of problem (9.5.4) is to include equality constraints as

$$\begin{aligned} &\underset{\mathbf{x}}{\text{minimize}}\ f_0 + \mathbf{c}^{\mathrm{T}}\mathbf{x} \\ &\text{subject to } \mathbf{A}_1^{\mathrm{T}}\mathbf{x} \geqslant \mathbf{b}_1 \\ &\qquad\qquad\ \mathbf{A}_2^{\mathrm{T}}\mathbf{x} = \mathbf{b}_2, \qquad \mathbf{x} \geqslant \mathbf{0}. \end{aligned} \tag{9.5.6}$$

This problem can be reduced to (9.5.4) by writing the equality constraints as $\mathbf{A}_2^{\mathrm{T}}\mathbf{x} \geqslant \mathbf{b}_2$ and $-\mathbf{A}_2^{\mathrm{T}}\mathbf{x} \geqslant -\mathbf{b}_2$. Introducing non-negative multipliers $\boldsymbol{\lambda}_1, \boldsymbol{\lambda}_2^+, \boldsymbol{\lambda}_2^-$, and $\boldsymbol{\pi}$, and defining $\boldsymbol{\lambda}_2 = \boldsymbol{\lambda}_2^+ - \boldsymbol{\lambda}_2^-$, $\boldsymbol{\lambda} = \begin{pmatrix}\boldsymbol{\lambda}_1\\ \boldsymbol{\lambda}_2\end{pmatrix}$, $\mathbf{b} = \begin{pmatrix}\mathbf{b}_1\\ \mathbf{b}_2\end{pmatrix}$, and $\mathbf{A} = [\mathbf{A}_1 : \mathbf{A}_2]$, then from (9.5.5) the dual can be written

$$\begin{aligned} &\underset{\boldsymbol{\lambda}}{\text{minimize}}\ f_0 + \mathbf{b}^{\mathrm{T}}\boldsymbol{\lambda} \\ &\text{subject to } \mathbf{A}\boldsymbol{\lambda} \leqslant \mathbf{c}, \qquad \boldsymbol{\lambda}_1 \geqslant \mathbf{0}. \end{aligned} \tag{9.5.7}$$

In general the strengthening of a primal linear inequality constraint to become an equality just causes the bound on the multiplier in the dual to be relaxed. It is readily observed that all these duals arising from linear programming have the symmetry property that the dual of the dual simplifies to give the primal problem.

Another useful application of the dual is to solve the primal quadratic programming (QP) problem

$$\begin{aligned}&\underset{\mathbf{x}}{\text{minimize}}\ \tfrac{1}{2}\mathbf{x}^{\mathrm{T}}\mathbf{G}\mathbf{x}+\mathbf{g}^{\mathrm{T}}\mathbf{x}\\&\text{subject to } \mathbf{A}^{\mathrm{T}}\mathbf{x}\geqslant\mathbf{b}\end{aligned}\tag{9.5.8}$$

in which $\mathbf{G}$ is positive definite. The assumptions of theorem 9.5.1 are again clearly satisfied, so from (9.5.1) the Wolfe dual is

$$\begin{aligned}&\underset{\mathbf{x},\boldsymbol{\lambda}}{\text{maximize}}\ \tfrac{1}{2}\mathbf{x}^{\mathrm{T}}\mathbf{G}\mathbf{x}+\mathbf{g}^{\mathrm{T}}\mathbf{x}-\boldsymbol{\lambda}^{\mathrm{T}}(\mathbf{A}^{\mathrm{T}}\mathbf{x}-\mathbf{b})\\&\text{subject to } \mathbf{G}\mathbf{x}+\mathbf{g}-\mathbf{A}\boldsymbol{\lambda}=\mathbf{0},\qquad \boldsymbol{\lambda}\geqslant\mathbf{0}.\end{aligned}$$

Using the constraints of this dual to eliminate $\mathbf{x}$ from the objective function gives the problem

$$\begin{aligned}&\underset{\boldsymbol{\lambda}}{\text{maximize}}\ -\tfrac{1}{2}\boldsymbol{\lambda}^{\mathrm{T}}(\mathbf{A}^{\mathrm{T}}\mathbf{G}^{-1}\mathbf{A})\boldsymbol{\lambda}+\boldsymbol{\lambda}^{\mathrm{T}}(\mathbf{b}-\mathbf{A}^{\mathrm{T}}\mathbf{G}^{-1}\mathbf{g})-\tfrac{1}{2}\mathbf{g}^{\mathrm{T}}\mathbf{G}^{-1}\mathbf{g}\\&\text{subject to } \boldsymbol{\lambda}\geqslant\mathbf{0}.\end{aligned}\tag{9.5.9}$$

This is again a quadratic programming problem in the multipliers $\boldsymbol{\lambda}$, but subject only to the bounds $\boldsymbol{\lambda}\geqslant\mathbf{0}$. As such this can make the problem easier to solve. Once the solution $\boldsymbol{\lambda}^*$ to (9.5.9) has been found then $\mathbf{x}^*$ is obtained by solving the equations $\mathbf{G}\mathbf{x}=\mathbf{A}\boldsymbol{\lambda}^*-\mathbf{g}$ used to eliminate $\mathbf{x}$. In a similar way to (9.5.6), the addition of equality constraints into (9.5.8) causes no significant difficulty.

An example of duality for a non-quadratic objective function is in the solution of maximum entropy problems in information theory (Eriksson, 1980). The primal problem is

$$\begin{aligned}&\underset{\mathbf{x}}{\text{minimize}}\ \sum_{i=1}^{n} x_i\log(x_i/c_i)\\&\text{subject to } \mathbf{A}^{\mathrm{T}}\mathbf{x}=\mathbf{b},\qquad \mathbf{x}\geqslant\mathbf{0}\end{aligned}\tag{9.5.10}$$

where the constants c_i are positive and $\mathbf{A}$ is $n\times m$ with $m\ll n$. One possible method of solution is to eliminate variables as described in Section 11.1. However since $m\ll n$ this does not reduce the size of the problem very much and it is more advantageous to consider solving the dual. For $c>0$, the function $f(x)=x\log(x/c)$ is convex on $x>0$ since $f''(x)>0$ and hence by continuity is convex on $x\geqslant 0$ $(0\log 0=0)$. Thus (9.5.10) is a convex programming problem. Introducing multipliers $\boldsymbol{\lambda}$ and $\boldsymbol{\pi}$ respectively, the condition $\nabla_x\mathscr{L}=\mathbf{0}$ in (9.5.1) becomes

$$\log(x_i/c_i)+1-\mathbf{e}_i^{\mathrm{T}}\mathbf{A}\boldsymbol{\lambda}-\pi_i=0\tag{9.5.11}$$

for $i=1,2,\ldots,n$. It follows that

$$x_i = c_i \exp(\mathbf{e}_i^T \mathbf{A} \boldsymbol{\lambda} + \pi_i - 1) \tag{9.5.12}$$

and hence that $x_i^* > 0 \ \forall \ i$. Thus the bounds $\mathbf{x} \geqslant \mathbf{0}$ are inactive with $\boldsymbol{\pi}^* = \mathbf{0}$ and can be ignored. After relaxing the bounds $\boldsymbol{\lambda} \geqslant \mathbf{0}$ to allow for equality constraints the Wolfe dual (9.5.1) thus becomes

$$\underset{\mathbf{x}, \boldsymbol{\lambda}}{\text{maximize}} \quad \Sigma x_i \log(x_i/c_i) - \boldsymbol{\lambda}^T(\mathbf{A}^T\mathbf{x} - \mathbf{b})$$

subject to (9.5.11). These constraints can be written as in (9.5.12) and used to eliminate x_i from the dual objective function. Thus the optimum multipliers $\boldsymbol{\lambda}^*$ can be found by solving

$$\underset{\boldsymbol{\lambda}}{\text{minimize}} \quad h(\boldsymbol{\lambda}) \triangleq \sum_i c_i \exp(\mathbf{e}_i^T \mathbf{A} \boldsymbol{\lambda} - 1) - \boldsymbol{\lambda}^T \mathbf{b}$$

without any constraints. This is a problem in only m variables. It is easy to show that

$$\nabla_\lambda h = \mathbf{A}^T\mathbf{x} - \mathbf{b}$$
$$\nabla_\lambda^2 h = \mathbf{A}^T[\operatorname{diag} x_i]\mathbf{A}$$

where x_i is dependent on $\boldsymbol{\lambda}$ through (9.5.12). Since these derivatives are available, and the Hessian is positive definite, Newton's method with line search (Volume 1, Section 3.1) can be used to solve the problem. For large sparse primal problems Eriksson (1980) has also investigated the use of conjugate gradient methods.

An example of the use of duality to solve convex programming problems in which both the objective and constraint functions are non-linear arises in the study of geometric programming. A description of this technique, and of how the Wolfe dual enables the problem to be solved efficiently, is given in detail in Section 13.2.

It is important to consider what happens if the primal has no solution. In particular it is desirable that the dual also does not have a solution, and it is shown that this is often but not always true. The primal problem can fail to have a solution in a number of ways and firstly the case is considered in which the primal is *unbounded* ($f(\mathbf{x}) \to -\infty$). A useful result is then as follows.

Theorem 9.5.2

If V is the infimum of $f(\mathbf{x})$ for feasible $\mathbf{x}$ in the primal problem (9.4.8), and v is the supremum of $\mathscr{L}(\mathbf{x}, \boldsymbol{\lambda})$ for feasible $\mathbf{x}, \boldsymbol{\lambda}$ in the dual, then $V \geqslant v$.

Proof

Let $\mathbf{x}'$ be primal feasible and $\mathbf{x}, \boldsymbol{\lambda}$ dual feasible. Then by convexity of f, dual feasibility, concavity of c_i, and non-negativity of c_i' and λ_i in turn, it follows that

$$f' - f \geqslant \mathbf{g}^T(\mathbf{x}' - \mathbf{x}) = \sum_i \lambda_i \mathbf{a}_i^T(\mathbf{x}' - \mathbf{x})$$
$$\geqslant \sum_i \lambda_i(c_i' - c_i) \geqslant -\sum_i \lambda_i c_i.$$

Hence $f' \geqslant f - \Sigma_i \lambda_i c_i = \mathscr{L}$, so taking the infimum over all $\mathbf{x}'$ and the supremum over all $\mathbf{x}, \boldsymbol{\lambda}$ it follows that $V \geqslant v$. □

If the primal problem is unbounded it follows that $V = v = -\infty$ and this is not possible if any feasible $\mathbf{x}, \boldsymbol{\lambda}$ exists. Thus an *unbounded primal implies an inconsistent dual.*

Next the case in which the primal constraints are inconsistent is considered. This result is implied by theorem 9.5.2 if the dual is unbounded; however the converse is not always true. An example of this is given in Question 9.24 in which although the primal is inconsistent yet the dual has a solution. However for linear constraints this possibility is excluded. In this case the constraints can be written $\mathbf{c}(\mathbf{x}) \triangleq \mathbf{A}^T\mathbf{x} - \mathbf{b} \geqslant \mathbf{0}$. Now the set $\{\mathbf{x} : \mathbf{A}^T\mathbf{x} \geqslant \mathbf{b}\}$ is empty if and only if there exists a vector $\boldsymbol{\lambda} \geqslant \mathbf{0}$ such that $\mathbf{A}\boldsymbol{\lambda} = \mathbf{0}$ and $\mathbf{b}^T\boldsymbol{\lambda} > 0$. (This result is similar to Farkas' lemma (lemma 9.2.4) and is proved in a similar way.) Now let $\mathbf{x}, \boldsymbol{\lambda}$ be dual feasible and $\boldsymbol{\lambda}'$ be the vector which exists above. Then $(\mathbf{x}, \boldsymbol{\lambda} + \alpha\boldsymbol{\lambda}')$ is dual feasible for all $\alpha \geqslant 0$ and

$$\mathscr{L}(\mathbf{x}, \boldsymbol{\lambda} + \alpha\boldsymbol{\lambda}') = \mathscr{L}(\mathbf{x}, \boldsymbol{\lambda}) + \alpha\boldsymbol{\lambda}'^T\mathbf{b}$$

which $\to \infty$ as $\alpha \to \infty$. Thus for linearly constrained problems, *if the primal is infeasible and the dual is feasible, then the dual is unbounded.* It is also possible that both the primal and the dual may be infeasible. Thus using the Wolfe dual for linear constraint problems is always satisfactory in that a failure to solve the dual always implies that the primal has no solution.

A final possibility which should also be considered (although it cannot occur for linear or quadratic programming problems) is that the primal (or dual) problem is bounded but has no solution. In this case there are open questions about the nature of solutions to the dual (or primal) problems and the situation is described in more detail by Wolfe (1961).

Questions for Chapter 9

1. Verify that the points $\mathbf{x}' = \begin{pmatrix} 1 \\ 0 \\ 0 \end{pmatrix}$ and $\mathbf{x}'' = \begin{pmatrix} -\frac{1}{3} \\ \frac{2}{3} \\ \frac{2}{3} \end{pmatrix}$ satisfy Kuhn–Tucker (first order necessary) conditions foı the problem

$$\begin{aligned} &\text{minimize} && x_2 + x_3 \\ &\text{subject to} && x_1 + x_2 + x_3 = 1 \\ &&& x_1^2 + x_2^2 + x_3^2 = 1 \end{aligned}$$

and evaluate the corresponding Lagrange multipliers.

2. By drawing a diagram of the feasible region and the contours of $f(\mathbf{x})$ determine the solution of the problem

$$\begin{aligned} &\text{minimize} && f(\mathbf{x}) \triangleq -x_1 + x_2 \\ &\text{subject to} && 0 \leqslant x_1 \leqslant a \\ &&& 0 \leqslant x_2 \leqslant 1 \\ &&& x_2 \geqslant x_1^2 \end{aligned}$$

where a is a fixed positive constant. Show that the set of active constraints at the solution differs according to whether or not a is greater than a certain fixed value $\bar{a}$, and determine $\bar{a}$. Obtain the Lagrange multipliers of the active constraints in both cases and verify that the KT conditions are satisfied.

3. A parcel has its longest side of length x_1 and its two other sides are of length x_2 and x_3. Postage regulations are that each dimension should be no greater than 42 in, and that the total girth (that is $2(x_2 + x_3)$) plus length should be no greater than 72 in. State the constrained minimization problem which determines the parcel of maximum volume which is permissible. Use the symmetry between x_2 and x_3 to eliminate x_3, draw a diagram of the feasible region, and show (approximately) how the objective function behaves. Identify two possibilities for the set of active constraints at the solution. Solve these (as if equality constraints) for $\mathbf{x}$ and $\boldsymbol{\lambda}$, and determine at which point the multipliers satisfy KT conditions for inequality constraints.
4. It is desired to build a warehouse of width x_1, height x_2, and length x_3 (in metres), with capacity 1500 m^3. Building costs per square metre are: walls £4, roof £6, floor plus land £12. For aesthetic reasons, the width should be twice the height. State the problem which determines the dimensions of the warehouse of minimum cost and write down the KT conditions. By eliminating x_1 and x_3, show that to the nearest metre, $x_2 = 10$ minimizes the cost, and hence find x_1 and x_3. Determine the optimum multipliers in the KT conditions.

   It can be shown that changing $c_i(\mathbf{x}) = 0$ to $c_i(\mathbf{x}) = \epsilon_i$ in the problem induces a change $\lambda_i \epsilon_i$ (to first order) in $f(\mathbf{x})$ at the resulting solution. Estimate the change in cost on reducing the required capacity by 10 per cent.
5. List all the stationary points of the function

$$f(\mathbf{x}) = -x_1^2 - 4x_2^2 - 16x_3^2,$$

subject to the constraint $c(\mathbf{x}) = 0$, where $c(\mathbf{x})$ is given in turn by

(i) $c(\mathbf{x}) = x_1 - 1$,
(ii) $c(\mathbf{x}) = x_1 x_2 - 1$,
(iii) $c(\mathbf{x}) = x_1 x_2 x_3 - 1$.

6. Solve the problem

$$\begin{aligned} &\text{minimize } f(x, y) \triangleq x^2 + y^2 + 3xy + 6x + 19y \\ &\text{subject to } 3y + x = 5. \end{aligned}$$

7. a, b, and c are positive constants. Find the least value of a sum of three positive numbers x, y, and z subject to the constraint

$$\frac{a}{x} + \frac{b}{y} + \frac{c}{z} = 1$$

by the method of Lagrange multipliers, assuming that the positivity conditions are not active.

8. Consider the problem

$$\underset{\mathbf{x}}{\text{minimize}} \ \tfrac{1}{2}\alpha(x_1 - 1)^2 - x_1 - x_2$$

$$\text{subject to } x_1 \geqslant x_2^2,\ x_1^2 + x_2^2 \leqslant 1.$$

Find all the points at which both the constraints are active. One of these points is $\mathbf{x}' = (0.618, 0.786)^{\mathrm{T}}$ to three decimal places. Working to this accuracy, find the range of values of α for which $\mathbf{x}'$ satisfies the KT conditions.

9. Consider the problem

$$\begin{aligned} &\text{maximize } x_2 \\ &\text{subject to } (3 - x_1)^3 - (x_2 - 2) \geqslant 0 \\ &\qquad\qquad 3x_1 + x_2 \geqslant 9. \end{aligned}$$

(i) Derive the KT conditions for this problem, and find all solutions of these.
(ii) Solve the problem graphically.
(iii) Repeat the analysis of (i) and (ii) for the same problem with the additional constraint

$$2x_1 - 3x_2 \geqslant 0.$$

10. Under what conditions on the problem are the KT conditions (a) necessary, (b) sufficient, (c) necessary *and* sufficient for the solution of an inequality constrained optimization problem?

Form the KT conditions for the problem

$$\begin{aligned} &\text{maximize } (x + 1)^2 + (y + 1)^2 \\ &\text{subject to } x^2 + y^2 \leqslant 2 \\ &\qquad\qquad y \leqslant 1 \end{aligned}$$

and hence determine the solution.

11. Find the point on the ellipse defined by the intersection of the surfaces $x + y = 1$ and $x^2 + 2y^2 + z^2 = 1$ which is nearest to the origin. Use (i) the method of Lagrange multipliers, (ii) direct elimination.

12. A bookmaker offers odds of $r_i : 1$ against each of n runners in a race. A punter bets a proportion x_i of his total stake t on each runner. Assume that only one runner can win and that $r_1 > r_2 > \cdots > r_n > 0$. Clearly the punter can guarantee not to lose money if and only if

$$r_i x_i \geqslant \sum_{j \neq i} x_j, \qquad i = 1, 2, \ldots, n.$$

Show that this situation can arise if $\Sigma_i\, 1/(r_i + 1) \leqslant 1$. (Consider $x_i = c/(r_i + 1)$ where $1/c = \Sigma_i\, 1/(r_i + 1)$.)

If this condition holds, and if it is equally likely that any runner can win, the expected profit to the punter is $(t\, \Sigma_i (r_i + 1)x_i) - t$. Show that the choice of the x_i which gives maximum expected profit, yet which guarantees no loss of money, can be posed as a constrained minimization problem involving only bounds on the variables and an equality constraint. By showing that the KT conditions are satisfied, verify that the solution is

$$x_1 = 1 - \sum_{i>1} \frac{1}{r_i + 1}$$

$$x_i = \frac{1}{r_i + 1}, \qquad i > 1.$$

13. Consider the relationship between the method of Lagrange multipliers in Section 9.1 and the direct elimination method in Section 7.2 and Question 7.5. In the latter notation, show that (9.1.3) can be rearranged to give $\boldsymbol{\lambda}^* = [\mathbf{A}_1^*]^{-1}\mathbf{g}_1^*$. Hence use the result of Question 7.5 to show that if $\mathbf{x}^*$ satisfies (9.1.3) then it is a stationary point of the function $\psi(\mathbf{x}_2)$ in (7.2.2).
14. Attempt to solve the problem in Question 7.4 by the method of Lagrange multipliers. Show that either $y = 0$ or $\lambda = -1$ and both of these imply a contradiction, so no solution of (9.1.5) exists. Explain this fact in terms of the regularity assumption (9.2.4) and the independence of the vectors $\mathbf{a}_i^*$, $i \in E$.
15. Show that if the matrix $\mathbf{A}^*$ in (9.1.3) has full rank then the Lagrange multipliers $\boldsymbol{\lambda}^*$ are uniquely defined by $\boldsymbol{\lambda}^* = \mathbf{A}^{*+}\mathbf{g}^*$ or by solving any $m \times m$ subsystem $\mathbf{A}_1^*\boldsymbol{\lambda}^* = \mathbf{g}_1^*$ where $\mathbf{A}_1^*$ is non-singular (see Question 9.13). Computationally the former is most stable if the matrix factors $\mathbf{A}^* = \mathbf{QR}$ ($\mathbf{Q}$ orthogonal, $\mathbf{R}$ upper triangular) are calculated and $\boldsymbol{\lambda}^*$ is obtained by back substitution in $\mathbf{R}\boldsymbol{\lambda}^* = \mathbf{Q}^T\mathbf{g}^*$.
16. By examining second order conditions, determine whether or not each of the points $\mathbf{x}'$ and $\mathbf{x}''$ are local solutions to the problem in Question 9.1.
17. By examining second order conditions, determine the nature (maximizer, minimizer, or saddle point) of each of the stationary points obtained in Question 9.5.
18. Given an optimal b.f.s. to an LP, show that the reduced costs $\hat{\mathbf{c}}_N$ are the Lagrange multipliers to the LP after having used the equations $\mathbf{Ax} = \mathbf{b}$ to eliminate the basic variables.
19. For the LP in Question 8.1, illustrate (for non-negative $\mathbf{x}$) the plane $3x_1 + x_2 + x_3 = 12$ and the line along which it intersects the plane $x_1 - x_2 + x_3 = -8$. Hence show that the feasible region is a convex set and give its extreme points. Do the same for the LP which results from deleting the condition $x_1 - x_2 + x_3 = -8$.
20. For the LP (8.1.1) in standard form, prove that

(i) $\exists$ a solution $\Rightarrow$ $\exists$ an extreme point which is a solution,
(ii) $\mathbf{x}$ is an extreme point $\Rightarrow$ $\mathbf{x}$ has $p \leqslant m$ positive components,
(iii) $\mathbf{x}$ is an extreme point and $\mathbf{A}$ has full rank $\Rightarrow$ $\exists$ a b.f.s. at $\mathbf{x}$ (degenerate iff $p < m$),
(iv) $\exists$ a b.f.s. at $\mathbf{x}$ $\Rightarrow$ $\mathbf{x}$ is an extreme point.

(By a b.f.s. at $\mathbf{x}$ is meant that there exists a partition into B and N variables such that $\mathbf{x}_B \geqslant \mathbf{0}$, $\mathbf{x}_N = \mathbf{0}$, and $\mathbf{A}_B$ is non-singular (see Section 8.2).) Part (i) follows from the convexity of S in theorem 9.4.1, the result of Question 9.21, and the fact that an extreme point of S must be an extreme point of (8.1.1) by the linearity of f. The remaining results follow from Question 9.22. Part (ii) is true since if $\mathbf{A}_P$ has rank p then $p \leqslant m$. Part (iii) is derived by including all positive components of $\mathbf{x}$ in B. If $p < m$, other independent columns of $\mathbf{A}$ exist which can augment $\mathbf{A}_P$ to become a square non-singular matrix, and the corresponding variables (with zero value, hence degeneracy) are added to B. In part (iv) the non-singularity of $\mathbf{A}_B$ implies that $\mathbf{A}_P$ has full rank and hence $\mathbf{x}$ is extreme.

21. Consider a closed convex set $K \subset \mathbb{R}^{\oplus} = \{\mathbf{x} : \mathbf{x} \geqslant \mathbf{0}\}$. Show that K has an extreme point in any supporting hyperplane $S = \{\mathbf{x} : \mathbf{c}^T\mathbf{x} = z\}$. Show that $T \triangleq K \cap S$ is not empty. Since $T \subset \mathbb{R}^{\oplus}$, construct an extreme point in the following way. Choose the set $T_1 \subset T$ such that the first component t_1 of all vectors $\mathbf{t} \in T_1$ is smallest. Then choose a set $T_2 \subset T_1$ with the smallest second component, and so on. Show that a unique point $\mathbf{t}^*$ is ultimately determined which is extreme in T. Hence show that $\mathbf{t}^*$ is extreme in K (Hadley, 1962, chapter 2, theorem III).
22. Let $\mathbf{x}$ be any feasible point in (8.1.1), with p positive components, let $\mathbf{A}_P$ be the matrix with columns $\mathbf{a}_i \ \forall i : x_i > 0$ and $\mathbf{x}_P$ likewise. Show that $\mathbf{x}$ is an extreme point of (8.1.1) iff $\mathbf{A}_P$ has full rank, in the following way. If $\mathbf{A}_P$ has not full rank there exists $\mathbf{u} \neq \mathbf{0}$ such that $\mathbf{A}_P\mathbf{u} = \mathbf{0}$. By examining $\mathbf{x}_P \pm \epsilon\mathbf{u}$ show that $\mathbf{x}$ is not extreme. If $\mathbf{A}_P$ has full rank, show that $\mathbf{x}_P = \mathbf{A}_P^+\mathbf{b}$ is uniquely defined. Let $\mathbf{x} = (1-\theta)\mathbf{x}_0 + \theta\mathbf{x}_1$, $\theta \in (0, 1)$, $\mathbf{x}_0, \mathbf{x}_1$ feasible, and show that the zero components of $\mathbf{x}$ must be zero in $\mathbf{x}_0$ and $\mathbf{x}_1$. Hence show that the remaining components are defined by $\mathbf{A}_P^+\mathbf{b}$ and hence $\mathbf{x}_0 = \mathbf{x}_1 = \mathbf{x}$, so that $\mathbf{x}$ is extreme.
23. Show that the dual of the problem

$$\begin{aligned}&\text{minimize } \tfrac{1}{2}\sigma x_1^2 + \tfrac{1}{2}x_2^2 + x_1\\&\text{subject to } x_1 \geqslant 0\end{aligned}$$

is a maximization problem in terms of a Lagrange multiplier λ. For the cases $\sigma = +1$ and $\sigma = -1$, investigate whether the local solution of the dual gives the multiplier λ^* which exists at the local solution to the primal, and explain the difference between the two cases.
24. Consider the problem

$$\begin{aligned}&\text{minimize } f(x) \triangleq 0\\&\text{subject to } c(x) \triangleq -e^x \geqslant 0.\end{aligned}$$

Verify that the constraint is concave but inconsistent, so that the feasible region is empty. Set up the dual problem and show that it is solved by $\lambda = 0$ and any x.

Chapter 10

Quadratic Programming

10.1 Equality Constraints

Like linear programming problems, another optimization problem which can be solved in a finite number of steps is a *quadratic programming (QP) problem*. In terms of (7.1.1) this is a problem in which the objective function $f(\mathbf{x})$ is quadratic and the constraint functions $c_i(\mathbf{x})$ are linear. Thus the problem is to find a solution $\mathbf{x}^*$ to

$$\begin{aligned} &\underset{\mathbf{x}}{\text{minimize}} \ q(\mathbf{x}) \triangleq \tfrac{1}{2}\mathbf{x}^{\mathrm{T}}\mathbf{G}\mathbf{x} + \mathbf{g}^{\mathrm{T}}\mathbf{x} \\ &\text{subject to } \mathbf{a}_i^{\mathrm{T}}\mathbf{x} = b_i, \qquad i \in E \\ &\qquad\qquad\ \mathbf{a}_i^{\mathrm{T}}\mathbf{x} \geqslant b_i, \qquad i \in I, \end{aligned} \tag{10.1.1}$$

where it is always possible to arrange that the matrix G is symmetric. As in linear programming, the problem may be infeasible or the solution may be unbounded; however these possibilities are readily detected in the algorithms, so for the most part it is assumed that a solution $\mathbf{x}^*$ exists. If the Hessian matrix $\mathbf{G}$ is positive semi-definite, $\mathbf{x}^*$ is a global solution, and if $\mathbf{G}$ is positive definite, $\mathbf{x}^*$ is also unique. These results follow from the (strict) convexity of $q(\mathbf{x})$, so that (10.1.1) is a convex programming problem and theorem 9.4.1 and its corollary apply. When the Hessian $\mathbf{G}$ is indefinite then local solutions which are not global can occur, and the computation of any such local solution is of interest (see Section 10.4). A modern computer code for QP needs to be quite sophisticated, so a simplified account of the basic structure is given first, and is amplified or qualified later. In the first case it is shown in Sections 10.1 and 10.2 how equality constraint problems can be treated. The generalization of these ideas to handle inequality constraints is by means of an active set strategy and is described in Section 10.3. More advanced features of a QP algorithm which handle an indefinite Hessian and allow a sequence of problems to be solved efficiently are given in Section 10.4. QP problems with a special structure such as having only bounded constraints, or such as arise from least squares problems, are considered in Section 10.5. Early work on QP was often presented as a modification of the tableau form of the simplex method for linear programming. This work is described in Section 10.6 and has come to be expressed in terms of a *linear complementarity problem*. The equivalence between these methods and the active set methods is described and reasons for preferring the latter derivation are put forward.

Quadratic programming differs from linear programming in that it is possible to have meaningful problems (other than an $n \times n$ system of equations) in which there are no inequality constraints. This section therefore studies how to find a solution $\mathbf{x}^*$ to the equality constraint problem

$$\begin{aligned} &\underset{\mathbf{x}}{\text{minimize}} \quad q(\mathbf{x}) \triangleq \tfrac{1}{2}\mathbf{x}^{\mathrm{T}}\mathbf{G}\mathbf{x} + \mathbf{g}^{\mathrm{T}}\mathbf{x} \\ &\text{subject to } \mathbf{A}^{\mathrm{T}}\mathbf{x} = \mathbf{b}. \end{aligned} \tag{10.1.2}$$

It is assumed that there are $m \leqslant n$ constraints so that $\mathbf{b} \in \mathbb{R}^m$, and $\mathbf{A}$ is $n \times m$ and collects the column vectors $\mathbf{a}_i$, $i \in E$, in (10.1.1). It is assumed that $\mathbf{A}$ has rank m; if the constraints are consistent this can always be achieved by removing dependent constraints, although there may be numerical difficulties in recognizing this situation. This assumption also ensures that unique Lagrange multipliers $\boldsymbol{\lambda}^*$ exist, and calculation of these quantities, which may be required for sensitivity analysis or for use in an active set method, is also considered in this section.

A straightforward way of solving (10.1.2) is to use the constraints to eliminate variables. If the partitions

$$\mathbf{x} = \begin{pmatrix} \mathbf{x}_1 \\ \mathbf{x}_2 \end{pmatrix}, \qquad \mathbf{A} = \begin{pmatrix} \mathbf{A}_1 \\ \mathbf{A}_2 \end{pmatrix}, \qquad \mathbf{g} = \begin{pmatrix} \mathbf{g}_1 \\ \mathbf{g}_2 \end{pmatrix}, \qquad \mathbf{G} = \begin{bmatrix} \mathbf{G}_{11} & \mathbf{G}_{12} \\ \mathbf{G}_{21} & \mathbf{G}_{22} \end{bmatrix}$$

are defined, where $\mathbf{x}_1 \in \mathbb{R}^m$ and $\mathbf{x}_2 \in \mathbb{R}^{n-m}$, etc., then the equations in (10.1.2) become $\mathbf{A}_1^{\mathrm{T}}\mathbf{x}_1 + \mathbf{A}_2^{\mathrm{T}}\mathbf{x}_2 = \mathbf{b}$ and are readily solved (by Gaussian elimination, say) to give $\mathbf{x}_1$ in terms of $\mathbf{x}_2$; this is conveniently written

$$\mathbf{x}_1 = \mathbf{A}_1^{-\mathrm{T}}(\mathbf{b} - \mathbf{A}_2^{\mathrm{T}}\mathbf{x}_2). \tag{10.1.3}$$

Substituting into $q(\mathbf{x})$ gives the problem: minimize $\psi(\mathbf{x}_2)$, $\mathbf{x}_2 \in \mathbb{R}^{n-m}$ where $\psi(\mathbf{x}_2)$ is the quadratic function

$$\begin{aligned} \psi(\mathbf{x}_2) = {} & \tfrac{1}{2}\mathbf{x}_2^{\mathrm{T}}(\mathbf{G}_{22} - \mathbf{G}_{21}\mathbf{A}_1^{-\mathrm{T}}\mathbf{A}_2^{\mathrm{T}} - \mathbf{A}_2\mathbf{A}_1^{-1}\mathbf{G}_{12} + \mathbf{A}_2\mathbf{A}_1^{-1}\mathbf{G}_{11}\mathbf{A}_1^{-\mathrm{T}}\mathbf{A}_2^{\mathrm{T}})\mathbf{x}_2 \\ & + \mathbf{x}_2^{\mathrm{T}}(\mathbf{G}_{21} - \mathbf{A}_2\mathbf{A}_1^{-1}\mathbf{G}_{11})\mathbf{A}_1^{-\mathrm{T}}\mathbf{b} + \tfrac{1}{2}\mathbf{b}^{\mathrm{T}}\mathbf{A}_1^{-1}\mathbf{G}_{11}\mathbf{A}_1^{-\mathrm{T}}\mathbf{b} \\ & + \mathbf{x}_2^{\mathrm{T}}(\mathbf{g}_2 - \mathbf{A}_2\mathbf{A}_1^{-1}\mathbf{g}_1) + \mathbf{g}_1^{\mathrm{T}}\mathbf{A}_1^{-\mathrm{T}}\mathbf{b}. \end{aligned} \tag{10.1.4}$$

A unique minimizer $\mathbf{x}_2^*$ exists if the Hessian $\boldsymbol{\nabla}^2\psi$ in the quadratic term is positive definite, in which case $\mathbf{x}_2^*$ is obtained by solving the linear system $\boldsymbol{\nabla}\psi(\mathbf{x}_2) = \mathbf{0}$. Then $\mathbf{x}_1^*$ is found by substitution in (10.1.3). The Lagrange multiplier vector $\boldsymbol{\lambda}^*$ is defined by $\mathbf{g}^* = \mathbf{A}\boldsymbol{\lambda}^*$ (Section 9.1) where $\mathbf{g}^* = \boldsymbol{\nabla}q(\mathbf{x}^*)$, and can be calculated by solving the first partition $\mathbf{g}_1^* = \mathbf{A}_1\boldsymbol{\lambda}^*$. By definition of $q(\mathbf{x})$ in (10.1.2), $\mathbf{g}^* = \mathbf{g} + \mathbf{G}\mathbf{x}^*$ so an explicit expression for $\boldsymbol{\lambda}^*$ is

$$\boldsymbol{\lambda}^* = \mathbf{A}_1^{-1}(\mathbf{g}_1 + \mathbf{G}_{11}\mathbf{x}_1^* + \mathbf{G}_{12}\mathbf{x}_2^*). \tag{10.1.5}$$

An example of the method is given to solve the problem

$$\begin{aligned} &\underset{\mathbf{x}}{\text{minimize}} \quad q(\mathbf{x}) \triangleq x_1^2 + x_2^2 + x_3^2 \\ &\text{subject to } x_1 + 2x_2 - x_3 = 4 \\ &\qquad\qquad\; x_1 - x_2 + x_3 \quad -2. \end{aligned} \tag{10.1.6}$$

To eliminate x_3, the equations are written

$$x_1 + 2x_2 = 4 + x_3$$

$$x_1 - x_2 = -2 - x_3$$

and are readily solved by Gaussian elimination giving

$$x_1 = 0 - \tfrac{1}{3}x_3, \qquad x_2 = 2 + \tfrac{2}{3}x_3. \tag{10.1.7}$$

It is easily verified that this solution corresponds to (10.1.3) with $\mathbf{x}_1 = \begin{pmatrix} x_1 \\ x_2 \end{pmatrix}$, $\mathbf{x}_2 = (x_3)$, $\mathbf{A}_1 = \begin{bmatrix} 1 & 1 \\ 2 & -1 \end{bmatrix}$, and $\mathbf{A}_2 = [-1 \quad 1]$. Substituting (10.1.7) into $q(\mathbf{x})$ in (10.1.6) gives

$$\psi(x_3) = \tfrac{14}{9}x_3^2 + \tfrac{8}{3}x_3 + 4 \tag{10.1.8}$$

which corresponds to (10.1.4) with $\mathbf{G}_{11} = \begin{bmatrix} 2 & 0 \\ 0 & 2 \end{bmatrix}$, $\mathbf{G}_{12} = \mathbf{G}_{21}^{\mathrm{T}} = \mathbf{0}$, $\mathbf{G}_{22} = [2]$, and $\mathbf{g} = \mathbf{0}$. The Hessian matrix in (10.1.8) is $[\frac{28}{9}]$ which is positive definite, so the minimizer is obtained by setting $\nabla\psi = \mathbf{0}$ and is $x_3^* = -\frac{6}{7}$. Back substitution in (10.1.6) gives $x_1^* = \frac{2}{7}$ and $x_2^* = \frac{10}{7}$. The system $\mathbf{g}^* = \mathbf{A}\boldsymbol{\lambda}^*$ becomes

$$\tfrac{2}{7}\begin{pmatrix} 2 \\ 10 \\ -6 \end{pmatrix} = \begin{bmatrix} 1 & 1 \\ 2 & -1 \\ -1 & 1 \end{bmatrix} \begin{pmatrix} \lambda_1^* \\ \lambda_2^* \end{pmatrix}$$

and solving the first two rows gives $\lambda_1^* = \frac{8}{7}$ and $\lambda_2^* = -\frac{4}{7}$, and this is consistent with the third row.

Direct elimination of variables is not the only way of solving (10.1.2) nor may it be the best. A *generalized elimination method* is possible in which essentially a linear transformation of variables is made initially. Let $\mathbf{S}$ and $\mathbf{Z}$ be $n \times m$ and $n \times (n - m)$ matrices respectively such that $[\mathbf{S} : \mathbf{Z}]$ is non-singular, and in addition let $\mathbf{A}^{\mathrm{T}}\mathbf{S} = \mathbf{I}$ and $\mathbf{A}^{\mathrm{T}}\mathbf{Z} = \mathbf{0}$. $\mathbf{S}^{\mathrm{T}}$ can be regarded as a left generalized inverse for $\mathbf{A}$ so that a solution of $\mathbf{A}^{\mathrm{T}}\mathbf{x} = \mathbf{b}$ is given by $\mathbf{x} = \mathbf{S}\mathbf{b}$. However this solution is non-unique in general and other feasible points are given by $\mathbf{x} = \mathbf{S}\mathbf{b} + \boldsymbol{\delta}$ where $\boldsymbol{\delta}$ is in the null column space of $\mathbf{A}$. This is the linear space

$$\{\boldsymbol{\delta} : \mathbf{A}^{\mathrm{T}}\boldsymbol{\delta} = \mathbf{0}\} \tag{10.1.9}$$

which has dimension $n - m$. The purpose of the matrix $\mathbf{Z}$ is that its columns $\mathbf{z}_1, \mathbf{z}_2, \ldots, \mathbf{z}_{n-m}$ act as basis vectors (or reduced coordinate directions) for this null space. That is to say, at any feasible point $\mathbf{x}$ any feasible correction $\boldsymbol{\delta}$ can be

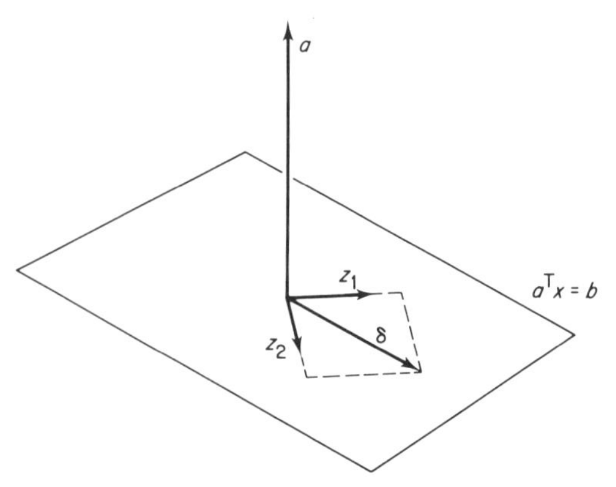

Figure 10.1.1 Reduced coordinates for the feasible region

written as

$$\boldsymbol{\delta} = \mathbf{Zy} = \sum_{i=1}^{n-m} \mathbf{z}_i y_i \tag{10.1.10}$$

where $y_1, y_2, \ldots, y_{n-m}$ are the components (or reduced variables) in each reduced coordinate direction (see Figure 10.1.1). Thus any feasible point $\mathbf{x}$ can be written

$$\mathbf{x} = \mathbf{Sb} + \mathbf{Zy}. \tag{10.1.11}$$

This can be interpreted (Figure 10.1.2) as a step from the origin to the feasible point $\mathbf{Sb}$, followed by a feasible correction $\mathbf{Zy}$ to reach the point $\mathbf{x}$. Thus (10.1.11) provides a way of eliminating the constraints $\mathbf{A}^{\mathrm{T}}\mathbf{x} = \mathbf{b}$ in terms of the vector of reduced variables $\mathbf{y}$ which has $n - m$ elements, and is therefore a generalization of (10.1.3). Substituting into $q(\mathbf{x})$ gives the reduced quadratic function

$$\psi(\mathbf{y}) = \tfrac{1}{2}\mathbf{y}^{\mathrm{T}}\mathbf{Z}^{\mathrm{T}}\mathbf{GZy} + (\mathbf{g} + \mathbf{GSb})^{\mathrm{T}}\mathbf{Zy} + \tfrac{1}{2}(\mathbf{g} + \mathbf{GSb})^{\mathrm{T}}\mathbf{Sb}. \tag{10.1.12}$$

If $\mathbf{Z}^{\mathrm{T}}\mathbf{GZ}$ is positive definite then a unique minimizer $\mathbf{y}^*$ exists which (from $\nabla\psi(\mathbf{y}) = \mathbf{0}$) solves the linear system

$$(\mathbf{Z}^{\mathrm{T}}\mathbf{GZ})\mathbf{y} = -\mathbf{Z}^{\mathrm{T}}(\mathbf{g} + \mathbf{GSb}). \tag{10.1.13}$$

The solution is best achieved by computing $\mathbf{LL}^{\mathrm{T}}$ or $\mathbf{LDL}^{\mathrm{T}}$ factors of $\mathbf{Z}^{\mathrm{T}}\mathbf{GZ}$ which also enables the positive definite condition to be checked. Then $\mathbf{x}^*$ is obtained by substituting into (10.1.11). The matrix $\mathbf{Z}^{\mathrm{T}}\mathbf{GZ}$ in (10.1.12) is often referred to as the *reduced Hessian matrix* and the vector $\mathbf{Z}^{\mathrm{T}}(\mathbf{g} + \mathbf{GSb})$ as the *reduced gradient vector*. Notice that $\mathbf{g} + \mathbf{GSb} = \nabla q(\mathbf{Sb})$ is the gradient vector of $q(\mathbf{x})$ at $\mathbf{x} = \mathbf{Sb}$ (just as $\mathbf{g}$ is $\nabla q(\mathbf{0})$), so reduced derivatives are obtained by a matrix operation with $\mathbf{Z}^{\mathrm{T}}$. In addition, premultiplying by $\mathbf{S}^{\mathrm{T}}$ in $\mathbf{g}^* = \mathbf{A}\boldsymbol{\lambda}^*$ gives the equation

$$\boldsymbol{\lambda}^* = \mathbf{S}^{\mathrm{T}}\mathbf{g}^* \tag{10.1.14}$$

which can be used to calculate Lagrange multipliers. Explicit expressions for $\mathbf{x}^*$ and

λ^* in terms of the original data are

$$\mathbf{x}^* = \mathbf{Sb} - \mathbf{Z}(\mathbf{Z}^T\mathbf{GZ})^{-1}\mathbf{Z}^T(\mathbf{g} + \mathbf{GSb}) \tag{10.1.15}$$

and

$$\begin{aligned}\lambda^* &= \mathbf{S}^T(\mathbf{g} + \mathbf{Gx}^*)\\ &= (\mathbf{S}^T - \mathbf{S}^T\mathbf{GZ}(\mathbf{Z}^T\mathbf{GZ})^{-1}\mathbf{Z}^T)\mathbf{g} + \mathbf{S}^T(\mathbf{G} - \mathbf{GZ}(\mathbf{Z}^T\mathbf{GZ})^{-1}\mathbf{Z}^T\mathbf{G})\mathbf{Sb}.\end{aligned} \tag{10.1.16}$$

These formulae would not be used directly in computation but are useful in showing the relationship with the alternative method of Lagrange multipliers for solving equality constraint QP problems, considered in Section 10.2.

Depending on the choice of **S** and **Z**, a number of methods exist which can be interpreted in this way, and a general procedure is described below for constructing matrices **S** and **Z** with the correct properties. However one choice of particular importance is obtained by way of any QR factorization (Householder's, for example) of the matrix **A**. This can be written

$$\mathbf{A} = \mathbf{Q}\begin{bmatrix}\mathbf{R}\\ \mathbf{0}\end{bmatrix} = [\mathbf{Q}_1 \;\; \mathbf{Q}_2]\begin{bmatrix}\mathbf{R}\\ \mathbf{0}\end{bmatrix} = \mathbf{Q}_1\mathbf{R} \tag{10.1.17}$$

where **Q** is $n \times n$ and orthogonal, **R** is $m \times m$ and upper triangular, and $\mathbf{Q}_1$ and $\mathbf{Q}_2$ are $n \times m$ and $n \times (n - m)$ respectively. The choices

$$\mathbf{S} = \mathbf{A}^{+T} = \mathbf{Q}_1\mathbf{R}^{-T}, \qquad \mathbf{Z} = \mathbf{Q}_2 \tag{10.1.18}$$

are readily observed to have the correct properties. Moreover the vector **Sb** in (10.1.11) and Figure 10.1.2 is observed to be orthogonal to the constraint manifold, and the reduced coordinate directions $\mathbf{z}_i$ are also mutually orthogonal. The solution $\mathbf{x}^*$ is obtained as above by setting up and solving (10.1.13) for $\mathbf{y}^*$ and then substituting into (10.1.11). **Sb** is calculated by forward substitution in $\mathbf{R}^T\mathbf{v} = \mathbf{b}$ followed by forming $\mathbf{Sb} = \mathbf{Q}_1\mathbf{v}$. The multipliers λ^* in (10.1.14) are calculated by backward substitution in

$$\mathbf{R}\lambda^* = \mathbf{Q}_1^T\mathbf{g}^*. \tag{10.1.19}$$

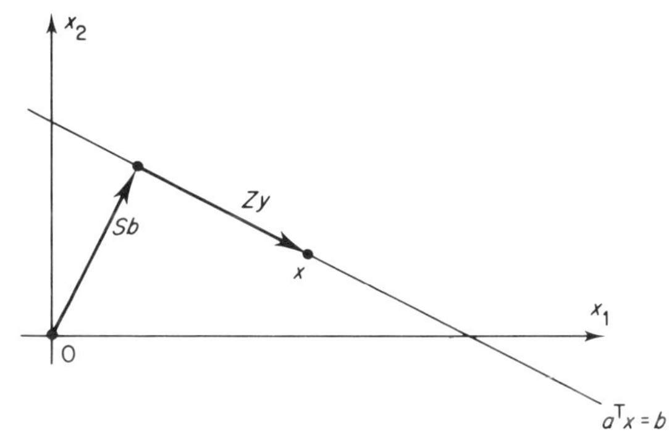

Figure 10.1.2 Generalized elimination in the special case (10.1.18)

This scheme is due to Gill and Murray (1974a) and I shall refer to it as the *orthogonal factorization method*. It is recommended for non-sparse problems because it is stable in regard to propagation of round-off errors (see below).

An example is given of the method applied to problem (10.1.6). The matrices involved are

$$\mathbf{S} = \tfrac{1}{14}\begin{bmatrix} 5 & 8 \\ 4 & -2 \\ -1 & 4 \end{bmatrix}, \qquad \mathbf{Z} = \begin{bmatrix} 1 \\ -2 \\ -3 \end{bmatrix}$$

(unnormalized) and these can be represented by the QR decomposition

$$\mathbf{A} = \begin{bmatrix} 1/\sqrt{6} & 4/\sqrt{21} & 1/\sqrt{14} \\ 2/\sqrt{6} & -1/\sqrt{21} & -2/\sqrt{14} \\ 1/\sqrt{6} & 2/\sqrt{21} & -3/\sqrt{14} \end{bmatrix}\begin{bmatrix} \sqrt{6} & -\sqrt{6}/3 \\ 0 & \sqrt{21}/3 \\ 0 & 0 \end{bmatrix}.$$

The vector $\mathbf{Sb} = \frac{1}{7}(2, 10, -6)^{\mathrm{T}}$, and since $\mathbf{g} = \mathbf{0}$ and $\mathbf{GSb} = 2\mathbf{Sb}$ it follows that $\mathbf{Z}^{\mathrm{T}}(\mathbf{g} + \mathbf{GSb}) = \mathbf{0}$. Hence $\mathbf{y}^* = \mathbf{0}$ and $\mathbf{x}^* = \mathbf{Sb} = \frac{1}{7}(2, 10, -6)^{\mathrm{T}}$. Also $\mathbf{g}^* = \mathbf{g} + \mathbf{GSb} = \frac{2}{7}(2, 10, -6)^{\mathrm{T}}$ so $\boldsymbol{\lambda}^* = (\frac{8}{7}, -\frac{4}{7})^{\mathrm{T}}$ and these results agree with those obtained by direct elimination.

A general scheme for computing suitable $\mathbf{S}$ and $\mathbf{Z}$ matrices is the following. Choose any $n \times (n - m)$ matrix $\mathbf{V}$ such that $[\mathbf{A} : \mathbf{V}]$ is non-singular. Let the inverse be expressed in partitioned form

$$[\mathbf{A} : \mathbf{V}]^{-1} = \begin{bmatrix} \mathbf{S}^{\mathrm{T}} \\ \mathbf{Z}^{\mathrm{T}} \end{bmatrix} \tag{10.1.20}$$

where $\mathbf{S}$ and $\mathbf{Z}$ are $n \times m$ and $n \times (n - m)$ respectively. Then it follows that $\mathbf{S}^{\mathrm{T}}\mathbf{A} = \mathbf{I}$ and $\mathbf{Z}^{\mathrm{T}}\mathbf{A} = \mathbf{0}$ so these matrices are suitable for use in the generalized elimination method. The resulting method can also be interpreted as one which makes a linear transformation with the matrix $[\mathbf{A} : \mathbf{V}]$, as described at the start of Section 12.4. The methods described earlier can all be identified as special cases of this scheme. If

$$\mathbf{V} = \begin{bmatrix} \mathbf{0} \\ \mathbf{I} \end{bmatrix} \tag{10.1.21}$$

is chosen, then the identity

$$\begin{bmatrix} \mathbf{A}_1 & \mathbf{0} \\ \mathbf{A}_2 & \mathbf{I} \end{bmatrix}^{-1} = \begin{bmatrix} \mathbf{A}_1^{-1} & \mathbf{0} \\ -\mathbf{A}_2\mathbf{A}_1^{-1} & \mathbf{I} \end{bmatrix} = \begin{bmatrix} \mathbf{S}^{\mathrm{T}} \\ \mathbf{Z}^{\mathrm{T}} \end{bmatrix} \tag{10.1.22}$$

gives expressions for $\mathbf{S}$ and $\mathbf{Z}$. It can easily be verified that in this case the method reduces to the direct elimination method. Alternatively if the choice

$$\mathbf{V} = \mathbf{Q}_2 \tag{10.1.23}$$

is made, where $\mathbf{Q}_2$ is defined by (10.1.17), then the above orthogonal factorization method is obtained. This follows by virtue of the identity

$$[\mathbf{A} : \mathbf{V}]^{-1} = [\mathbf{Q}_1\mathbf{R} : \mathbf{Q}_2]^{-1} = \begin{bmatrix} \mathbf{R}^{-1}\mathbf{Q}_1^{\mathrm{T}} \\ \mathbf{Q}_2^{\mathrm{T}} \end{bmatrix}, \tag{10.1.24}$$

which can also be expressed as

$$[\mathbf{A} : \mathbf{V}]^{-1} = \begin{bmatrix} \mathbf{A}^+ \\ \mathbf{V}^+ \end{bmatrix}, \tag{10.1.25}$$

where $\mathbf{A}^+ = (\mathbf{A}^T\mathbf{A})^{-1}\mathbf{A}^T$ is the full rank Penrose generalized inverse matrix, and (10.1.14) can be written

$$\boldsymbol{\lambda}^* = \mathbf{A}^+\mathbf{g}^* \tag{10.1.26}$$

in this case. In fact for any given $\mathbf{S}$ and $\mathbf{Z}$, a unique $\mathbf{V}$ must exist by virtue of (10.1.20). However if $\mathbf{Z}$ is given but $\mathbf{S}$ is arbitrary then some freedom is possible in the choice of $\mathbf{V}$; this point is discussed further in Question 10.6. The generalization expressed by (10.1.20) is not entirely an academic one; it may be preferable not to form $\mathbf{S}$ and $\mathbf{Z}$ but to perform any convenient stable factorization of $[\mathbf{A} : \mathbf{V}]$ and use this to generate $\mathbf{S}$ and $\mathbf{Z}$ indirectly: in particular the simple form of (10.1.21) and (10.1.22) could be advantageous for larger problems or for large sparse problems.

Another method which can be described within this framework is the *reduced gradient method* of Wolfe (1963a). In terms of the active set method (Section 10.3) the matrix $\mathbf{V}$ is formed from normal vectors $\mathbf{a}_i$ of *inactive* constraints which have previously been active. Thus when a constraint becomes inactive the column $\mathbf{a}_i$ in $\mathbf{A}$ is transferred to $\mathbf{V}$ so that $[\mathbf{A} : \mathbf{V}]^{-1}$ need not be recomputed and only the partition line is repositioned. When an inactive constraint becomes active, the incoming vector $\mathbf{a}_i$ replaces one column of $\mathbf{V}$. The choice of column is arbitrary and can be made so as to keep $[\mathbf{A} : \mathbf{V}]$ well conditioned. This exchange of columns is analogous to that in linear programming (see equation (8.2.14)) and developments such as those described in Section 8.5 for taking account of sparsity or maintaining stability can be taken over into this method.

Other methods for QP also can be interpreted in these terms. Murray (1971) gives a method in which a matrix $\mathbf{Z}$ is constructed for which

$$\mathbf{Z}^T\mathbf{G}\mathbf{Z} = \mathbf{D} \tag{10.1.27}$$

where $\mathbf{D}$ is diagonal. In this method columns of $\mathbf{Z}$ can be interpreted as conjugate directions, and are related to the matrix ($\hat{\mathbf{Z}}$ say) in the orthogonal factorization method by $\mathbf{Z} = \hat{\mathbf{Z}}\mathbf{L}^{-1}$, where $\hat{\mathbf{Z}}^T\mathbf{G}\hat{\mathbf{Z}}$ has factors $\mathbf{L}\mathbf{D}\mathbf{L}^T$. Because of this the method essentially represents inverse matrix information and may be doubtful on grounds of stability. However it does have the advantage that the solution of (10.1.13) becomes trivial and no storage of factors of $\mathbf{Z}^T\mathbf{G}\mathbf{Z}$ is required. It is also possible to consider Beale's (1959) method (see Section 10.6) partly within this framework since it can be interpreted as selecting conjugate search directions from the rows of $[\mathbf{A} : \mathbf{V}]^{-1}$. In this case the matrix $\mathbf{V}$ is formed from gradient differences $\boldsymbol{\gamma}^{(j)}$ on previous iterations. However the method also has some disadvantageous features.

It is not easy to give a unique best choice amongst these many methods, because this depends on the type of problem and also on whether the method is to be considered as part of an active set method for inequality constraints (Section 10.3). Moreover there are other methods described in Section 10.2, some of which are of interest. However the orthogonal factorization method is advantageous in that calculating $\mathbf{Z}$ involves operations with elementary orthogonal matrices which are very stable numerically. Also this choice ($\mathbf{Z} = \mathbf{Q}_2$) gives the best possible bound

$$\kappa(\mathbf{Z}^T\mathbf{G}\mathbf{Z}) \leqslant \kappa(\mathbf{G}) \tag{10.1.28}$$

on the condition number $\kappa(\mathbf{Z}^T\mathbf{G}\mathbf{Z})$. This is not to say that it gives the best $\kappa(\mathbf{Z}^T\mathbf{G}\mathbf{Z})$ itself, as Gill and Murray (1974a) erroneously claim. Indeed a trivial modification to Murray's (1971) method above with $\mathbf{D} = \mathbf{I}$ enables a $\mathbf{Z}$ matrix to be calculated for which $\kappa(\mathbf{Z}^T\mathbf{G}\mathbf{Z}) = 1$. In fact the relevance of the condition number is doubtful in that it can be changed by symmetric row and column scaling in $\mathbf{Z}^T\mathbf{G}\mathbf{Z}$ without changing the propagation of errors. Insofar as (10.1.28) does have some meaning it indicates that $\kappa(\mathbf{Z}^T\mathbf{G}\mathbf{Z})$ cannot become arbitrarily large. However this conclusion can also be obtained in regard to other methods if careful attention is given to pivoting, and I see no reason why these methods could not be implemented in a reasonably stable way. On the other hand, an arbitrary choice of $\mathbf{V}$ which makes the columns of $\mathbf{Z}$ very close to being dependent might be expected to induce unpleasantly large growth of round-off errors.

10.2 Lagrangian Methods

An alternative way of deriving the solution $\mathbf{x}^*$ to (10.1.2) and the associated multipliers $\boldsymbol{\lambda}^*$ is by the *method of Lagrange multipliers* (9.1.5). The Lagrangian function (9.1.6) becomes

$$\mathscr{L}(\mathbf{x}, \boldsymbol{\lambda}) = \tfrac{1}{2}\mathbf{x}^T\mathbf{G}\mathbf{x} + \mathbf{g}^T\mathbf{x} - \boldsymbol{\lambda}^T(\mathbf{A}^T\mathbf{x} - \mathbf{b}) \tag{10.2.1}$$

and the stationary point condition (9.1.7) yields the equations

$$\begin{aligned} \nabla_x \mathscr{L} = \mathbf{0}: &\quad \mathbf{G}\mathbf{x} + \mathbf{g} - \mathbf{A}\boldsymbol{\lambda} = \mathbf{0} \\ \nabla_\lambda \mathscr{L} = \mathbf{0}: &\quad \mathbf{A}^T\mathbf{x} - \mathbf{b} \qquad = \mathbf{0} \end{aligned}$$

which can be rearranged to give the linear system

$$\begin{bmatrix} \mathbf{G} & -\mathbf{A} \\ -\mathbf{A}^T & \mathbf{0} \end{bmatrix} \begin{pmatrix} \mathbf{x} \\ \boldsymbol{\lambda} \end{pmatrix} = -\begin{pmatrix} \mathbf{g} \\ \mathbf{b} \end{pmatrix}. \tag{10.2.2}$$

The coefficient matrix is referred to as the *Lagrangian matrix* and is symmetric but not positive definite. If the inverse exists and is expressed as

$$\begin{bmatrix} \mathbf{G} & -\mathbf{A} \\ -\mathbf{A}^T & \mathbf{0} \end{bmatrix}^{-1} = \begin{bmatrix} \mathbf{H} & -\mathbf{T} \\ -\mathbf{T}^T & \mathbf{U} \end{bmatrix} \tag{10.2.3}$$

then the solution to (10.2.2) can be written

$$\mathbf{x}^* = -\mathbf{H}\mathbf{g} + \mathbf{T}\mathbf{b} \tag{10.2.4}$$

$$\boldsymbol{\lambda}^* = \mathbf{T}^T\mathbf{g} - \mathbf{U}\mathbf{b}. \tag{10.2.5}$$

These relationships are used by Fletcher (1971) to solve the equality constraint problem (10.3.1) which arises in the active set method. Since $\mathbf{b} = \mathbf{0}$ in that case only the matrices $\mathbf{H}$ and $\mathbf{T}$ need to be stored. Explicit expressions for $\mathbf{H}$, $\mathbf{T}$, and $\mathbf{U}$ when $\mathbf{G}^{-1}$ exists are

$$\begin{aligned}\mathbf{H} &= \mathbf{G}^{-1} - \mathbf{G}^{-1}\mathbf{A}(\mathbf{A}^{\mathrm{T}}\mathbf{G}^{-1}\mathbf{A})^{-1}\mathbf{A}^{\mathrm{T}}\mathbf{G}^{-1}\\ \mathbf{T} &= \mathbf{G}^{-1}\mathbf{A}(\mathbf{A}^{\mathrm{T}}\mathbf{G}^{-1}\mathbf{A})^{-1}\\ \mathbf{U} &= -(\mathbf{A}^{\mathrm{T}}\mathbf{G}^{-1}\mathbf{A})^{-1}\end{aligned} \tag{10.2.6}$$

and are readily verified by multiplying out the Lagrangian matrix and its inverse. Murtagh and Sargent (1969) suggest methods for linear constraints which use this representation by storing $(\mathbf{A}^{\mathrm{T}}\mathbf{G}^{-1}\mathbf{A})^{-1}$ and $\mathbf{G}^{-1}$. However it is not necessary for $\mathbf{G}^{-1}$ to exist for the Lagrangian matrix to be non-singular. Neither of the above methods are recommended in practice since they represent inverses directly and so have potential stability problems. In fact if $\mathbf{S}$ and $\mathbf{Z}$ are defined by (10.1.20) then an alternative representation of the inverse Lagrangian matrix is

$$\begin{aligned}\mathbf{H} &= \mathbf{Z}(\mathbf{Z}^{\mathrm{T}}\mathbf{G}\mathbf{Z})^{-1}\mathbf{Z}^{\mathrm{T}}\\ \mathbf{T} &= \mathbf{S} - \mathbf{Z}(\mathbf{Z}^{\mathrm{T}}\mathbf{G}\mathbf{Z})^{-1}\mathbf{Z}^{\mathrm{T}}\mathbf{G}\mathbf{S}\\ \mathbf{U} &= \mathbf{S}^{\mathrm{T}}\mathbf{G}\mathbf{Z}(\mathbf{Z}^{\mathrm{T}}\mathbf{G}\mathbf{Z})^{-1}\mathbf{Z}^{\mathrm{T}}\mathbf{G}\mathbf{S} - \mathbf{S}^{\mathrm{T}}\mathbf{G}\mathbf{S}.\end{aligned} \tag{10.2.7}$$

This can be verified by using relationships derived from (10.1.20) (see Question 10.7). In this case it follows that (10.2.4) and (10.2.5) are identical to the expressions (10.1.15) and (10.1.16) derived in the generalized elimination method. Thus it can be regarded that the computation of $\mathbf{S}$, $\mathbf{Z}$, and the $\mathbf{L}\mathbf{L}^{\mathrm{T}}$ factors of $\mathbf{Z}^{\mathrm{T}}\mathbf{G}\mathbf{Z}$ in any of the elimination methods is essentially a subtle way of factorizing the Lagrangian matrix in a Lagrangian method. In particular the orthogonal factorization method is stable and therefore valuable. Equations (10.2.7) prove that the Lagrangian matrix is non-singular if and only if there exists $\mathbf{Z}$ such that the matrix $\mathbf{Z}^{\mathrm{T}}\mathbf{G}\mathbf{Z}$ is non-singular (see Question 10.11). Furthermore $\mathbf{x}^*$ is a unique local minimizer if and only if $\mathbf{Z}^{\mathrm{T}}\mathbf{G}\mathbf{Z}$ is positive definite by virtue of the second order conditions (Section 2.1) applied to the quadratic function (10.1.12).

Some other observations on the structure of the inverse in (10.2.3) are the following. Firstly $\mathbf{T}^{\mathrm{T}}\mathbf{A} = \mathbf{I}$ so $\mathbf{T}^{\mathrm{T}}$ is a left generalized inverse for $\mathbf{A}$. If $\mathbf{Z}^{\mathrm{T}}\mathbf{G}\mathbf{Z}$ is positive definite, $\mathbf{H}$ is positive semi-definite with rank $n - m$. It also satisfies $\mathbf{H}\mathbf{A} = \mathbf{0}$ so projects any vector $\mathbf{v}$ into the constraint manifold (since $\mathbf{A}^{\mathrm{T}}\mathbf{H}\mathbf{v} = \mathbf{0}$). If $\mathbf{x}^{(k)}$ is any feasible point then it follows that $\mathbf{A}^{\mathrm{T}}\mathbf{x}^{(k)} = \mathbf{b}$, and $\mathbf{g}^{(k)} = \mathbf{g} + \mathbf{G}\mathbf{x}^{(k)}$ is the gradient vector of $q(\mathbf{x})$ at $\mathbf{x}^{(k)}$. Then using (10.2.6) it follows that (10.2.4) and (10.2.5) can be rearranged as

$$\mathbf{x}^* = \mathbf{x}^{(k)} - \mathbf{H}\mathbf{g}^{(k)} \tag{10.2.8}$$

$$\boldsymbol{\lambda}^* = \mathbf{T}^{\mathrm{T}}\mathbf{g}^{(k)}. \tag{10.2.9}$$

Equation (10.2.8) has a close relationship with Newton's method and shows that $\mathbf{H}$ contains the correct curvature information for the feasible region and so can be regarded as a reduced inverse Hessian matrix. Therefore these formulae can be considered to define a projection method with respect to a metric $\mathbf{G}$ (if positive definite); that is if $\mathbf{G}$ is replaced by $\mathbf{I}$ then $\mathbf{H} = \mathbf{A}\mathbf{A}^{+} = \mathbf{P}$ which is an orthogonal projection matrix, and the gradient projection method of Section 11.1 is obtained. In fact it can be established that $\mathbf{H} = (\mathbf{PGP})^+$, the generalized inverse of the projected Hessian matrix.

There are methods for factorizing a general non-singular symmetric matrix which is not positive definite, initially due to Bunch and Parlett (1971), which can also be used. This involves calculating $\mathbf{LDL}^T$ factors of the Lagrangian matrix (with some symmetric pivoting) in which $\mathbf{D}$ is block diagonal with a mixture of 1×1 and 2×2 blocks. This therefore provides a stable means of solving (10.2.2), especially for problems in which $\mathbf{G}$ is indefinite. However the method does ignore the zero matrix which exists in (10.2.2) and so does not take full advantage of the structure. Also since pivoting is involved the factors are not very convenient for being updated in an active set method. On the other hand, the Lagrangian matrix is non-singular if and only if $\mathbf{A}$ has full rank, so generalized elimination methods which form $\mathbf{S}$ and $\mathbf{Z}$ matrices in a reasonably stable way when $\mathbf{A}$ has full rank cannot be unstable as a means of factorizing the Lagrangian matrix as long as $\mathbf{Z}^T\mathbf{GZ}$ is positive definite. Thus the stable generalized elimination methods seem to be preferable.

Yet another way of solving the Lagrangian system (10.2.2) is to factorize the Lagrangian matrix forward. Firstly $\mathbf{LDL}^T$ factors of $\mathbf{G}$ are calculated which is stable so long as $\mathbf{G}$ is positive definite. Then these factors are used to eliminate the off-diagonal partitions ($-\mathbf{A}$ and $-\mathbf{A}^T$) in the Lagrangian matrix. The $\mathbf{0}$ partition then becomes changed to $-\mathbf{A}^T\mathbf{G}^{-1}\mathbf{A}$ which is negative definite and so can be factorized as $-\hat{\mathbf{L}}\hat{\mathbf{D}}\hat{\mathbf{L}}^T$ with $\hat{\mathbf{D}} > \mathbf{0}$. The resulting factors $\bar{\mathbf{L}}\bar{\mathbf{D}}\bar{\mathbf{L}}^T$ of the Lagrangian matrix are therefore given by

$$\bar{\mathbf{L}} = \begin{bmatrix} \mathbf{L} & \\ \mathbf{B}^T & \hat{\mathbf{L}} \end{bmatrix}, \qquad \bar{\mathbf{D}} = \begin{bmatrix} \mathbf{D} & \\ & -\hat{\mathbf{D}} \end{bmatrix} \tag{10.2.10}$$

where $\mathbf{B}$ is defined by $\mathbf{LDB} = -\mathbf{A}$ and is readily computed by forward substitution. To avoid loss of precision in forming $\mathbf{A}^T\mathbf{G}^{-1}\mathbf{A}$ (the 'squaring' effect), $\hat{\mathbf{L}}$ and $\hat{\mathbf{D}}$ are best calculated by forming $\mathbf{D}^{1/2}\mathbf{L}^T\mathbf{A}$ and using the QR method to factorize this matrix into $\mathbf{Q}\hat{\mathbf{D}}^{1/2}\hat{\mathbf{L}}^T$. This method is most advantageous when $\mathbf{G}$ has some structure (for example a band matrix) which can be exploited and when the number of constraints m is small (see Question 10.3). It is not entirely obvious that the method is stable with regard to round-off error, especially when $\mathbf{G}$ is nearly singular, but the method has been used successfully. The requirement that $\mathbf{G}$ is positive definite is important, however, in that otherwise computation of the $\mathbf{LDL}^T$ factors may not be possible or may be unstable. The method is closely related to the dual transformation described in (9.5.8) and (9.5.9) in the case that the constraints are equalities.

10.3 The Active Set Method

Most QP problems involve inequality constraints and so can be expressed in the form given in (10.1.1). This section describes how methods for solving equality constraints can be generalized to handle the inequality problem by means of an *active set method* similar to that described for LP problems in Section 8.3. It is convenient to consider first the case in which the Hessian matrix $\mathbf{G}$ is positive definite which ensures that any solution is a unique global minimizer, and that some potential difficulties are avoided. However it is possible with care to handle the more general case which is described in Section 10.4.

In the active set method certain constraints, indexed by the *active set* $\mathscr{A}$, are regarded as equalities whilst the rest are temporarily disregarded, and the method adjusts this set in order to identify the correct active constraints at the solution to (10.1.1). Because the objective function is quadratic there may be m $(0 \leqslant m \leqslant n)$ active constraints, in contrast to linear programming when $m = n$. On iteration k a feasible point $\mathbf{x}^{(k)}$ is known which satisfies the active constraints as equalities, that is $\mathbf{a}_i^{\mathrm{T}}\mathbf{x}^{(k)} = b_i$, $i \in \mathscr{A}$. Also except in degenerate cases $\mathbf{a}_i^{\mathrm{T}}\mathbf{x}^{(k)} > b_i$, $i \notin \mathscr{A}$, so that the current active set $\mathscr{A}$ is equivalent to the set of active constraints $\mathscr{A}^{(k)}$ defined in (7.1.2). Each iteration attempts to locate the solution to an *equality problem* (EP) in which only the active constraints occur. This is most conveniently done by shifting the origin to $\mathbf{x}^{(k)}$ and looking for a correction $\boldsymbol{\delta}^{(k)}$ which solves

$$\begin{aligned} &\underset{\boldsymbol{\delta}}{\text{minimize}}\ \tfrac{1}{2}\boldsymbol{\delta}^{\mathrm{T}}\mathbf{G}\boldsymbol{\delta} + \boldsymbol{\delta}^{\mathrm{T}}\mathbf{g}^{(k)} \\ &\text{subject to } \mathbf{a}_i^{\mathrm{T}}\boldsymbol{\delta} = 0, \qquad i \in \mathscr{A} \end{aligned} \tag{10.3.1}$$

where $\mathbf{g}^{(k)}$ is defined by $\mathbf{g}^{(k)} = \mathbf{g} + \mathbf{G}\mathbf{x}^{(k)}$ and is $\nabla q(\mathbf{x}^{(k)})$ for the function $q(\mathbf{x})$ defined in (10.1.1). This of course is basically in the form of (10.1.2) so can be solved by any of the methods of Sections 10.1 and 10.2. If $\boldsymbol{\delta}^{(k)}$ is feasible with regard to the constraints not in $\mathscr{A}$, then the next iterate is taken as $\mathbf{x}^{(k+1)} = \mathbf{x}^{(k)} + \boldsymbol{\delta}^{(k)}$. If not then a line search is made in the direction of $\boldsymbol{\delta}^{(k)}$ to find the best feasible point as described in (8.3.6). This can be expressed by defining the solution of (10.3.1) as a search direction $\mathbf{s}^{(k)}$, and choosing the step $\alpha^{(k)}$ to solve

$$\alpha = \min\left(1, \min_{\substack{i:i\notin\mathscr{A},\\ \mathbf{a}_i^{\mathrm{T}}\mathbf{s}^{(k)}<0}} \frac{b_i - \mathbf{a}_i^{\mathrm{T}}\mathbf{x}^{(k)}}{\mathbf{a}_i^{\mathrm{T}}\mathbf{s}^{(k)}}\right) \tag{10.3.2}$$

so that $\mathbf{x}^{(k+1)} = \mathbf{x}^{(k)} + \alpha^{(k)}\mathbf{s}^{(k)}$. If $\alpha^{(k)} < 1$ in (10.3.2) then a new constraint becomes active, defined by the index (p say) which achieves the min in (10.3.2), and this index is added to the active set $\mathscr{A}$.

If $\mathbf{x}^{(k)}$ (that is $\boldsymbol{\delta} = \mathbf{0}$) solves the current EP (10.3.1) then it is possible to compute multipliers ($\boldsymbol{\lambda}^{(k)}$ say) for the active constraints as described in Sections 10.1 and 10.2. The vectors $\mathbf{x}^{(k)}$ and $\boldsymbol{\lambda}^{(k)}$ then satisfy all the first order conditions (9.1.16) for the inequality constraint problem except possibly that $\lambda_i \geqslant 0$, $i \in I$. Thus a test is made to determine whether $\lambda_i^{(k)} \geqslant 0\ \forall\, i \in \mathscr{A}^{(k)} \cap I$; if so then the first order conditions are satisfied and these are sufficient to ensure a global solution since $q(\mathbf{x})$ is convex. Otherwise there exists an index, q say, $q \in \mathscr{A}^{(k)} \cap I$, such that $\lambda_q^{(k)} < 0$. In this case, following the discussion in Section 9.1, it is possible to reduce $q(\mathbf{x})$ by allowing the constraint q to become inactive. Thus q is removed from $\mathscr{A}$ and the algorithm continues as before by solving the resulting EP (10.3.1). If there is more than one index for which $\lambda_q^{(k)} < 0$ then it is usual to select q to solve

$$\min_{i \in \mathscr{A} \cap I} \lambda_i^{(k)}. \tag{10.3.3}$$

This selection works quite well and is convenient, so is usually used. One slight disadvantage is that it is not invariant to scaling of the constraints so some attention to scaling may be required. An invariant but more complex test can be devised based on the expected reduction in $q(\mathbf{x})$ (Fletcher and Jackson, 1974). To summarize the algorithm, therefore, if $\mathbf{x}^{(1)}$ is a given feasible point and $\mathscr{A} = \mathscr{A}^{(1)}$ is the corresponding active set then the active set method is defined as follows.

(a) Given $\mathbf{x}^{(1)}$ and $\mathscr{A}$, set $k = 1$.
(b) If $\boldsymbol{\delta} = \mathbf{0}$ does not solve (10.3.1), go to (d).
(c) Compute Lagrange multipliers $\boldsymbol{\lambda}^{(k)}$ and solve (10.3.3); if $\lambda_q^{(k)} \geqslant 0$ terminate with $\mathbf{x}^* = \mathbf{x}^{(k)}$, otherwise remove q from $\mathscr{A}$.
(d) Solve (10.3.1) for $\mathbf{s}^{(k)}$. (10.3.4)
(e) Find $\alpha^{(k)}$ to solve (10.3.2) and set $\mathbf{x}^{(k+1)} = \mathbf{x}^{(k)} + \alpha^{(k)}\mathbf{s}^{(k)}$.
(f) If $\alpha^{(k)} < 1$, add p to $\mathscr{A}$.
(g) Set $k = k + 1$ and go to (b).

An illustration of the method for a simple QP problem is shown in Figure 10.3.1. In this QP the constraints are the bounds $\mathbf{x} \geqslant \mathbf{0}$ and a general constraint $\mathbf{a}^T\mathbf{x} \geqslant b$. At $\mathbf{x}^{(1)}$ the bounds $x_1 \geqslant 0$ and $x_2 \geqslant 0$ are both active and so $\mathbf{x}^{(1)}$ (that is $\boldsymbol{\delta} = \mathbf{0}$) solves the current EP (10.3.2). Calculating $\boldsymbol{\lambda}^{(1)}$ shows that the constraint $x_2 \geqslant 0$ has negative multiplier and so becomes inactive, so that only $x_1 \geqslant 0$ is active. The corresponding EP is now solved by $\mathbf{x}^{(2)}$ (or $\boldsymbol{\delta} = \mathbf{x}^{(2)} - \mathbf{x}^{(1)}$) which is feasible and so becomes the next iterate. Calculating $\boldsymbol{\lambda}^{(2)}$ indicates that the constraint $x_1 \geqslant 0$ has a negative multiplier so this too becomes inactive leaving $\mathscr{A}$ empty. The corresponding EP is now solved by $\mathbf{x}'$ which is infeasible, so the search direction $\mathbf{s}^{(2)} = \mathbf{x}' - \mathbf{x}^{(2)}$ is calculated from (10.3.1) and a line search along $\mathbf{s}^{(2)}$ by solving (10.3.2) gives $\mathbf{x}^{(3)}$ as the best feasible point. The general constraint $\mathbf{a}^T\mathbf{x} \geqslant b$ becomes active at $\mathbf{x}^{(3)}$ so is added to $\mathscr{A}$. Since $\mathbf{x}^{(3)}$ does not solve the current EP, multipliers are not calculated, but instead the EP is solved, yielding $\mathbf{x}^{(4)}$ as the next

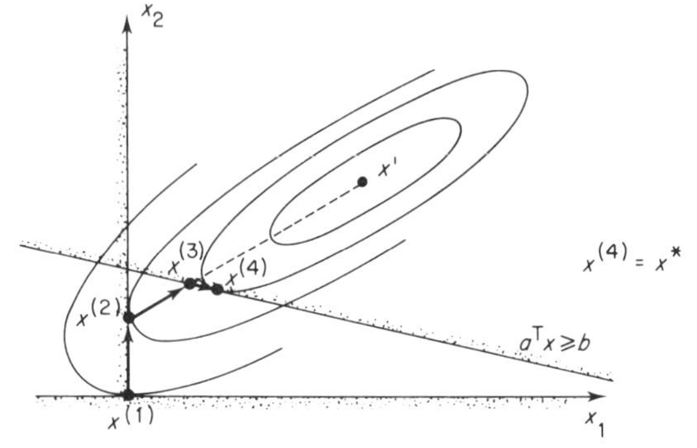

Figure 10.3.1 The active set method for a simple QP problem

iterate. Calculation of multipliers at $\mathbf{x}^{(4)}$ indicates that $\mathbf{x}^{(4)}$ is the solution of the inequality QP problem, and the algorithm terminates. A numerical example with similar properties is given in Question 10.4.

Termination of the algorithm in general can be proved if each step $\alpha^{(k)} \neq 0$, in which case $q(\mathbf{x})$ is reduced on each iteration. It is a consequence of (10.3.2) (see Question 10.10) that the vectors $\mathbf{a}_i$, $i \in \mathscr{A}$, arc independent so the EP in (10.3.1) is always well defined. The termination proof relies on the fact that there is a subsequence $\{\mathbf{x}^{(k)}\}$ of iterates which solve the current EP. (Only when $\alpha^{(k)} < 1$ in (10.3.2) is $\mathbf{x}^{(k+1)}$ not a solution to the EP. In this case an index p is added to $\mathscr{A}$. This can happen at most n times in which case $\mathbf{x}^{(k)}$ is then a vertex, so solves the corresponding EP.) Since the number of possible EPs is finite, since each $\mathbf{x}^{(k)}$ in the subsequence is the unique global solution of an EP, and since $q(\mathbf{x}^{(k)})$ is monotonically decreasing, it follows that termination must occur. This proof fails when steps of zero length are taken and $q(\mathbf{x}^{(k)})$ does not decrease, in which case the algorithm can *cycle* by returning to a previous active set in the sequence. This is caused by *degeneracy* in the constraint set and the situation is similar to that described in Section 8.6. There is the additional possibility of a tie occurring when the value $\alpha^{(k)} = 1$ (for which $\mathbf{s}^{(k)}$ solves (10.3.1)) is also the value for which inactive constraints become active in (8.3.6). It is possible to give perturbation results which enable these ties to be broken in such a way as to theoretically avoid cycling, as in Section 8.6, but in practice there are some difficulties (again see Section 8.6) and most current algorithms ignore the fact that degeneracy can occur.

An important feature of an active set method concerns efficient solution of the EP (10.3.1) when changes are made to $\mathscr{A}$. As in linear programming it is possible to avoid refactorizing the Lagrangian matrix on each iteration (requiring $O(n^3)$ housekeeping operations) and instead to *update* the factors whenever a change is made to $\mathscr{A}$ ($O(n^2)$ operations). Some care has to be taken in the different methods in choosing how the various factors are represented to ensure that this updating is possible and efficient. These computations are quite intricate to write down and explain, even for just one method, so I shall not give details. Different ideas are used in different methods; amongst these are simplex-type updates and the idea for representing **QR** factors of **A** without requiring **Q**, described in Section 8.5. Use of rank one updates and use of elementary Givens rotations often occur (Fletcher and Powell, 1975; Gill and Murray, 1978a). Permutations (interchange or cyclic) in the constraint or variable orderings are often required to achieve a favourable structure (for example Fletcher and Jackson, 1974). A review of a number of these techniques is given by Gill and Murray (1978a). A description of the details in the case that the orthogonal factorization method in Section 10.1 is used to solve the EP is given by Gill and Murray (1978b). Another important feature of a similar nature is that the algorithms should treat bounds (when $\mathbf{a}_i = \pm\mathbf{e}_i$) in an efficient way. For instance if the first p ($\leqslant m$) active constraints are lower bounds on the variables 1 to p, then the matrices **A**, **S**, and **Z** have the structure

$$\mathbf{A} = \underset{\displaystyle\quad p \qquad m-p}{\begin{bmatrix} \mathbf{I} & \mathbf{A}_1 \\ \mathbf{0} & \mathbf{A}_2 \end{bmatrix}} \begin{matrix} p \\ n-p \end{matrix}, \qquad \mathbf{S} = \underset{\displaystyle\quad p \qquad\quad m-p}{\begin{bmatrix} \mathbf{I} & \mathbf{0} \\ -\mathbf{S}_2^T\mathbf{A}_1^T & \mathbf{S}_2^T \end{bmatrix}}, \qquad \mathbf{Z} = \underset{\displaystyle n-m}{\begin{bmatrix} \mathbf{0} \\ \mathbf{Z}_2 \end{bmatrix}}$$

where $\mathbf{S}_2^T$ is any left generalized inverse for $\mathbf{A}_2$ ($\mathbf{S}_2^T\mathbf{A}_2 = \mathbf{I}$). This structure cuts down storage and operations requirements but makes the updating more complicated as there are four cases to consider (adding or removing either a bound or a general constraint to or from $\mathscr{A}$).

The active set method requires an initial feasible point $\mathbf{x}^{(1)}$; this can be calculated by the techniques described in Section 8.4, solving the auxiliary problem (8.4.6). This calculates a feasible vertex (possibly including some pseudoconstraints), which becomes the required feasible point $\mathbf{x}^{(1)}$. Factors of the $\mathbf{A}$ matrix for the active set can also be passed on to be used in the main algorithm. An alternative possibility which avoids the feasible point requirement is to bias the phase I cost function by adding to it a multiple $\nu q(\mathbf{x})$ of the quadratic function. For sufficiently small ν, the problem can then be solved in a single phase. This leads to a study of QP-like methods for the problem

$$\text{minimize } \psi(\mathbf{x}) \triangleq q(x) + \| \boldsymbol{\ell}(\mathbf{x})^+ \|_1 \tag{10.3.5}$$

(for inequality constraints $\boldsymbol{\ell}(\mathbf{x}) \leqslant \mathbf{0}$), where $\boldsymbol{\ell}(\mathbf{x}) = \mathbf{A}^T\mathbf{x} - \mathbf{b}$ and $a^+ = \max(a, 0)$. An active set method is also possible which differs from (10.3.4) in a number of ways. One is that the line search (10.3.2) is changed to search for a minimizer of (10.3.5): if a constraint $\ell_p(\mathbf{x})$ becomes active (zero) then it is added to $\mathscr{A}$ as before. Another difference is that the first order conditions for (10.3.5) are that multipliers $\boldsymbol{\lambda}$ exist which satisfy $0 \leqslant \lambda_i \leqslant 1$ (see (14.3.8) and later in Section 14.3). Thus test (10.3.3) is changed in an obvious way to choose λ_q as that multiplier which violates these conditions the most. Likewise inactive constraints have their corresponding multiplier either 0 or 1 according to whether or not $\ell_i(\mathbf{x}) < 0$ or > 0. The EP which is solved is

$$\begin{aligned} &\underset{\mathbf{x}}{\text{minimize}} \quad q(\mathbf{x}) + \sum_{i \notin \mathscr{A}} \lambda_i \ell_i(\mathbf{x}) \\ &\text{subject to } \ell_i(\mathbf{x}) = 0, \qquad i \in \mathscr{A}, \end{aligned} \tag{10.3.6}$$

and the techniques of Sections 10.1 and 10.2 are again relevant, together with suitable updating methods.

The relationship between the active set method (10.3.4) and some methods derived as extensions of linear programming is described in Section 10.6. Possible variations on (10.3.4) which have been suggested include the removal of more than one constraint (with $\lambda_i^{(k)} < 0$) from $\mathscr{A}$ in step (c) (Goldfarb, 1972), or the modification of step (b) so as to accept an approximate solution to the EP. Both these possibilities tend to induce smaller active sets. Goldfarb (1972) argues that this is advantageous, although no extensive experience is available and I would expect the amount of improvement to be small. The modification in which more than one constraint is removed also creates a potential difficulty for indefinite QP associated with the need to maintain stable factors of the curvature matrix $\mathbf{Z}^T\mathbf{G}\mathbf{Z}$.

10.4 Advanced Features

A modern code for QP should be of wider application than has so far been considered; it should be able to handle an indefinite QP problem, and it should enable the user to carry forward efficiently information from a previous QP calculation when a sequence of problems is being solved. These extensions are considered in this section. Indefinite QP refers to the case in which the Hessian $\mathbf{G}$ is not required to be positive definite, so will often be indefinite, although the negative and positive semi-definite cases are also included. For equality constrained QP no difficulties arise and a unique global solution exists if and only if the reduced Hessian matrix $\mathbf{Z}^T\mathbf{G}\mathbf{Z}$ is positive definite. However when inequalities are present, and excluding the positive semi-definite case, it is possible for local solutions to exist which are not global. The problem of global optimization, here as elsewhere in the book, is usually very difficult so the algorithms only aim to find a local solution. In fact a modification of the active set method is used which only guarantees to find a KT point for which the reduced Hessian $\mathbf{Z}^T\mathbf{G}\mathbf{Z}$ is positive definite. If $\lambda_i^* > 0$, $i \in I^*$, then these conditions are sufficient for a local solution. However if there exists $\lambda_i^* = 0$, $i \in I^*$, then it is possible that the method will find a KT point which is not a local solution. In this case there usually exist arbitrarily small perturbations of the problem, (a) which make the KT point a local solution or (b) which make the point no longer a KT point so that the active set method continues to find a better local solution. Both these features are illustrated in Question 10.9. However although it is theoretically possible for these situations to arise, in practice (especially with round-off errors) it is unlikely and the algorithm is usually successful in finding a local (and often global) solution.

The main difficulty in solving indefinite QP problems arises in a different way. The possibility exists for an arbitrary choice of $\mathscr{A}$ that the resulting reduced Hessian matrix $\mathbf{Z}^T\mathbf{G}\mathbf{Z}$ may be indefinite so that its $\mathbf{L}\mathbf{D}\mathbf{L}^T$ factors may not exist or may be unstable in regard to round-off propagation. (Of course if $\mathbf{G}$ is positive definite then so is $\mathbf{Z}^T\mathbf{G}\mathbf{Z}$ and no problems arise.) To some extent the active set method avoids the difficulty. Initially $\mathbf{x}^{(1)}$ is a vertex so the null space (10.1.9) has dimension zero and the problem does not arise. Also if $\mathbf{Z}^T\mathbf{G}\mathbf{Z}$ is positive definite for some given $\mathscr{A}$ and a constraint is added to $\mathscr{A}$, then the new null space is a subset of the previous one, so the new matrix $\mathbf{Z}^T\mathbf{G}\mathbf{Z}$ of one less dimension is also positive definite.

The difficulty arises therefore in step (c) of (10.3.4). In this case $\mathbf{x}^{(k)}$ solves the current EP and $\mathbf{Z}^T\mathbf{G}\mathbf{Z}$ is positive definite. However when a constraint index q is removed from $\mathscr{A}$, the dimension of the null space is increased and an extra column is adjoined to $\mathbf{Z}$. This causes an extra row and column to be adjoined to $\mathbf{Z}^T\mathbf{G}\mathbf{Z}$, and it is possible that the resulting matrix is not positive definite. Computing the new $\mathbf{L}\mathbf{D}\mathbf{L}^T$ factors is just the usual step for extending a Choleski factorization, but when $\mathbf{Z}^T\mathbf{G}\mathbf{Z}$ is not positive definite the new element of $\mathbf{D}$ is either negative or zero. (The latter case corresponds to a singular Lagrangian matrix.) This implies that the correction $\boldsymbol{\delta}^{(k)}$ which is a stationary point of (10.3.1) is no longer a minimizer, so some thought must be given to the choice of the search direction $\mathbf{s}^{(k)}$ in step (d) of (10.3.4). Since $\mathbf{x}^{(k)}$ is not a KT point, feasible descent directions exist and any such direction can be chosen for $\mathbf{s}^{(k)}$. One possibility is to consider the line caused by increasing the slack on constraint q at $\mathbf{x}^{(k)}$. Thus b_q is changed to $b_q + \alpha$ and from (10.2.4) the resulting search direction is $\mathbf{s}^{(k)} = \mathbf{T}\mathbf{e}_q$, the column of $\mathbf{T}$ corresponding to constraint q. In fact when the updated $\mathbf{D}$ has a negative element this is equivalent to choosing $\mathbf{s}^{(k)} = -\boldsymbol{\delta}^{(k)}$ where $\boldsymbol{\delta}^{(k)}$ is the stationary point of (10.3.1) for the updated problem. In this case the unit step $\alpha^{(k)} = 1$ in (10.3.2) no longer has significance. Thus the required modification to (10.3.4) when $\mathbf{D}$ has a negative element is to choose $\mathbf{s}^{(k)} = -\boldsymbol{\delta}^{(k)}$ in step (d), and to solve (8.3.6) rather than (10.3.2) to obtain $\alpha^{(k)}$ in step (e). If $\mathbf{D}$ has a zero element it can be replaced by $-\epsilon$ where $\epsilon > 0$ is negligible. This causes $\mathbf{s}^{(k)}$ to be very large but the length of $\mathbf{s}^{(k)}$ is unimportant in solving (8.3.6). It may be that (8.3.6) has no solution; this is the way in which an unbounded solution of an indefinite QP is indicated.

It can be seen therefore that in exact arithmetic, with the above changes, there is no real difficulty at all. The situation in which $\mathbf{D}$ has one negative element is only temporary since the algorithm proceeds to add constraints, reducing the dimension of the null space, until the solution of an EP is found, in which case the matrix $\mathbf{Z}^T\mathbf{G}\mathbf{Z}$ and hence $\mathbf{D}$ is positive definite. No further forward extension of the $\mathbf{L}\mathbf{D}\mathbf{L}^T$ factorization is made which might break down. The only adverse possibility is that there is no bound on the possible size of the negative element of $\mathbf{D}$. Thus when this element is computed the resulting round-off error can be significant. Therefore Gill and Murray (1978b) suggest that this possibility is monitored, and if a large negative element is observed, then the factors are recomputed at a later stage.

An alternative possibility which avoids this round-off problem is the following. When a constraint q is determined as in step (c), then it can be left in $\mathscr{A}$ as a *pseudo-constraint*. The right hand side b_q is allowed to increase (implicitly) to $b_q + \alpha$ and the search direction $\mathbf{s}^{(k)} = \mathbf{T}\mathbf{e}_q$ described above is followed. If a new constraint p becomes active in the line search, then p is added to $\mathscr{A}$ and the process is repeated. Only when the solution of an EP is located in a subsequent line search is the index q removed from $\mathscr{A}$. This approach has the advantage that all the reduced Hessian matrices $\mathbf{Z}^T\mathbf{G}\mathbf{Z}$ which are computed are positive definite. The disadvantage is that additional degenerate situations arise when the incoming vector $\mathbf{a}_p$ in step (f) of (10.3.4) is dependent on the vectors $\mathbf{a}_i$, $i \in \mathscr{A}' \cup q$. This arises most obviously when moving from one vertex to another, as in linear programming. A possible remedy is to have formulae for updating the factors of the Lagrangian matrix when an *exchange* of constraint indices in $\mathscr{A}$ is made (as in the simplex method for LP). An idea of this type is used by Fletcher (1971) and should be practicable with more stable methods but has not, I believe, been investigated.

Another important feature of a QP code is that it should allow information to be passed forward from a previous QP calculation when the problem has only been perturbed by a small amount. This is an example of *parametric programming* (see Section 8.3 for the LP case). In this case it can be advantageous to solve an initial EP with the active set $\mathscr{A}$ from the previous QP calculation. The resulting point $\mathbf{x}^{(1)}$ will often solve the new QP problem. If $\mathbf{A}$ is unchanged then $\mathbf{Z}$ need not be computed in most methods, and if $\mathbf{G}$ is also unchanged then $\mathbf{Z}^T\mathbf{G}\mathbf{Z}$ does not change so its factors need not be recomputed. There are, however, some potential difficulties. If $\mathbf{A}$ changes it may be that it comes close to losing rank, so that round-off difficulties occur. In indefinite QP problems, if $\mathbf{A}$ or $\mathbf{G}$ change it may be that $\mathbf{Z}^T\mathbf{G}\mathbf{Z}$ is no longer positive definite. In both these cases the option of using the previous active set should be abandoned. Another possibility is that the initial solution $\mathbf{x}^{(1)}$ of the EP is infeasible with respect to certain constraints $i \in \mathscr{A}$. In this case it is preferable not to have to resort to trying to find a feasible vertex as in Section 8.4. A better idea is to replace the right hand sides b_i of the violated constraints by $b_i + \theta$ and to trace the solution of the QP problem as $\theta \downarrow 0$. An alternative approach is based on using an exact penalty function as described at the end of Section 8.4 and in (10.3.5).

Another aspect of parametric programming is in finding the sensitivity of the solution with respect to changes in $\mathbf{g}$ and $\mathbf{b}$. The resulting rate of change of $\mathbf{x}^*$ and $\boldsymbol{\lambda}^*$ can be obtained by substituting in (10.2.4) and (10.2.5), and the effect on $q(\mathbf{x}^*)$ by using the definition of $q(\mathbf{x})$ in (10.1.1). Notice however that dq/db_i is approximated by λ_i^* by virtue of (9.1.10). These estimates may be invalid if there exist multipliers for which $\lambda_i^* = 0, i \in I^*$.

10.5 Special QP Problems

In some cases a QP problem has a special structure which can be utilized in order to gain improved efficiency or stability. A number of these possibilities are described in this section. An important case arises when the only constraints in the problem are the bounds

$$\ell_i \leqslant x_i \leqslant u_i, \qquad i = 1, 2, \ldots, n \tag{10.5.1}$$

where any u_i or $-\ell_i$ may be ∞ and $\ell_i \leqslant u_i$. Thus the vectors $\mathbf{a}_i$ are $\pm\mathbf{e}_i$ and it is possible to express the EP (after permuting the variables so that the active bounds are those on the first m variables) with

$$\mathbf{A} = \mathbf{S} = \begin{bmatrix} \mathbf{I} \\ \mathbf{0} \end{bmatrix} \begin{matrix} m \\ n-m \end{matrix}, \qquad \mathbf{Z} = \mathbf{V} = \begin{bmatrix} \mathbf{0} \\ \mathbf{I} \end{bmatrix} \begin{matrix} m \\ n-m \end{matrix}, \tag{10.5.2}$$

and $\mathbf{Z}^{\mathrm{T}}\mathbf{G}\mathbf{Z} = \mathbf{G}_{22}$ (see the definition before (10.1.3)). Thus no calculations on $\mathbf{Z}$ are required and this can be exploited in a special purpose method. The only updating which is required is to the $\mathbf{LDL}^{\mathrm{T}}$ factors of $\mathbf{Z}^{\mathrm{T}}\mathbf{G}\mathbf{Z}$. Also from (10.1.14) the multipliers $\boldsymbol{\lambda}^{(k)}$ at the solution to an EP are just the corresponding elements of $\mathbf{g}^{(k)}$ by virtue of the above $\mathbf{S}$. Another feature of this problem is that degeneracy in the constraints cannot arise. Also there are no difficulties in dealing with indefinite QP using the second method (based on pseudoconstraints) described in Section 10.4. The only exchange formula which is needed is that for replacing a lower bound on x_i by its upper bound, or vice versa, which is trivial. Pseudoconstraints (Section 8.4) are also important for calculating an initial vertex for the problem, especially if u_i and $-\ell_i$ are large. Methods for this problem are given by Gill and Murray (1976b) and by Fletcher and Jackson (1974). The latter method updates a partial factorization of the matrix $\begin{bmatrix} \mathbf{G}_{22} & \mathbf{G}_{12} \\ \mathbf{G}_{21} & \mathbf{G}_{11} \end{bmatrix}$; it now seems preferable to me to update merely the $\mathbf{LDL}^{\mathrm{T}}$ factors of $\mathbf{G}_{22}$ as mentioned above.

Another special case of QP arises when $\mathbf{G}$ is a positive definite diagonal matrix. By scaling the variables the objective function can be written with $\mathbf{G} = \mathbf{I}$ as

$$q(\mathbf{x}) = \tfrac{1}{2}\mathbf{x}^{\mathrm{T}}\mathbf{x} + \mathbf{g}^{\mathrm{T}}\mathbf{x}. \tag{10.5.3}$$

The resulting problem is known as a *least distance problem* since the solution $\mathbf{x}^*$ in (10.1.1) can be interpreted as the feasible point which is the least distance (in L_2) from the point $\mathbf{x} = -\mathbf{g}$ (after writing (10.5.3) as $\frac{1}{2}\|\mathbf{x}+\mathbf{g}\|_2^2$ and ignoring a constant term). Again a special purpose method is appropriate since the choice $\mathbf{Z} = \mathbf{Q}_2$ is advantageous in that $\mathbf{Z}^T\mathbf{G}\mathbf{Z} = \mathbf{Z}^T\mathbf{Z} = \mathbf{I}$. Thus no special provision need be made for updating factors of $\mathbf{Z}^T\mathbf{G}\mathbf{Z}$ and only the updating operations on $\mathbf{Z}$ are required, for which the formulation of Gill and Murray (1978b) can be used. Since $\mathbf{G}$ is positive definite, no special difficulties arise.

Yet another special case of QP occurs when the problem is to find a *best least squares solution* of the linear equations

$$\mathbf{B}\mathbf{x} = \mathbf{c} \tag{10.5.4}$$

where $\mathbf{B}$ is $p \times n$ and $\mathbf{c} \in \mathbb{R}^p$, subject to the constraints in (10.1.1). In this case the objective function is the sum of squares of residuals defined by

$$q(\mathbf{x}) = (\mathbf{B}\mathbf{x} - \mathbf{c})^T(\mathbf{B}\mathbf{x} - \mathbf{c}). \tag{10.5.5}$$

It is possible to proceed as in the standard method with $\mathbf{G} = \mathbf{B}^T\mathbf{B}$ and $\mathbf{g} = -\mathbf{B}^T\mathbf{c}$, after dividing by 2 and omitting the constant term. However it is well known that to form $\mathbf{B}^T\mathbf{B}$ explicitly in a linear least squares calculation causes the 'squaring effect' to occur which can cause the loss of twice as many significant figures as are necessary. It is therefore again advantageous to have a special purpose routine which avoids the difficulty. This can be done by updating QR factors of the matrix $\mathbf{B}\mathbf{Z}$. That is to say,

$$\mathbf{B}\mathbf{Z} = \mathbf{P}\begin{bmatrix}\mathbf{U}\\ \mathbf{0}\end{bmatrix} = \mathbf{P}_1\mathbf{U} \tag{10.5.6}$$

is defined where $\mathbf{P}$ is $p \times p$ orthogonal, $\mathbf{U}$ is $(n-m)\times(n-m)$ upper triangular, and $\mathbf{P}_1$ is the first $n-m$ columns of $\mathbf{P}$. It follows that $\mathbf{U}^T\mathbf{U}$ is the Choleski factorization of $\mathbf{Z}^T\mathbf{G}\mathbf{Z}$ $(= \mathbf{Z}^T\mathbf{B}^T\mathbf{B}\mathbf{Z})$ and can thus be used to solve (10.1.3) or (10.3.1). An algorithm is described by Mifflin (1978).

Finally the special case of least squares QP subject only to bounds on the variables is considered. It is possible to introduce a further special purpose method, similar to those described above in which, using (10.5.2) and (10.5.6), $\mathbf{B}$ is written $[\mathbf{B}_1 : \mathbf{B}_2]$ and $\mathbf{P}_1\mathbf{U}$ factors of $\mathbf{B}_2$ are used to solve the EP. An alternative possibility is to transform the problem into a least distance problem using the Wolfe dual (theorem 9.5.1). For example the QP

$$\begin{aligned}&\underset{\mathbf{x}}{\text{minimize}}\ \tfrac{1}{2}\mathbf{x}^T\mathbf{B}^T\mathbf{B}\mathbf{x} + \mathbf{g}^T\mathbf{x}\\ &\text{subject to } \mathbf{x} \geqslant \mathbf{0}\end{aligned} \tag{10.5.7}$$

becomes

$$\underset{\mathbf{x},\boldsymbol{\lambda}}{\text{maximize}}\ \tfrac{1}{2}\mathbf{x}^T\mathbf{B}^T\mathbf{B}\mathbf{x} + \mathbf{g}^T\mathbf{x} - \boldsymbol{\lambda}^T\mathbf{x} \tag{10.5.8}$$

$$\text{subject to } \mathbf{B}^T\mathbf{B}\mathbf{x} + \mathbf{g} - \boldsymbol{\lambda} = \mathbf{0} \tag{10.5.9}$$

$$\boldsymbol{\lambda} \geqslant \mathbf{0} \tag{10.5.10}$$

using (9.5.1). Eliminating $\boldsymbol{\lambda}$ from (10.5.8) using (10.5.9) and writing $\mathbf{Bx} = \mathbf{u}$ yields the least distance problem

$$\begin{aligned} &\underset{\mathbf{u}}{\text{minimize}}\ \tfrac{1}{2}\mathbf{u}^{\mathrm{T}}\mathbf{u} \\ &\text{subject to } \mathbf{B}^{\mathrm{T}}\mathbf{u} \geqslant -\mathbf{g}. \end{aligned} \tag{10.5.11}$$

The system can be solved by the method described above giving a solution $\mathbf{u}^*$. The solution $\mathbf{x}^*$ of (10.5.7) can then be recovered in the following way. By applying the dual transformation (theorem 9.5.1) to (10.5.11) it is easily seen that the dual of (10.5.11) is (10.5.7) so that the vector $\mathbf{x}$ is the multiplier vector for the constraints in (10.5.11). Thus $x_i^* = 0$ if i corresponds to an inactive constraint at the solution of (10.5.11). Furthermore a QR factorization of the columns of $\mathbf{B}$ corresponding to the active constraints is available (see (10.1.17)) so the remaining elements of $\mathbf{x}^*$ can be determined from $\mathbf{Bx}^* = \mathbf{u}^*$. Thus bounded least squares problems can be reduced to least distance problems, and vice versa. This result is a special case of the duality between (9.5.8) and (9.5.9) given in Section 9.5.

10.6 Complementary Pivoting and Other Methods

A number of methods for QP have been suggested as extensions of the simplex method for LP. The earliest of these is probably the Dantzig–Wolfe algorithm (Dantzig, 1963) which solves the KT conditions for the QP

$$\begin{aligned} &\underset{\mathbf{x}}{\text{minimize}}\ \tfrac{1}{2}\mathbf{x}^{\mathrm{T}}\mathbf{Gx} + \mathbf{g}^{\mathrm{T}}\mathbf{x} \\ &\text{subject to } \mathbf{A}^{\mathrm{T}}\mathbf{x} \geqslant \mathbf{b}, \qquad \mathbf{x} \geqslant \mathbf{0}. \end{aligned} \tag{10.6.1}$$

Introducing multipliers $\mathbf{y}$ for the constraints $\mathbf{A}^{\mathrm{T}}\mathbf{x} \geqslant \mathbf{b}$ and $\mathbf{u}$ for the bounds $\mathbf{x} \geqslant \mathbf{0}$ gives the Lagrangian function

$$\mathscr{L}(\mathbf{x}, \mathbf{y}, \mathbf{u}) = \tfrac{1}{2}\mathbf{x}^{\mathrm{T}}\mathbf{Gx} + \mathbf{g}^{\mathrm{T}}\mathbf{x} - \mathbf{y}^{\mathrm{T}}(\mathbf{A}^{\mathrm{T}}\mathbf{x} - \mathbf{b}) - \mathbf{u}^{\mathrm{T}}\mathbf{x}. \tag{10.6.2}$$

Defining slack variables $\mathbf{v} = \mathbf{A}^{\mathrm{T}}\mathbf{x} - \mathbf{b}$, the KT conditions (9.1.16) become

$$\begin{aligned} &\mathbf{u} - \mathbf{Gx} + \mathbf{Ay} = \mathbf{g} \\ &\mathbf{v} - \mathbf{A}^{\mathrm{T}}\mathbf{x} = -\mathbf{b} \\ &\mathbf{u}, \mathbf{v}, \mathbf{x}, \mathbf{y} \geqslant \mathbf{0}, \qquad \mathbf{u}^{\mathrm{T}}\mathbf{v} = 0, \qquad \mathbf{x}^{\mathrm{T}}\mathbf{y} = 0. \end{aligned} \tag{10.6.3}$$

The method assumes that $\mathbf{G}$ is positive definite which is a sufficient condition that any solution to (10.6.3) solves (10.6.1). The system (10.6.3) can be expressed in the form

$$\begin{aligned} &\mathbf{w} - \mathbf{Mz} = \mathbf{q} \\ &\mathbf{w} \geqslant \mathbf{0}, \qquad \mathbf{z} \geqslant \mathbf{0}, \qquad \mathbf{w}^{\mathrm{T}}\mathbf{z} = 0 \end{aligned} \tag{10.6.4}$$

where

$$\mathbf{w} = \begin{pmatrix} \mathbf{u} \\ \mathbf{v} \end{pmatrix}, \qquad \mathbf{z} = \begin{pmatrix} \mathbf{x} \\ \mathbf{y} \end{pmatrix}, \qquad \mathbf{M} = \begin{bmatrix} \mathbf{G} & -\mathbf{A} \\ \mathbf{A}^{\mathrm{T}} & \mathbf{0} \end{bmatrix}, \qquad \mathbf{q} = \begin{pmatrix} \mathbf{g} \\ -\mathbf{b} \end{pmatrix}. \tag{10.6.5}$$

This has come to be referred to as a *linear complementarity problem* and there are other problems in game theory and boundary value calculations which can also be expressed in this way. The Dantzig–Wolfe method for solving (10.6.3) has become known as the *principal pivoting method* for solving (10.6.5) which is outlined below. However all methods for solving (10.6.4) have some features in common. They carry out row operations on the equations in (10.6.4) in a closely related way to LP (Section 8.2). Thus the variables $\mathbf{w}$ and $\mathbf{z}$ together are rearranged into basic (B) and nonbasic (N) variables $\mathbf{x}_B$ and $\mathbf{x}_N$. At a general stage, a tableau representation

$$[\mathbf{I} : \hat{\mathbf{A}}] \begin{pmatrix} \mathbf{x}_B \\ \mathbf{x}_N \end{pmatrix} = \hat{\mathbf{b}} \tag{10.6.6}$$

of the equations in (10.6.4) is available and the variables have the values $\mathbf{x}_B = \hat{\mathbf{b}}$ and $\mathbf{x}_N = \mathbf{0}$. A variable x_q, $q \in N$, is chosen to be increased and the effect on $\mathbf{x}_B$ from (10.6.6) is given by

$$\mathbf{x}_B = \hat{\mathbf{b}} - \hat{\mathbf{a}}_q x_q \tag{10.6.7}$$

where $\hat{\mathbf{a}}_q$ is the column of $\hat{\mathbf{A}}$ corresponding to variable x_q. An element x_i, $i \in B$, becomes zero therefore when $x_q = \hat{b}_i / \hat{a}_{iq}$, and this is used in a test similar to (8.2.11). This determines some variable x_p which has become zero, and so p and q are interchanged between B and N as in LP. The tableau is then rearranged by making row operations which reduce the old x_q column of the tableau to a unit vector (the *pivot step* of LP – see Section 8.2 and Table 8.2.1). The algorithms terminate when $\hat{\mathbf{b}} \geqslant \mathbf{0}$ and when the solution is complementary, that is $z_i \in B \Leftrightarrow w_i \in N$ and $z_i \in N \Leftrightarrow w_i \in B$, for all i.

The principal pivoting method is initialized by $\mathbf{x}_B = \mathbf{w}$, $\mathbf{x}_N = \mathbf{z}$, $\hat{\mathbf{A}} = -\mathbf{M}$, and $\hat{\mathbf{b}} = \mathbf{q}$. This is complementary so the algorithm terminates if $\hat{\mathbf{b}} \geqslant \mathbf{0}$. If not, there is an element $w_q \in B$ for which $\hat{b}_q < 0$ (the most negative is chosen if more than one exists). The complementary variable $z_q \in N$ is then chosen to be increased. The effect on the basic variables is observed by virtue of (10.6.7). As long as any z_i variables in B stay non-negative, then z_q is increased until $w_q \uparrow 0$. Then z_q and w_q are interchanged in B and N and the tableau is updated; the resulting tableau is again complementary so the iteration can be repeated. It may be, however, that when z_q is increased then a basic variable $z_p \downarrow 0$. In this case a pivot interchange is made to give a non-complementary tableau. On the next iteration the complement w_p is chosen to be increased. If this causes $w_q \uparrow 0$ then complementarity is restored and the algorithm can proceed as described above. However a different z_p may become zero in which case the same operation of increasing the complement is repeated. Since there are only a finite number of z_i, $i \in B$, ultimately complementarity is always restored. An example of the method applied to the problem of Question 10.4 is given in Table 10.6.1. It is not difficult along the lines of Question 8.17 to show that the principal pivoting (or Dantzig–Wolfe method) is equivalent to the active set method (10.3.4). A complementary solution is equivalent to the solution of an EP, and choosing the most negative w_q is analogous to (10.3.3). The case when $w_q \uparrow 0$ is the same as choosing $\alpha^{(k)} = 1$

in (10.3.2), and the case $z_p \downarrow 0$ is equivalent to $\alpha^{(k)} < 1$ when an inactive constraint becomes active in the line search. These features can all be observed in Table 10.6.1. Solving the problem in the tableau form has the advantage of simplicity for a small problem and can give a compact code for microcomputer applications. Otherwise, however, the tableau form is disadvantageous; it does not exploit efficient factorizations of the Lagrangian matrix and instead updates essentially an inverse matrix which can be numerically unstable. Slack variables for all the inequalities are required so that the tableau can be large. Only a restricted form (10.6.1) of the QP problem is solved and the method may fail for indefinite QP. Thus in general a modern code for the active set method is preferable.

A different algorithm for solving the linear complementarity problem (10.6.4) is due to Lemke (see the review of Cottle and Dantzig, 1968). In this method the tableau is extended by adding an extra variable z_0 and a corresponding extra column $-\mathbf{e} = -(1, 1, \ldots, 1)^{\mathrm{T}}$ to $\hat{\mathbf{A}}$. Problem (10.6.4) is initialized by $\mathbf{x}_B = \mathbf{w}$, $\mathbf{x}_N = (z_0, \mathbf{z})$, $\hat{\mathbf{A}} = (-\mathbf{e}, -\mathbf{M})$, $\hat{\mathbf{b}} = \mathbf{q}$, and the values of the variables are $\mathbf{x}_N = \mathbf{0}$ and $\mathbf{x}_B = \hat{\mathbf{b}}$. If $\hat{\mathbf{b}} \geqslant \mathbf{0}$ then this is the solution; otherwise z_0 is increased in (10.6.6) until all the variables $\mathbf{w}$ are non-negative. The algorithm now attempts to reduce z_0 down to zero whilst retaining feasibility in (10.6.4). The effect of the initial step is to drive a variable,

Table 10.6.1 Tableaux for the principal pivoting (Dantzig–Wolfe) method

w_1	1			(-2)	1	-1	-3
w_2		1		1	-2	-1	0
w_3			1	1	1	0	2
				z_1	z_2	z_3	$\hat{b}$

z_1	1			$-\frac{1}{2}$	$-\frac{1}{2}$	$\frac{1}{2}$	$\frac{3}{2}$
w_2		1		$\frac{1}{2}$	$-\frac{3}{2}$	$-\frac{3}{2}$	$-\frac{3}{2}$
w_3			1	$\frac{1}{2}$	$(\frac{3}{2})$	$-\frac{1}{2}$	$\frac{1}{2}$
				w_1	z_2	z_3	$\hat{b}$

z_1	1			$-\frac{1}{3}$	$\frac{1}{3}$	$\frac{1}{3}$	$\frac{5}{3}$
w_2		1		1	1	(-2)	-1
z_2			1	$\frac{1}{3}$	$\frac{2}{3}$	$-\frac{1}{3}$	$\frac{1}{3}$
				w_1	w_2	z_3	$\hat{b}$

z_1	1			$-\frac{1}{6}$	$\frac{1}{2}$	$\frac{1}{6}$	$\frac{3}{2}$
z_3		1		$-\frac{1}{2}$	$-\frac{1}{2}$	$-\frac{1}{2}$	$\frac{1}{2}$
z_2			1	$\frac{1}{6}$	$\frac{1}{2}$	$-\frac{1}{6}$	$\frac{1}{2}$
				w_1	w_3	w_2	$\hat{b}$

w_p say, to zero (corresponding to the most negative q_i). Then w_p and z_0 are interchanged in the pivot step and the complement z_p is increased at the start of the second iteration. In general the variable being increased and the variable which is driven to zero are interchanged at the pivot step. The next variable to be increased is then the complement of the one which is driven to zero. The algorithm terminates when z_0 becomes zero, which gives a solution to (10.6.4). An example of the method applied to the problem in Question 10.4 is given in Table 10.6.2. An observation about the method is that including the z_0 term makes each tableau in the method a solution of a modified form of problem (10.6.1) in which $\mathbf{g}$ is replaced by $\mathbf{g} + z_0\mathbf{e}$ and $\mathbf{b}$ by $\mathbf{b} - z_0\mathbf{e}$. Thus Lemke's method can be regarded as a parametric programming solution to (10.6.1) in which the solution is traced out as the parameter $z_0 \downarrow 0$. A step in which $w_p \downarrow 0$ and then z_p is increased corresponds to an active constraint becoming inactive as the parameter changes. Vice versa, a step in which $z_p \downarrow 0$ and then w_p is increased corresponds to an inactive constraint becoming active. A form of Lemke's algorithm without the z_0 variable arises if there is some column $\mathbf{m}_q \geqslant \mathbf{0}$ in $\mathbf{M}$ such that $m_{jq} > 0$ when $q_j < 0$. Then an initial

Table 10.6.2 Tableaux for the Lemke pivoting method

w_1	1			(−1)	−2	1	−1	−3
w_2		1		−1	1	−2	−1	0
w_3			1	−1	1	1	0	2
				z_0	z_1	z_2	z_3	$\hat{b}$

z_0	1			−1	2	−1	1	3
w_2		1		−1	(3)	−3	0	3
w_3			1	−1	3	0	1	5
				w_1	z_1	z_2	z_3	$\hat{b}$

z_0	1			$-\frac{1}{3}$	$-\frac{2}{3}$	1	1	1
z_1		1		$-\frac{1}{3}$	$\frac{1}{3}$	−1	0	1
w_3			1	0	−1	(3)	1	2
				w_1	w_2	z_2	z_3	$\hat{b}$

z_0	1			$-\frac{1}{3}$	$-\frac{1}{3}$	$-\frac{1}{3}$	($\frac{2}{3}$)	$\frac{1}{3}$
z_1		1		$-\frac{1}{3}$	0	$\frac{1}{3}$	$\frac{1}{3}$	$\frac{5}{3}$
z_2			1	0	$-\frac{1}{3}$	$\frac{1}{3}$	$\frac{1}{3}$	$\frac{2}{3}$
				w_1	w_2	w_3	z_3	$\hat{b}$

z_3	1			$-\frac{1}{2}$	$-\frac{1}{2}$	$-\frac{1}{2}$	$\frac{3}{2}$	$\frac{1}{2}$
z_1		1		$-\frac{1}{6}$	$\frac{1}{6}$	$\frac{1}{2}$	$-\frac{1}{2}$	$\frac{3}{2}$
z_2			1	$\frac{1}{6}$	$-\frac{1}{6}$	$\frac{1}{2}$	$-\frac{1}{2}$	$\frac{1}{2}$
				w_1	w_2	w_3	z_0	$\hat{b}$

step to increase z_q is possible which from (10.6.6) makes $\mathbf{w} \geqslant \mathbf{0}$. Then the algorithm proceeds as above and terminates when complementarity is achieved (see Question 10.13).

Another popular early technique was *Beale's method* (Beale, 1959) which was developed as an extension of linear programming. However the method can also be described briefly as an active set method (Goldfarb, private communication) in the following way. Consider the active set method for LP in Section 8.3 applied to minimize a quadratic function. If the kth line search is terminated by reaching an unconstrained minimum, then the difference in gradients $\boldsymbol{\gamma}^{(k)} = \mathbf{g}^{(k+1)} - \mathbf{g}^{(k)}$ is exchanged with the outgoing constraint normal in the matrix $\mathbf{A}$ in (8.3.3). After a sequence of steps in which constraints are removed from $\mathscr{A}$ the effect is that the search direction $\mathbf{s}^{(k+1)}$ is orthogonal to the vectors $\boldsymbol{\gamma}^{(1)}, \ldots, \boldsymbol{\gamma}^{(k)}$ and hence conjugate to $\mathbf{s}^{(1)}, \ldots, \mathbf{s}^{(k)}$. Thus the possibility of quadratic termination is introduced (see Volume 1, Section 2.5). Unfortunately if a previously inactive constraint then becomes active, the sequence of conjugate directions is broken. To recover from this Beale's method requires some unproductive iterations in which the now irrelevant $\boldsymbol{\gamma}^{(j)}$ vectors are removed from the $\mathbf{A}$ matrix. The possibility of rearranging the tableau in order to avoid these iterations has been studied by Benveniste (1979). The method is then likely to be closely related to Murray's (1971) method which also recurs conjugate directions. Since essentially an inverse matrix is updated, the more stable factorizations described in Section 10.1 are preferred. The relationship between the active set method, the Dantzig–Wolfe method, Beale's method, and others is discussed more extensively by Goldfarb (1972).

Questions for Chapter 10

1. Consider the equality quadratic programming problem

$$\text{minimize} \quad q(\mathbf{x}) \triangleq \tfrac{1}{2}\mathbf{x}^{\mathrm{T}} \begin{bmatrix} 3 & -1 & 0 \\ -1 & 2 & -1 \\ 0 & -1 & 1 \end{bmatrix} \mathbf{x} + \begin{pmatrix} 1 \\ 1 \\ 1 \end{pmatrix}^{\mathrm{T}} \mathbf{x}$$

$$\text{subject to } x_1 + 2x_2 + x_3 = 4.$$

Eliminate x_1 and express the resulting function in the form $\tfrac{1}{2}\mathbf{y}^{\mathrm{T}}\mathbf{U}\mathbf{y} + \mathbf{v}^{\mathrm{T}}\mathbf{y}$, where $\mathbf{y} = \begin{pmatrix} x_2 \\ x_3 \end{pmatrix}$, $\mathbf{U}$ is a constant symmetric matrix, and $\mathbf{v}$ is a constant vector. Hence find the solution $\mathbf{x}^*$ to the QP. Find the Lagrange multiplier λ^* of the equality constraint. Does $\mathbf{x}^*$ solve the QP

$$\begin{aligned} &\text{minimize} \quad q(\mathbf{x}) \\ &\text{subject to } x_1 + 2x_2 + x_3 \geqslant 4, \qquad \mathbf{x} \geqslant \mathbf{0}? \end{aligned}$$

2. Solve the equality constraint QP problem

$$\underset{\boldsymbol{\delta}}{\text{minimize}} \quad \delta_1^2 - \delta_1\delta_2 + \delta_2^2 - \delta_2\delta_3 + \delta_3^2 + 2\delta_1 - \delta_2$$

$$\begin{aligned} \text{subject to } 3\delta_1 - \delta_2 + \delta_3 &= 0 \\ 2\delta_1 - \delta_2 - \delta_3 &= 0 \end{aligned}$$

by eliminating the variables δ_1 and δ_2. Exhibit the matrices $\mathbf{A}$, $\mathbf{A}_1$, $\mathbf{A}_2$ and $\mathbf{Z}$, and a matrix $\mathbf{S}$ such that $\mathbf{S}^T\mathbf{A} = \mathbf{I}$. Hence find the multipliers of the two constraints at the solution.

3. Solve the equality constraint QP

$$\text{minimize} \quad \sum_{i=1}^{n} ix_i^2$$

$$\text{subject to} \quad \sum_{i=1}^{n} x_i = K \qquad (K > 0)$$

by the method of Lagrange multipliers. Does this solution minimize the function subject to the constraints

$$x_i \geqslant 0 \quad i = 1, 2, \ldots, n \quad \text{and} \quad \sum_{i=1}^{n} x_i \geqslant K?$$

Give the explicit solution of the equality constraint problem when $n = 3$, $K = 10$.

4. Starting from $\mathbf{x}^{(1)} = \mathbf{0}$, illustrate the steps taken by the active set method to solve the QP problem

$$\begin{aligned} \underset{\mathbf{x}}{\text{minimize}} \quad & x_1^2 - x_1x_2 + x_2^2 - 3x_1 \\ \text{subject to} \quad & x_1 \geqslant 0 \\ & x_2 \geqslant 0 \\ & -x_1 - x_2 \geqslant -2. \end{aligned}$$

For simplicity, solve each EP without making a shift of variables. Verify that the method is equivalent to the Dantzig–Wolfe method (Table 10.6.1).

5. Illustrate the matrices $\mathbf{V}$, $\mathbf{S}$, and $\mathbf{Z}$ when solving the QP problem in Question 9.6 by (a) direct elimination, (b) orthogonal factorization (Gill and Murray, 1974a).

6. Given matrices $\mathbf{A}$ and $\mathbf{Z}$ which are $n \times m$ and $n \times (n-m)$ respectively $(1 < m < n)$ consider how to choose an $n \times (n-m)$ matrix $\mathbf{V}$ such that (10.1.20) holds. Let the matrices be partitioned after the mth row as

$$\mathbf{A} = \begin{bmatrix} \mathbf{A}_1 \\ \mathbf{A}_2 \end{bmatrix}, \qquad \mathbf{V} = \begin{bmatrix} \mathbf{V}_1 \\ \mathbf{V}_2 \end{bmatrix}, \qquad \mathbf{S} = \begin{bmatrix} \mathbf{S}_1 \\ \mathbf{S}_2 \end{bmatrix}, \qquad \mathbf{Z} = \begin{bmatrix} \mathbf{Z}_1 \\ \mathbf{Z}_2 \end{bmatrix}.$$

Show that $\mathbf{S}_1 = \mathbf{A}_1^{-T}(\mathbf{I} - \mathbf{A}_2^T\mathbf{S}_2)$, $\mathbf{V}_1 = -\mathbf{A}_1\mathbf{S}_2^T\mathbf{Z}_2^{-T}$ and $\mathbf{V}_2 = \mathbf{Z}_2^{-T}(\mathbf{I} - \mathbf{Z}_1^T\mathbf{V}_1)$ define $\mathbf{S}_1$, $\mathbf{V}_1$, and hence $\mathbf{V}_2$ in terms of $\mathbf{S}_2$. It follows that $\mathbf{V}$ is underdetermined to the extent of an arbitrary choice of the $(n-m) \times m$ matrix $\mathbf{S}_2$. Investigate the choice $\mathbf{S}_2 = \mathbf{0}$.

7. Let matrices $\mathbf{G}$, $\mathbf{A}$, $\mathbf{S}$, and $\mathbf{Z}$ be $n \times n$, $n \times m$, $n \times m$, and $n \times (n-m)$ respectively $(m \leqslant n)$, let $\mathbf{S}^T\mathbf{A} = \mathbf{I}$, $\mathbf{Z}^T\mathbf{A} = \mathbf{0}$, and let the matrix $[\mathbf{S} : \mathbf{Z}]$ be non-singular. Show that there exists a unique $n \times (n-m)$ matrix $\mathbf{V}$ such that $[\mathbf{A} : \mathbf{V}]^{-1} = [\mathbf{S} : \mathbf{Z}]^T$, and show that $\mathbf{S}^T\mathbf{V} = \mathbf{0}$, $\mathbf{Z}^T\mathbf{V} = \mathbf{I}$, and $\mathbf{A}\mathbf{S}^T + \mathbf{V}\mathbf{Z}^T = \mathbf{I}$.

Hence show that (10.2.3) holds where $\mathbf{H}$, $\mathbf{T}$, and $\mathbf{U}$ are defined by (10.2.7). Show how the problem

$$\underset{\mathbf{x}}{\text{minimize}}\ \tfrac{1}{2}\mathbf{x}^{\mathrm{T}}\mathbf{G}\mathbf{x} + \mathbf{g}^{\mathrm{T}}\mathbf{x}$$

$$\text{subject to } \mathbf{A}^{\mathrm{T}}\mathbf{x} = \mathbf{b}$$

can be solved by the method of Lagrange multipliers, and use (10.2.3) and (10.2.7) to obtain explicit expressions for the solution $\mathbf{x}^*$ and the associated Lagrange multiplier $\boldsymbol{\lambda}^*$. Verify that these reduce to (10.1.15) and (10.1.16) respectively.

8. Establish equations (10.2.8) and (10.2.9) and show that the matrix $\mathbf{H}$ in (10.2.3) satisfies $\mathbf{H} = (\mathbf{PGP})^+$ where $\mathbf{P} = \mathbf{I} - \mathbf{AA}^+$.
9. Consider the indefinite QP problem

$$\underset{\mathbf{x}}{\text{minimize}}\ -x_1x_2$$

$$\begin{aligned}\text{subject to } 2 \geqslant x_1 + x_2 &\geqslant 0\\ 2 \geqslant x_1 - x_2 &\geqslant -2\end{aligned}$$

and show that if the active set method starts at the vertex $\mathbf{x}^{(1)} = (1, -1)^{\mathrm{T}}$ then the method terminates at a point $\mathbf{x}'$ which is not a local solution to the problem. Show that there exist arbitrary small perturbations to the constraints such that (a) $\mathbf{x}'$ is a local solution, (b) the algorithm does not terminate at $\mathbf{x}'$ but finds the global solution.

10. In the active set method assume that the initial active set $\mathscr{A}^{(1)}$ has the property that the set of vectors $\mathbf{a}_i$, $i \in \mathscr{A}^{(1)}$, are independent. Show using equation (10.3.2) that any vector $\mathbf{a}_p$ which is added to the set is not dependent on the other vectors in the set, and hence prove inductively that the independence condition is retained.
11. If the Lagrangian matrix in (10.2.2) is non-singular, show that $\mathbf{A}$ has full rank and hence that a matrix $\mathbf{Z}$ can be chosen as in (10.1.18) with orthogonal columns. If $\mathbf{Z}^{\mathrm{T}}\mathbf{G}\mathbf{Z}$ is singular, consider $\mathbf{G} + \nu\mathbf{I}$ as $\nu \to 0$ and show that (10.2.7) becomes unbounded whereas (10.2.3) does not. This is a contradiction and proves that $\mathbf{Z}^{\mathrm{T}}\mathbf{G}\mathbf{Z}$ is non-singular.
12. Show that the least distance to a fixed point $\mathbf{p}$ of any point $\mathbf{x}$ which satisfies the constraints $\mathbf{A}^{\mathrm{T}}\mathbf{x} \geqslant \mathbf{b}$ can be found by solving a certain QP problem. If all the constraints are active, show by the method of Lagrange multipliers that the problem reduces to the solution of a system of linear equations. In general, when there may be inactive constraints, use the Wolfe dual (theorem 9.5.1) to show that the optimum multipliers satisfy a QP problem with simple bounds only. How does the solution of this problem determine the least distance solution?
13. State the linear complementarity problem arising from Question 10.4 and show that it satisfies the conditions for the modified form of Lemke's method (Section 10.6) to be applicable, without introducing the variable z_0. Hence solve the problem in this way.

14. Show that the problem of finding the point of least Euclidean distance from the origin (in $\mathbb{R}^3$) subject to

$$x_1 + 2x_2 - x_3 \geqslant 4$$
$$-x_1 + x_2 - x_3 \leqslant 2$$

can be formulated as a quadratic programming problem. By eliminating x_1 and x_2, calculate a solution to the equality problem in which both constraints are active. Is this the solution to the least distance quadratic programming problem?

Chapter 11

General Linearly Constrained Optimization

11.1 Equality Constraints

The next class of problem which it is convenient to consider is that in which the objective function is general but the constraints remain linear, that is

$$\begin{aligned} &\underset{\mathbf{x}}{\text{minimize}}\ f(\mathbf{x}) \\ &\text{subject to } \mathbf{a}_i^T\mathbf{x} = b_i, \qquad i \in E \\ &\qquad\qquad\ \ \mathbf{a}_i^T\mathbf{x} \geqslant b_i, \qquad i \in I. \end{aligned} \tag{11.1.1}$$

The methods for handling the linear constraints are largely those used in quadratic programming, but the non-quadratic objective function introduces an extra level of difficulty. One aspect of this is that in general the problem can no longer be solved finitely so the solution $\mathbf{x}^*$ is obtained in the limit of some iterative sequence $\{\mathbf{x}^{(k)}\}$. The equality constraint problem ($I = \emptyset$) is considered in this section and is readily handled by generalized elimination, so that the main features are essentially those relating to unconstrained optimization as described in Volume 1. Methods for calculating Lagrange multipliers at the solution of an equality constraint problem are also considered at the end of this section. Inequality constraints can be handled by an active set method (Section 11.2) which is a generalization of that used in quadratic programming. A new decision which occurs is how accurately to solve each equality problem which arises: a poor decision in this respect can lead to the phenomenon of *zigzagging* which can slow down the rate of convergence appreciably. Methods for overcoming this problem are described in Section 11.3. In addition, some alternative possibilities for handling inequality constraints by way of a reduction to a sequence of quadratic programming problems are described at the end of Section 11.2.

The rest of this section is devoted to finding the solution $\mathbf{x}^*$ of the equality constraint problem

$$\begin{aligned} &\underset{\mathbf{x}}{\text{minimize}}\ f(\mathbf{x}) \\ &\text{subject to } \mathbf{A}^T\mathbf{x} = \mathbf{b} \end{aligned} \tag{11.1.2}$$

in which $\mathbf{A}$ is $n \times m$ and $\mathbf{b} \in \mathbb{R}^m$. As in Section 10.1 it is assumed that $\text{rank}(\mathbf{A}) = m$.

The *generalized elimination method* of Section 10.1 is used to provide a reduction to an unconstrained problem and includes both the elimination of variables method and the orthogonal factorization method, amongst others. Thus matrices $\mathbf{S}$ and $\mathbf{Z}$ are introduced such that $\mathbf{S}^T\mathbf{A} = \mathbf{I}$, $\mathbf{Z}^T\mathbf{A} = \mathbf{0}$, and $[\mathbf{S} : \mathbf{Z}]$ is non-singular. If the current iterate is the feasible point $\mathbf{x}^{(k)}$ then a general feasible point can be expressed as

$$\mathbf{x} = \mathbf{x}^{(k)} + \boldsymbol{\delta} = \mathbf{x}^{(k)} + \mathbf{Z}\mathbf{y} \tag{11.1.3}$$

where $\boldsymbol{\delta} = \mathbf{Z}\mathbf{y}$ is a feasible correction in the null space of $\mathbf{A}$ (see (10.1.9) and (10.1.10)). Thus an equivalent form of (11.1.2) is to solve the reduced unconstrained problem

$$\underset{\mathbf{y}}{\text{minimize}}\ \psi(\mathbf{y}) \triangleq f(\mathbf{x}^{(k)} + \mathbf{Z}\mathbf{y}). \tag{11.1.4}$$

For computational convenience and stability it is usually best to define a new reduced function on each iteration as (11.1.4) indicates. By using the chain rule in (11.1.3) it follows that

$$\nabla_y = \mathbf{Z}^T\nabla_x \tag{11.1.5}$$

and so derivatives of (11.1.4) are given by

$$\nabla_y\psi(\mathbf{y}) = \mathbf{Z}^T\mathbf{g}(\mathbf{x}) \tag{11.1.6}$$

which is the *reduced gradient vector*, and by

$$\nabla_y^2\psi(\mathbf{y}) = \mathbf{Z}^T[\mathbf{G}(\mathbf{x})]\mathbf{Z} \tag{11.1.7}$$

which is the *reduced Hessian matrix*.

It is possible to apply any appropriate technique for unconstrained optimization described in Volume 1 to solve the reduced problem (11.1.4) and most of the remarks in Chapter 2 about the structure of methods apply here also. From (11.1.6) and (11.1.7), sufficient conditions for $\mathbf{x}^*$ to be optimal are that $\mathbf{Z}^T\mathbf{g}^* = \mathbf{0}$ and $\mathbf{Z}^T\mathbf{G}^*\mathbf{Z}$ is positive definite, and necessary conditions are that $\mathbf{Z}^T\mathbf{g}^* = \mathbf{0}$ or $\mathbf{Z}^T\mathbf{G}^*\mathbf{Z}$ is positive semi-definite. These conditions are readily shown to be equivalent to those described in Section 9.1 and 9.3 (see Question 11.1). The initial feasible point $\mathbf{x}^{(1)}$ can be $\mathbf{x}^{(1)} = \mathbf{S}\mathbf{b}$, which is the closest feasible point to the origin in the orthogonal factorization method, or more generally $\mathbf{x}^{(1)} = \mathbf{x}' + \mathbf{S}(\mathbf{b} - \mathbf{A}^T\mathbf{x}')$ which is the closest feasible point to any given point $\mathbf{x}'$. As in Chapter 3 it is important to start by considering *Newton's method* which is based on the quadratic model obtained by truncating the Taylor series for $\psi(\mathbf{y})$ about $\mathbf{y} = \mathbf{0}$, that is

$$\psi(\mathbf{y}) \approx q^{(k)}(\mathbf{y}) = f^{(k)} + \mathbf{y}^T\mathbf{Z}^T\mathbf{g}^{(k)} + \tfrac{1}{2}\mathbf{y}^T\mathbf{Z}^T\mathbf{G}^{(k)}\mathbf{Z}\mathbf{y}. \tag{11.1.8}$$

The origin in $\mathbf{y}$-space corresponds to the point $\mathbf{x}^{(k)}$ in $\mathbf{x}$-space (see (11.1.3)) and so $f^{(k)}$, $\mathbf{g}^{(k)}$, and $\mathbf{G}^{(k)}$ refer to quantities evaluated at $\mathbf{x}^{(k)}$. The model quadratic $q^{(k)}(\mathbf{y})$ refers to the function obtained on iteration k. Thus the basic Newton's method chooses $\mathbf{y}^{(k)}$ to minimize $q^{(k)}(\mathbf{y})$. A unique minimizer exists if and only if $\mathbf{Z}^T\mathbf{G}^{(k)}\mathbf{Z}$ is positive definite and is obtained by making $\nabla q^{(k)} = \mathbf{0}$, that is by solving

$$(\mathbf{Z}^{\mathrm{T}}\mathbf{G}^{(k)}\mathbf{Z})\mathbf{y} = -\mathbf{Z}^{\mathrm{T}}\mathbf{g}^{(k)} \tag{11.1.9}$$

for $\mathbf{y} = \mathbf{y}^{(k)}$, in which case the next iterate is $\mathbf{x}^{(k+1)} = \mathbf{x}^{(k)} + \mathbf{Z}\mathbf{y}^{(k)}$ from (11.1.3). If any $\mathbf{x}^{(k)}$ is sufficiently close to $\mathbf{x}^*$ then the method converges and the rate is second order. However the method can fail to converge, and moreover is undefined when $\mathbf{Z}^{\mathrm{T}}\mathbf{G}^{(k)}\mathbf{Z}$ is not positive definite. Changes to correct these disadvantages include the use of a line search and the possibility of modifying the Hessian matrix to become positive definite. The latter includes the ideas of restricted step and Levenberg–Marquardt methods. All of this development is given in Volume 1, Section 3.1 and Chapter 5, and the reader is referred there for further details. One point of interest is that a step restriction $\| \mathbf{y} \|_2 \leqslant h^{(k)}$ can be handled by a modification $\mathbf{Z}^{\mathrm{T}}\mathbf{G}^{(k)}\mathbf{Z} + \nu\mathbf{I}$ for some $\nu \geqslant 0$ as in Section 5.2 and corresponds to a restriction $\| \mathbf{x} - \mathbf{x}^{(k)} \|_2 \leqslant h^{(k)}$ if the orthogonal factorization method is used to reduce the problem. However for other generalized elimination methods this correspondence does not hold.

It is often the case that the user is unable or does not wish to supply formulae for second derivatives so it is important to consider methods which require only first derivative information or even no derivatives at all. When first derivatives are available, one possibility is a finite difference Newton method (see Section 3.2). In this case the differences are best taken in $\mathbf{y}$-space so that the ith column of the reduced Hessian matrix is defined by

$$\mathbf{Z}^{\mathrm{T}}(\mathbf{g}(\mathbf{x}^{(k)} + \mathbf{z}_i h) - \mathbf{g}^{(k)})/h \tag{11.1.10}$$

for some small h, and after symmetrization this matrix is used to replace $\mathbf{Z}^{\mathrm{T}}\mathbf{G}^{(k)}\mathbf{Z}$ in (11.1.9). Only $n - m$ additional gradient evaluations are required per iteration so the method might be useful when $n - m$ is small, especially as a way to estimate an initial reduced Hessian approximation.

However for the reasons described in Section 3.2 it is usually preferable to select a quasi-Newton method to solve the reduced problem (11.1.4). Various suggestions (see below) have been made but the most satisfactory in view of the development given here is due to Gill and Murray (1974c). The reduced Hessian matrix is approximated by a positive definite matrix $\mathbf{B}^{(k)}$ given in factored form

$$\mathbf{Z}^{\mathrm{T}}\mathbf{G}^{(k)}\mathbf{Z} \simeq \mathbf{B}^{(k)} = \mathbf{L}^{(k)}\mathbf{D}^{(k)}\mathbf{L}^{(k)\mathrm{T}}. \tag{11.1.11}$$

The search direction in $\mathbf{y}$-space is then defined by solving

$$\mathbf{B}^{(k)}\mathbf{p} = -\mathbf{Z}^{\mathrm{T}}\mathbf{g}^{(k)} \tag{11.1.12}$$

for $\mathbf{p} = \mathbf{p}^{(k)}$, by analogy with (11.1.9), and $f(\mathbf{x})$ is minimized approximately by a search along the line $\mathbf{x}^{(k)} + \alpha\mathbf{s}^{(k)}$, where $\mathbf{s}^{(k)} = \mathbf{Z}\mathbf{p}^{(k)}$. The conditions for terminating the line search are those described in Sections 2.4 and 2.6. More in keeping with the 'old-fashioned' development of Section 3.2 is to represent $\mathbf{B}^{(k)}$ by the inverse reduced Hessian approximation

$$\mathbf{H}^{(k)} = [\mathbf{B}^{(k)}]^{-1} \approx [\mathbf{Z}^{\mathrm{T}}\mathbf{G}^{(k)}\mathbf{Z}]^{-1} \tag{11.1.13}$$

and to compute $\mathbf{s}^{(k)}$ from

$$\mathbf{s}^{(k)} = -\mathbf{Z}\mathbf{H}^{(k)}\mathbf{Z}^{\mathrm{T}}\mathbf{g}^{(k)}. \tag{11.1.14}$$

In view of the remarks at the end of Section 3.2 it may be that (11.1.13) is no less stable than (11.1.11) whilst being somewhat more convenient to use. After each iteration $\mathbf{H}^{(k)}$ (or the factorization of $\mathbf{B}^{(k)}$) is updated to include the additional curvature information obtained in the line search. For the reasons described in Chapter 3 the BFGS formula is currently preferred. The required quantities $\boldsymbol{\gamma}^{(k)}$ and $\boldsymbol{\delta}^{(k)}$ (in (3.2.12) say) become differences in reduced gradients and variables, that is $\boldsymbol{\gamma}^{(k)} = \mathbf{Z}^{\mathrm{T}}(\mathbf{g}^{(k+1)} - \mathbf{g}^{(k)})$ and $\boldsymbol{\delta}^{(k)} = \mathbf{y}^{(k)}$. The initial choice $\mathbf{H}^{(1)}$ can be any positive definite matrix; $\mathbf{H}^{(1)} = \mathbf{I}$ is usually chosen in the absence of any other information. The properties described in Section 3.2 relating amongst others to the maintaining of positive definite matrices $\mathbf{H}^{(k)}$ and to quadratic termination (here in $n - m$ iterations) also hold good.

Another possibility for a suitable algorithm is to apply a conjugate gradient method (Section 4.1) to the reduced problem. This is likely to be preferable only for large problems in which the matrix $\mathbf{H}^{(k)}$ cannot conveniently be stored. In this case it is also important to give some thought to storage for the matrices $\mathbf{S}$ and $\mathbf{Z}$. The best possibility is to define these implicitly by factorizing the matrix $[\mathbf{A} : \mathbf{V}]$ in such a way as to retain sparsity as much as possible. Likewise a choice of $\mathbf{V}$ which retains sparsity should be made, for example that given in (10.1.21) corresponding to elimination of variables. For no-derivative problems the experience of Section 3.2 again suggests using a quasi-Newton method with difference approximations to derivatives. These are best calculated in the space of the reduced variables so that the ith element of $\mathbf{Z}^{\mathrm{T}}\mathbf{g}^{(k)}$ is estimated by

$$(f(\mathbf{x}^{(k)} + \mathbf{z}_i h) - f^{(k)})/h \tag{11.1.15}$$

for some small interval h, which requires only $n - m$ additional function evaluations. Near the solution it is preferable to use the corresponding central difference formula

$$\tfrac{1}{2}(f(\mathbf{x}^{(k)} + \mathbf{z}_i h) - f(\mathbf{x}^{(k)} - \mathbf{z}_i h))/h. \tag{11.1.16}$$

Conjugate direction set methods as in Section 4.2 are also possible (Buckley, 1975) but there is no evidence to suggest that they are preferable. Least squares problems ($f(\mathbf{x}) \triangleq \mathbf{r}(\mathbf{x})^{\mathrm{T}}\mathbf{r}(\mathbf{x})$) with linear constraints can be handled readily: the Jacobian matrix in the reduced coordinates is $\boldsymbol{\nabla}_y(\mathbf{r}^{(k)^{\mathrm{T}}}) = \mathbf{Z}^{\mathrm{T}}\mathbf{A}^{(k)}$ where $\mathbf{A} = \boldsymbol{\nabla}_x \mathbf{r}^{\mathrm{T}}$, and analogues of the Gauss–Newton method, etc., are obtained by using the estimate $\mathbf{Z}^{\mathrm{T}}\mathbf{G}^{(k)}\mathbf{Z} = 2\mathbf{Z}^{\mathrm{T}}\mathbf{A}^{(k)}\mathbf{A}^{(k)^{\mathrm{T}}}\mathbf{Z}$ in (11.1.9); the resulting properties are the same as those described in Section 6.1.

Historically the earliest methods for solving (11.1.2) used steepest descent searches as typified by Rosen's (1960) *gradient projection method*. This is equivalent to choosing a search direction $\mathbf{p}^{(k)} = -\boldsymbol{\nabla}_y \psi^{(k)} = -\mathbf{Z}^{\mathrm{T}}\mathbf{g}^{(k)}$ as the steepest descent vector in the reduced coordinates and hence $\mathbf{s}^{(k)} = -\mathbf{Z}\mathbf{Z}^{\mathrm{T}}\mathbf{g}^{(k)}$ as the search direction in $\mathbf{x}$-space. When $\mathbf{Z}$ is defined by the orthogonal factorization method it follows that $\mathbf{s}^{(k)} = -\mathbf{P}\mathbf{g}^{(k)}$ where $\mathbf{P} = \mathbf{Z}\mathbf{Z}^{\mathrm{T}} = \mathbf{I} - \mathbf{A}\mathbf{A}^{+}$ is a projection matrix which projects into the null space of $\mathbf{A}$. Rosen suggests calculating this matrix although it is preferable to use the implicit definition $\mathbf{P} = \mathbf{Z}\mathbf{Z}^{\mathrm{T}}$. The idea for using an active set method also derives from Rosen (1960). Quasi-Newton methods for linearly constrained problems were first considered by Goldfarb (1969) although he updates the $n \times n$ matrix

$$\bar{\mathbf{H}}^{(k)} = \mathbf{Z}\mathbf{H}^{(k)}\mathbf{Z}^{\mathrm{T}} \tag{11.1.17}$$

which is positive semi-definite with rank $n - m$ and is related to the $\mathbf{H}$ matrix in (10.2.3). It is easy to show that updating $\mathbf{H}^{(k)}$ by any formula in the Broyden family, using reduced differences as above, is equivalent to updating $\bar{\mathbf{H}}^{(k)}$ by the same formula with unreduced differences (Question 11.2). The choice $\mathbf{H}^{(1)} = \mathbf{I}$ is equivalent to choosing $\bar{\mathbf{H}}^{(1)} = \mathbf{Z}\mathbf{Z}^{\mathrm{T}}$ which is the projection matrix $\mathbf{P}$ when the orthogonal factorization method defines $\mathbf{Z}$. However computing the search direction using $\mathbf{s}^{(k)} = -\bar{\mathbf{H}}^{(k)}\mathbf{g}^{(k)}$ causes problems due to round-off errors in that $\mathbf{s}^{(k)}$ no longer remains in the null space of $\mathbf{A}$ when $\mathbf{x}^{(k)}$ is close to $\mathbf{x}^*$, and so the implicit representation through (11.1.14) is preferable. Another early quasi-Newton method is due to Murtagh and Sargent (1969), and updates estimates of $\mathbf{G}^{-1}$ and $(\mathbf{A}^{\mathrm{T}}\mathbf{G}^{-1}\mathbf{A})^{-1}$ using the rank one formula, and uses the representation (10.2.6) to solve the quadratic/linear approximating subproblem. However neither matrix needs to be positive definite and the rank one formula is potentially unstable, so again this method would not be recommended. The observation that a variety of methods can be classified as generalized elimination methods is due to Fletcher (1972a) and the implementation via $\mathbf{S}$ and $\mathbf{Z}$ matrices to Gill and Murray (1974a).

Finally it is of interest to consider the computation of the Lagrange multiplier vector $\boldsymbol{\lambda}^*$ at the solution to (11.1.2). This information is required when the equality problem is a subproblem in the active set method, but can also be useful in a sensitivity analysis as described in Section 9.1. Essentially $\boldsymbol{\lambda}^*$ is defined by $\mathbf{g}^* = \mathbf{A}^*\boldsymbol{\lambda}^*$ and is readily computed by (10.1.14) or (10.1.19). There is the additional complication however that $\mathbf{x}^*$ and hence $\mathbf{g}^*$ cannot be computed exactly due to the non-finite nature of methods for solving (11.1.2). Thus the effect of calculating approximate multipliers $\boldsymbol{\lambda}^{(k)}$ at a point $\mathbf{x}^{(k)}$ which approximates $\mathbf{x}^*$ is considered. In particular in the active set method $\mathbf{x}^{(k)}$ might be quite a poor approximation to $\mathbf{x}^*$ so the errors involved may not be negligible. The obvious possibility from (10.1.14) is to compute $\boldsymbol{\lambda}^{(k)}$ from

$$\boldsymbol{\lambda}^{(k)} = \mathbf{S}^{\mathrm{T}}\mathbf{g}^{(k)} \tag{11.1.18}$$

which can be regarded as a first order estimate of $\boldsymbol{\lambda}^*$ since $\mathbf{g}^{(k)} - \mathbf{g}^* = O(\mathbf{x}^{(k)} - \mathbf{x}^*)$. Unlike (10.1.14) the resulting $\boldsymbol{\lambda}^{(k)}$ depends on how $\mathbf{S}$ is chosen because the equations $\mathbf{g}^{(k)} = \mathbf{A}\boldsymbol{\lambda}^{(k)}$ are no longer consistent. The orthogonal factorization method in which $\mathbf{S}^{\mathrm{T}} = \mathbf{A}^+$ has a nice interpretation in that it gives the best least squares solution of these equations. Another left inverse for $\mathbf{A}$ with special properties is the matrix $\mathbf{T}^{\mathrm{T}}$ defined in (10.2.3). Thus another possibility is to compute $\boldsymbol{\lambda}^{(k)}$ from

$$\boldsymbol{\lambda}^{(k)} = \mathbf{T}^{\mathrm{T}}\mathbf{g}^{(k)}. \tag{11.1.19}$$

$\mathbf{T}$ includes curvature information about the problem and it can be shown that the resulting $\boldsymbol{\lambda}^{(k)}$ is a second order estimate of $\boldsymbol{\lambda}^*$ (see (10.2.9) and Question 11.3). Given that $\mathbf{T}$ is available and that $\mathbf{x}^{(k)}$ is sufficiently close to $\mathbf{x}^*$, then (11.1.19) gives a more accurate estimate of $\boldsymbol{\lambda}^*$. However for reasons of convenience or numerical stability it may be preferable to use (11.1.18).

For no-derivative problems, if differencing is carried out as in (11.1.15) or (11.1.16) to estimate reduced derivatives, then no information is available to compute Lagrange multipliers. In this case an additional m function evaluations are required to estimate the Lagrange multipliers $\lambda_i^{(k)}$ by using

$$\lambda_i^{(k)} \approx (f(\mathbf{x}^{(k)} + \mathbf{s}_i h) - f^{(k)})/h \qquad (11.1.20)$$

where $\mathbf{s}_i$ is the corresponding column of the matrix $\mathbf{S}$.

11.2 Inequality Constraints

Most problems of interest involve inequality constraints and so can be expressed as in (11.1.1). This section describes how the methods of the previous section can be generalized to handle this problem by means of an active set method similar to that described in Section 10.3. Some alternative possibilities are also discussed at the end of the section. In the active set method certain constraints indexed by the active set $\mathscr{A}$ are regarded as equalities and the rest are temporarily disregarded as described in Section 10.3. Each iterate $\mathbf{x}^{(k)}$ is a feasible point and each iteration attempts to locate the solution of the EP

$$\begin{aligned} &\underset{\boldsymbol{\delta}}{\text{minimize}} \ f(\mathbf{x}^{(k)} + \boldsymbol{\delta}) \\ &\text{subject to } \mathbf{a}_i^{\mathrm{T}} \boldsymbol{\delta} = 0, \qquad i \in \mathscr{A} \end{aligned} \qquad (11.2.1)$$

obtained by shifting the origin to the point $\mathbf{x}^{(k)}$. The initial point $\mathbf{x}^{(1)}$ can be found by the methods of Sections 8.4 or 10.3. The solution of (11.2.1) yields a search direction $\mathbf{s}^{(k)}$ according to the method used, and $\mathbf{x}^{(k+1)}$ is taken (ideally) as the best feasible point on the line $\mathbf{x}^{(k)} + \alpha \mathbf{s}^{(k)}$. In quadratic programming there is a clear distinction either that $\mathbf{x}^{(k+1)}$ solves the EP or that a previously inactive constraint becomes active in the line search. This is not so in the more general case (11.1.1) and the minimizer of the EP is only located in the limit of a sequence of iterations with the same $\mathscr{A}$. Thus on each iteration a decision must be taken as to whether $\mathbf{x}^{(k)}$ (that is $\boldsymbol{\delta} = \mathbf{0}$) is an acceptable solution of the EP; if not, then one or more iterations are carried out with the same active set. The definition of an acceptable solution must be made with some care and is considered in more detail in Section 11.3. If $\mathbf{x}^{(k)}$ is accepted as the solution to the EP, then Lagrange multipliers $\boldsymbol{\lambda}^{(k)}$ are examined to determine whether or not $\mathbf{x}^{(k)}$ is a KT point. If so then the iteration terminates; otherwise the index q of an inequality constraint for which $\lambda_q^{(k)} < 0$ is least is removed from $\mathscr{A}$ as in (10.3.3).

The line search for finding the best feasible point in this more general case is also not a finite process and is therefore more complicated than (10.3.2). As in Sections 2.4 and 2.6 of Volume 1 it is necessary to choose conditions which define a range of acceptable α-values, and to locate such a value by a combination of interpolation and sectioning in such a way that the iteration is guaranteed to terminate. Exactly the same holds good here, but there is the additional complication that α must not exceed the value $\bar{\alpha}^{(k)}$, say, defined as in (8.3.6) by

$$\bar{\alpha}^{(k)} = \min_{\substack{i:i \notin \mathscr{A} \\ \mathbf{a}_i^T \mathbf{s}^{(k)} < 0}} \frac{b_i - \mathbf{a}_i^T \mathbf{x}^{(k)}}{\mathbf{a}_i^T \mathbf{s}^{(k)}}.$$

Thus the initial choice for α in the line search, or any α-values determined by extrapolation, must be reduced to $\bar{\alpha}^{(k)}$ if they would otherwise exceed this value. It may be that there are no acceptable α-values in the sense of Section 2.6 for which $\alpha \leqslant \bar{\alpha}^{(k)}$. In this case $\alpha^{(k)} = \bar{\alpha}^{(k)}$ is chosen. It is in this way that the value $\alpha^{(k)}$ in the line search in step (e) of (11.2.2) below is determined. Thus the active set method for solving (11.1.1) can be summarized as follows.

(a) Given $\mathbf{x}^{(1)}$ and $\mathscr{A} = \mathscr{A}(\mathbf{x}^{(1)})$, set $k = 1$.
(b) If $\boldsymbol{\delta} = \mathbf{0}$ is not an acceptable solution of (10.3.1), go to (d).
(c) Let $\lambda_q^{(k)}$ solve $\min \lambda_i^{(k)}$, $i \in I \cap \mathscr{A}$; if $\lambda_q^{(k)} \geqslant 0$, terminate with $\mathbf{x}^* = \mathbf{x}^{(k)}$, otherwise remove q from $\mathscr{A}$.
(d) Solve (11.1.1) for $\mathbf{s}^{(k)}$. (11.2.2)
(e) Choose $\mathbf{x}^{(k+1)} = \mathbf{x}^{(k)} + \alpha^{(k)}\mathbf{s}^{(k)}$ as a near best feasible point along the line $\mathbf{x}^{(k)} + \alpha\mathbf{s}^{(k)}$.
(f) If $\alpha^{(k)} = \bar{\alpha}^{(k)}$, add p to $\mathscr{A}$.
(g) Set $k = k + 1$ and go to (b).

Note that after an index q is removed from $\mathscr{A}$ in step (c) it is assumed that the search direction $\mathbf{s}^{(k)}$ which is calculated in step (d) is downhill ($\mathbf{s}^{(k)T}\mathbf{g}^{(k)} < 0$) and strictly feasible with respect to the deleted constraint ($\mathbf{s}^{(k)T}\mathbf{a}_q > 0$). This can be done in most common methods (see Question 11.7). The choice of method for solving (11.1.1) in step (d) is otherwise any of the methods considered in Section 11.1, and depends on what derivative information is available, amongst other things. Possible difficulties caused by degeneracy are no less severe than those described in Section 10.3 and elsewhere and it is usual, albeit somewhat unsatisfactory, to assume that it does not occur. In this case convergence of the algorithm depends on how the choice of an acceptable solution in step (b) is made, and this is considered in more detail in Section 11.3.

An important feature of an active set method concerns the efficient solution of the EP (11.2.1) when changes are made to $\mathscr{A}$. It is possible to avoid recalculating $\mathbf{S}$ and $\mathbf{Z}$ in $O(n^3)$ operations and instead to update these matrices in $O(n^2)$ operations, taking advantage of the fact that a column is either added to or removed from $\mathbf{A}$. Updating of other matrices may also be possible. Again details of these computations are not given but the references towards the end of Section 10.3 are again relevant. It is interesting however to review the changes to the reduced Hessian matrix $\mathbf{Z}^T\mathbf{G}^{(k)}\mathbf{Z}$ when a change is made to the active set $\mathscr{A}$, and hence to $\mathbf{Z}$. For the basic Newton method $\mathbf{G}^{(k)}$ also changes in a general way so no advantage can be taken of updating techniques. However the situation is more favourable when using a quasi-Newton method. First of all consider the case when $\alpha^{(k)} = \bar{\alpha}^{(k)}$ in step (e) of (11.2.2) so that a constraint becomes active. It is possible that the curvature estimate $\gamma^T\delta \leqslant 0$ may occur (see Section 3.2) in which case the matrix $\mathbf{H}^{(k)}$ must not be updated. The new $\mathbf{Z}$ matrix has one fewer column than the old matrix; by making a linear transformation of the columns of $\mathbf{Z}$ it is possible to arrange matters so that column $\mathbf{z}_{n-m}$ is removed from $\mathbf{Z}$. In this case the new matrix $\mathbf{B}^{(k)}$ in (11.1.11) is obtained by removing row and column $n - m$ from the old matrix $\mathbf{B}^{(k)}$ (after transformation). This is the same operation that is carried out in quadratic programming and details are given for example by Gill and Murray (1978b). A corresponding formula for updating $\mathbf{H}^{(k)}$ can also be obtained (Question 11.4); in terms of the matrix $\overline{\mathbf{H}}^{(k)}$ used in Goldfarb's method (see 11.1.17)) the recurrence relation is

$$\overline{\mathbf{H}} := \overline{\mathbf{H}} - \frac{\overline{\mathbf{H}}\mathbf{a}\mathbf{a}^T\overline{\mathbf{H}}}{\mathbf{a}^T\overline{\mathbf{H}}\mathbf{a}} \tag{11.2.3}$$

suppressing superscript k, where $\mathbf{a}$ is the column that is added to $\mathbf{A}$. When a constraint index is removed from $\mathscr{A}$ as in step (c) of (11.2.2), an extra column $\mathbf{z}_{n-m+1}$ is adjoined to $\mathbf{Z}$. This extends the space of the free variables to one higher dimension, and no curvature information is available in the new direction. Hence $\mathbf{B}^{(k)}$ must be extended in an arbitrary way: the most convenient way is to assign

$$\mathbf{B}^{(k)} := \begin{bmatrix} \mathbf{B}^{(k)} & \mathbf{0} \\ \mathbf{0}^T & 1 \end{bmatrix} \tag{11.2.4}$$

which is in the spirit of an initial choice of $\mathbf{B}^{(1)} = \mathbf{I}$. $\mathbf{H}^{(k)}$ is updated in an analogous way and corresponds to an update

$$\overline{\mathbf{H}}^{(k)} := \overline{\mathbf{H}}^{(k)} + \mathbf{z}_{n-m+1}\mathbf{z}_{n-m+1}^T \tag{11.2.5}$$

in $\overline{\mathbf{H}}^{(k)}$.

It is interesting to consider quadratic termination results for the active set/quasi-Newton methods when the above updating formulae are used. If the active set is changed c times in all, and m_i, $i = 0, 1, \ldots, c$, is the number of active constraints after the ith change, then an obvious bound for the total number t of exact line searches is

$$t \leqslant \sum_{i=0}^{c} (n - m_i).$$

However Powell (1972) shows that this is unduly pessimistic and that a better bound is

$$t \leqslant c + n,$$

which shows that introducing arbitrary information when using (11.2.4) does not increase the bound by more than one. This result also allows for additional inexact line searches to be mixed with exact line searches. It seems that the tighter bound

$$t \leqslant c + n - \max_{0 \leqslant i \leqslant c} m_i$$

also follows from Powell's paper, which can be important for large almost linear problems. These bounds all beg the question as to whether the algorithm does terminate, that is whether there exists a finite c. This is by no means obvious since the algorithm can return to a previous active set, so the finiteness of all possible active sets does not immediately imply termination.

Another observation relating to the active set method when using a Levenberg–Marquardt parameter is also of interest. It is described in Volume 1, Section 5.2, how it is possible to control the iteration in two ways, either using the parameter $h^{(k)}$ (algorithm (5.1.6)) or the parameter $\nu^{(k)}$ (algorithm (5.2.7)). In an active set method it is far preferable to control the algorithm using $h^{(k)}$ (Holt and Fletcher, 1979), since a suitable value of $h^{(k)}$ is likely to be little affected by changes in active set, whereas this is not true for the $\nu^{(k)}$ parameter. An interesting example of the potential difficulties is described in Question 11.5.

Finally a method for solving (11.1.1) is described which avoids using the active set method. The method is a version of Newton's method in which at an iterate $\mathbf{x}^{(k)}$, a quadratic Taylor series approximation

$$f(\mathbf{x}^{(k)}+\boldsymbol{\delta}) \approx q^{(k)}(\boldsymbol{\delta}) = f^{(k)} + \boldsymbol{\delta}^{\mathrm{T}}\mathbf{g}^{(k)} + \tfrac{1}{2}\boldsymbol{\delta}^{\mathrm{T}}\mathbf{G}^{(k)}\boldsymbol{\delta} \tag{11.2.6}$$

is made, and the subproblem

$$\begin{aligned} &\underset{\boldsymbol{\delta}}{\text{minimize}}\quad q^{(k)}(\boldsymbol{\delta}) \\ &\text{subject to } \mathbf{a}_i^{\mathrm{T}}\boldsymbol{\delta} = b_i - \mathbf{a}_i^{\mathrm{T}}\mathbf{x}^{(k)}, \qquad i \in E \\ &\qquad\qquad\;\; \mathbf{a}_i^{\mathrm{T}}\boldsymbol{\delta} \geqslant b_i - \mathbf{a}_i^{\mathrm{T}}\mathbf{x}^{(k)}, \qquad i \in I \end{aligned} \tag{11.2.7}$$

derived directly from (11.1.1) is solved. To ensure global convergence a step restriction

$$\|\boldsymbol{\delta}\|_{\infty} \leqslant h^{(k)} \tag{11.2.8}$$

is used in an algorithm like (5.1.6). The method therefore solves a sequence of quadratic programming subproblems with inequality constraints. The method is a special case of the SOLVER method described in Section 12.3. A disadvantage is that each iteration is relatively expensive and advantage cannot usually be taken of the updating techniques of quasi-Newton methods to reduce the overall operation count to $O(n^2)$ operations per iteration. However the solution of this subproblem does enable the correct active set to be determined quickly and also avoids the problem of zigzagging (see the remarks at the end of Section 11.3) so may converge more rapidly when remote from the solution. Certainly if the function and derivatives are expensive to compute, this approach can be preferable. A quasi-Newton version of the algorithm is given by Fletcher (1972b), although it might be better to use as an updating formula the version of the BFGS formula due to Powell (1978a) based on (12.3.18).

11.3 Zigzagging

A feature which can adversely affect the rate of convergence of any type of method for handling inequality constraints is known as *zigzagging*. Although the set of active constraints $\mathscr{A}^*$ at a local solution is well defined, it may be that the sets $\mathscr{A}^{(k)}$ (as defined in (7.1.2)) obtained at the iterates $\mathbf{x}^{(k)}$ in some method do not settle down (so that $\mathscr{A}^{(k)} = \mathscr{A}^*$ for all $k \geqslant K$ where K is sufficiently large) but oscillate between different subsets of the constraints in the problem. For linear constraints, this corresponds to zigzagging between different linear manifolds corresponding to the feasible region in (11.2.1) for different $\mathscr{A}$ (see Figure 11.3.1). If the active set does settle down (with $\mathscr{A}^{(k)} = \mathscr{A}^*$ $\forall\, k \geqslant K$) then the rate of convergence becomes that for an equality constraint method (Section 11.1) which is usually superlinear for most good methods. However if zigzagging occurs then the rate can degenerate to being linear with a rate constant which is not small, and in some cases the method can fail to converge to a solution.

For the active set method of algorithm (11.2.2), the likelihood of zigzagging is correlated with the test for an acceptable solution implied in step (b). If it is possible to solve any EP exactly in a finite number of steps, and if step (b) tests for an exact solution to the EP (as in quadratic programming, (10.3.4), step (b)), then it is not possible for zigzagging to occur (excluding degenerate cases). This is because the algorithm terminates for the reasons described in Section 10.3; once a constraint index is removed from the active set $\mathscr{A}^{(k)}$ in step (c), then the algorithm can no longer return to that active set by virtue of the optimality of $\mathbf{x}^{(k)}$ and the fact that $f^{(k)}$ is monotonically decreasing. However it is not possible to solve any EP exactly when $f(\mathbf{x})$ is a general function, and it is inefficient to find the solution to high accuracy because the solution to the EP may not correspond to the solution of (11.1.1). The opposite possibility therefore is to allow step (c) to be taken on every iteration. It is this strategy which is most likely to cause zigzagging, especially when the multiplier estimates are poor. An example is given by Wolfe (1972) in which zigzagging causes non-convergence to a solution (see Question 11.6). The example is somewhat pathological because the objective function is not $\mathbb{C}^2$ and the steepest descent method is used to determine $\mathbf{s}^{(k)}$ in (11.2.2), step (d). Nonetheless it has been observed in practice with quasi-Newton methods and smooth functions that this strategy can induce a slow rate of linear convergence. Therefore what is required on step (b) is to have some compromise between these

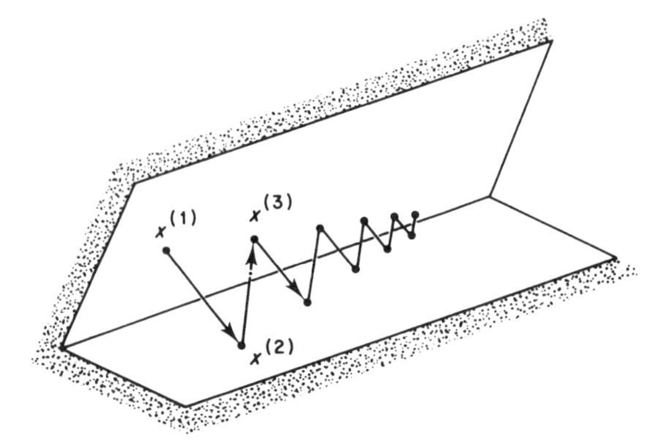

Figure 11.3.1 Zigzagging

two extremes which eliminates the possibility of zigzagging and yet which enables an active set $\mathscr{A}^{(k)}$ to be changed when it has become likely that it is not the same as $\mathscr{A}^*$.

Various ad hoc rules to prevent zigzagging are reviewed by Fletcher (1972a) and can work well enough in practice. However it seems preferable in view of the above discussion to try to eliminate zigzagging by looking at the accuracy to which the current EP has been solved. This is done by Rosen (1960), Goldfarb (1969), and Murtagh and Sargent (1969). They all suggest similar strategies which are typified by, although different to, the following (Fletcher, 1972a). At $\mathbf{x}^{(k)}$, a second order estimate of the reduction in $f(\mathbf{x})$ by keeping the same active set is given by

$$\Delta^{(k)} = \tfrac{1}{2}\mathbf{g}^{(k)^{\mathrm{T}}}\mathbf{Z}\mathbf{H}^{(k)}\mathbf{Z}^{\mathrm{T}}\mathbf{g}^{(k)} \tag{11.3.1}$$

by virtue of (11.1.8) and (11.1.9). $\mathbf{H}^{(k)}$ is positive definite and is either equal to or an approximation to the inverse reduced Hessian matrix, as in (11.1.13). If $\boldsymbol{\lambda}^{(k)}$ is a second order multiplier estimate, and $\lambda_i^{(k)} < 0$, then the additional function reduction obtained by removing the ith constraint index from $\mathscr{A}^{(k)}$ (assuming $(i \in I)$) is

$$\Delta_i^{(k)} = \frac{\tfrac{1}{2}\lambda_i^{(k)^2}}{-u_{ii}^{(k)}} \tag{11.3.2}$$

where $u_{ii}^{(k)}$ is the diagonal element of the matrix $\mathbf{U}^{(k)}$ defined by (10.2.3) at $\mathbf{x}^{(k)}$. This result is valid when $u_{ii}^{(k)} < 0$ which is implied by (10.2.6) when $\mathbf{G}^{(k)}$ is positive definite. Hence a possible strategy is to find the integer p which maximizes $\Delta_i^{(k)}$ over the set $\{i : \lambda_i^{(k)} < 0\}$ and then to define $\mathbf{x}^{(k)}$ as an acceptable solution (and hence remove the pth constraint) if $\Delta_p^{(k)} \geqslant \Delta^{(k)}$. The motivation for such a test is that a greater reduction in $f(x)$ is likely to occur by removing the constraint than by not doing so. For a quasi-Newton method a recurrence relation for the quantities $u_{ii}^{(k)}$ can be determined. An advantage of this test is that it is invariant under both linear transformations of the variables and scaling of the constraints. If $\mathbf{G}^{(k)}$ (or its approximation) is not positive definite, something has to be done to generalize the definitions of $\Delta^{(k)}$ and $\Delta_i^{(k)}$. One possibility is to define $\Delta^{(k)} = \infty$ if $\mathbf{H}^{(k)}$ is not positive definite and to have $\mathbf{x}^{(k)}$ as not being an acceptable solution. Otherwise, if $u_{ii}^{(k)} \geqslant 0$, then $\Delta_i^{(k)} = \infty$ is defined and p is chosen to minimize $\lambda_i^{(k)}$ over the set $\{i : \lambda_i^{(k)} < 0, \Delta_i^{(k)} = \infty\}$. Strategies based on comparing quantities like (11.3.1) and (11.3.2) have been used successfully in practice.

The dependence of the above test on the quantities $u_{ii}^{(k)}$ makes it somewhat unwieldy and it is preferable to have a more simple test for an acceptable solution with good invariance properties. It is also important to consider whether it can be proved that the resulting method converges to the solution of (11.1.1) and avoids zigzagging. A test which meets these criteria to a large extent can be derived in the following way. Let $\{\ell(k)\}$, $k = 1, 2, \ldots,$ be a sequence of integers such that $\ell(k)$ $(< k)$ is the greatest previous iteration index on which a constraint is removed from $\mathscr{A}$, or $\ell(k) = 1$ if no such index exists. Then a point $\mathbf{x}^{(k)}$ is acceptable in step (b) if

$$\Delta^{(k)} \leqslant f^{(\ell(k))} - f^{(k)}, \tag{11.3.3}$$

that is if the predicted function reduction for the current $\mathscr{A}$ is less than the total reduction since a constraint was last removed from $\mathscr{A}$. An alternative possibility is to define the right hand side of (11.3.3) as the total function reduction since $\mathscr{A}$ was last changed. The motivation for this test is that when zigzagging occurs the right hand side of (11.3.3) goes to zero, which ensures that $\Delta^{(k)}$ goes to zero, which usually ensures that convergence occurs for the subsequence of points with the same $\mathscr{A}$. If $\mathbf{x}^{(k)}$ is accepted as a solution of the EP then the decision as to which constraint to remove is that given in (11.2.2), step (c). If the constraints are prescaled then these tests are invariant under linear transformations of variables and constraint scalings. The test must also be incorporated with the termination test which is conveniently chosen to be $\Delta^{(k)} \leqslant \epsilon$ where $\epsilon > 0$ is some preset tolerance on the accuracy required in f^*. The detailed form of steps (b) and (c) is therefore

(b) if $\Delta^{(k)} > f^{(\ell(k))} - f^{(k)}$ then go to (d);
(c) (i) calculate $\boldsymbol{\lambda}^{(k)}$ from (11.1.18) or (11.1.19) and let $\lambda_q^{(k)}$ minimize $\lambda_i^{(k)}$, $i \in I \cap \mathscr{A}$; (11.3.4)
(ii) if $\lambda_q^{(k)} \geqslant 0$ then (if $\Delta^{(k)} \leqslant \epsilon$ then terminate else go to (d)) else remove q from $\mathscr{A}$.

In some cases it is possible to prove that convergence without zigzagging takes place when (11.3.4) is used in (11.2.2). Such a result is given below. To some extent this depends on the properties of the method used for finding the solution and multipliers of the EP. To make the result easy to state, it is assumed that the feasible region R is not empty, that $f(\mathbf{x})$ is $\mathbb{C}^2$ on R, and that the smallest eigenvalue μ_n of the Hessian matrix $\mathbf{G}(\mathbf{x})$ satisfies $\mu_n(\mathbf{x}) \geqslant a > 0$ on R, an assumption which implies strict convexity of f on R. Newton's method with line search is used and the search can be terminated by any of the conditions described in Section 2.4, so that convergence results like theorem 2.4.1 can be applied. It is also assumed for any $\mathbf{x} \in R$ that the vectors $\mathbf{a}_i$, $i \in \mathscr{A}(\mathbf{x})$, are independent. In this case either (11.1.18) or (11.1.19) can be used to calculate $\boldsymbol{\lambda}^{(k)}$; all that is required is that if $\mathbf{x}^{(k)} \to \mathbf{x}^*$ for some fixed $\mathscr{A}$ then $\boldsymbol{\lambda}^{(k)} \to \boldsymbol{\lambda}^*$.

Theorem 11.3.1

Under the above assumptions then the above method converges to the solution of (11.1.1) from any initial feasible point $\mathbf{x}^{(1)}$. *In addition if strict complementarity holds at* $\mathbf{x}^*$ *then no zigzagging can occur and* $\mathscr{A}^{(k)} = \mathscr{A}^*$ *for all k sufficiently large.*

Proof

If $\mathscr{A}^{(k)}$ is constant for all k sufficiently large then $\mathbf{x}^{(k)} \to \mathbf{x}^*$ by theorem 2.4.1 applied to the reduced problem. Otherwise the integers $\ell(k) \to \infty$; also the assumptions imply that $f(\mathbf{x})$ is bounded below so $f^{(k)} \downarrow f^\infty$ and $f^{(\ell(k))} - f^{(k)} \to 0$ which implies that $\Delta^{(k)} \to 0$ on the subsequence of iterations for which (11.3.3) holds. Let $\mathbf{x}' \neq \mathbf{x}''$ be two accumulation points ($f' = f'' = f^\infty$) such that $\mathbf{x}^{(k)} \to \mathbf{x}'$ and $\mathbf{x}^{(k+1)} \to \mathbf{x}''$ for a subsequence. Then $\mu_n^{(k)} \geqslant a$ and $\mathbf{s}^{(k)\mathrm{T}}\mathbf{g}^{(k)} < 0$ imply that $\mathbf{x}^{(k+1)}$

is not acceptable with respect to test (2.4.2) in the line search for some sufficiently large k, which is a contradiction. Thus the main sequence converges – say $\mathbf{x}^{(k)} \to \mathbf{x}^{\infty}$. Let K be sufficiently large such that any set $\mathscr{A}^{(k)}$, $k \geqslant K$, occurs infinitely often and restrict attention to $k \geqslant K$. For any such set $\mathscr{A}^{(k)}$, using positive definiteness and independence assumptions, it follows from $\Delta^{(k)} \to 0$ that $\mathbf{Z}^{\mathrm{T}}\mathbf{g}^{(k)} \to \mathbf{0}$ on a subsequence, and hence that $\mathbf{x}^{\infty}$ minimizes $f(\mathbf{x})$ on $\mathbf{a}_i^{\mathrm{T}}\mathbf{x} = b_i$, $i \in \mathscr{A}^{(k)}$, and that $\mathbf{g}^{\infty} = \Sigma_{i \in \mathscr{A}^{(k)}} \mathbf{a}_i^{\infty} \lambda_i^{\infty}$. Also $\lambda_i^{(k)} \to \lambda_i^{\infty}$, $i \in \mathscr{A}^{(k)}$. Therefore from the independence assumptions, $\mathbf{g}^{\infty} = \Sigma_{i \in \cap_k \mathscr{A}^{(k)}} \mathbf{a}_i^{\infty} \lambda_i^{\infty}$ and so $\lambda_i^{\infty} = 0$ for all $i \in \cup_k \mathscr{A}^{(k)} - \cap_k \mathscr{A}^{(k)}$. Also by continuity and since $\lambda_q^{(k)}$ is chosen as in (11.3.4), step (c)(i), it follows that $\lambda_i^{\infty} \geqslant 0$, $i \in I \cap_k \mathscr{A}^{(k)}$. Hence $\mathbf{x}^{\infty}$, $\boldsymbol{\lambda}^{\infty}$ is a KT point and by convexity solves (11.1.1). If strict complementarity $\lambda_i^* > 0$, $i \in \mathscr{A}^* \cap I$, holds then the set $\cup_k \mathscr{A}^{(k)} - \cap_k \mathscr{A}^{(k)}$ is empty and hence $\mathscr{A}^{(k)}$ is constant for sufficiently large k. Also $\mathscr{A}^{(k)} = \mathscr{A}^*$ follows because otherwise $\lambda_i^* = 0$, $i \in \mathscr{A}^* - \mathscr{A}^{(k)}$, is implied which is a contradiction. □

Exactly what can be proved about zigzagging of the active set method under weaker assumptions is a question of some interest.

It is also of interest to examine to what extent algorithms other than the active set method can exhibit zigzagging. In particular the algorithm defined by (11.2.7) and (11.2.8) can be shown to behave in a desirable way without requiring any anti-zigzagging precautions. The following is a brief sketch of this result. In a similar manner to theorems 5.1.1 and 5.1.2 it can be shown that the algorithm converges to the solution of (11.1.1) and that the restriction (11.2.8) becomes inactive. If $\boldsymbol{\lambda}^{(k+1)}$ are the multipliers of the linear constraints in (11.2.7) then by continuity it follows that $\boldsymbol{\lambda}^{(k+1)} \to \boldsymbol{\lambda}^*$; thus if $\lambda_i^* > 0$, $i \in \mathscr{A}^* \cap I$ it follows that $\mathscr{A}^{(k)} = \mathscr{A}^*$ for all k sufficiently large.

Questions for Chapter 11

1. Relate the conditions $\mathbf{Z}^{\mathrm{T}}\mathbf{g}^* = \mathbf{0}$ and $\mathbf{Z}^{\mathrm{T}}\mathbf{G}^*\mathbf{Z}$ positive definite arising from (11.1.6) and (11.1.7) to the first and second order conditions of Sections 9.1 and 9.3 applied to problem (11.1.2) when $\operatorname{rank}(\mathbf{A}) = m$. Show that $\mathbf{Z}^{\mathrm{T}}\mathbf{g}^* = \mathbf{0}$ is equivalent to $\mathbf{g}^* = \mathbf{A}\boldsymbol{\lambda}^*$. Also show that the columns of $\mathbf{Z}$ are a basis for the linear space (9.3.4) and hence that (9.3.5) is equivalent to the condition that $\mathbf{Z}^{\mathrm{T}}\mathbf{G}^*\mathbf{Z}$ is positive definite.
2. Consider any Broyden method with parameters $\phi^{(k)}$, $k = 1, 2, \ldots$, applied to solve the reduced problem (11.1.4) using the positive definite inverse reduced Hessian matrix $\mathbf{H}^{(k)}$ in (11.1.13). Show for the original problem (11.1.2) that an equivalent sequence $\{\mathbf{x}^{(k)}\}$ is obtained if the same $\phi^{(k)}$ are used in a Broyden family update of the positive semi-definite matrix $\overline{\mathbf{H}}^{(k)}$ defined in (11.1.17), when $\mathbf{s}^{(k)}$ is calculated from $\mathbf{s}^{(k)} = -\overline{\mathbf{H}}^{(k)}\mathbf{g}^{(k)}$. Assume that the initial matrices are related by $\overline{\mathbf{H}}^{(1)} = \mathbf{Z}\mathbf{H}^{(1)}\mathbf{Z}^{\mathrm{T}}$.

3. If $\mathbf{h}^{(k)} = \mathbf{x}^{(k)} - \mathbf{x}^*$ where $\mathbf{x}^*$ solves (11.1.2) show that the vector $\boldsymbol{\lambda}^{(k)}$ in (11.1.18) satisfies $\boldsymbol{\lambda}^{(k)} = \boldsymbol{\lambda}^* + O(\|\mathbf{h}^{(k)}\|)$ and is therefore a first order estimate of $\boldsymbol{\lambda}^*$. If $\boldsymbol{\lambda}^{(k)}$ is computed from (11.1.19) show that $\boldsymbol{\lambda}^{(k)} = \boldsymbol{\lambda}^* + O(\|\mathbf{h}^{(k)}\|^2)$. Use a result like (10.2.9) and assume that rank$(\mathbf{A}) = m$, that $\mathbf{Z}^{\mathrm{T}}\mathbf{G}^*\mathbf{Z}$ is positive definite, and that the expansion $\mathbf{g}^{(k)} = \mathbf{g}^* + \mathbf{G}^*\mathbf{h}^{(k)} + O(\|\mathbf{h}^{(k)}\|^2)$ is valid.
4. Assume that $n \times n$ matrices are related by

$$\hat{\mathbf{B}} = \begin{bmatrix} \mathbf{B} & \mathbf{b} \\ \mathbf{b}^{\mathrm{T}} & \beta \end{bmatrix}$$

and let $\hat{\mathbf{H}} = \hat{\mathbf{B}}^{-1}$ and $\mathbf{H} = \mathbf{B}^{-1}$. Show that $\hat{\mathbf{H}}$ and $\mathbf{H}$ are related by

$$\hat{\mathbf{H}} - \mu\mathbf{u}\mathbf{u}^{\mathrm{T}} = \begin{bmatrix} \mathbf{H} & \mathbf{0} \\ \mathbf{0}^{\mathrm{T}} & 0 \end{bmatrix}$$

where either $\mathbf{u}^{\mathrm{T}} = (\mathbf{b}^{\mathrm{T}}\mathbf{H} : -1)$ and $\mu = 1/(\beta - \mathbf{b}^{\mathrm{T}}\mathbf{H}\mathbf{b})$ or $\mathbf{u} = \hat{\mathbf{H}}\mathbf{e}_n$ and $\mu = u_n^{-1}$, depending whether it is $\hat{\mathbf{H}}$ or $\mathbf{H}$ that is known. Hence deduce (11.2.3).
5. For fixed $\mathbf{x}'$, show that adding a linear equality constraint to problem (11.1.2) cannot increase the condition number of the reduced Hessian matrix at $\mathbf{x}'$, assuming that $\mathbf{x}'$ is feasible in both problems. It might be thought in general therefore that adding linear constraints cannot degrade the conditioning of an optimization problem. This is not so as the following example shows. Consider best least squares data fitting with a sum of exponentials $ae^{\alpha t} + be^{\beta t}$. The unknown parameters a, α, b, β must be chosen to best fit some given data. The problem usually has a well-defined (albeit ill-conditioned) solution. Consider adding the linear constraint $\alpha = \beta$. Then the parameters a and b become under-determined and the reduced Hessian matrix is singular at any feasible point. Explain this apparent contradiction.
6. Consider the linear constraint problem

$$\underset{\mathbf{x}}{\text{minimize}}\ \tfrac{4}{3}(x_1^2 - x_1x_2 + x_1^2)^{3/4} - x_3$$

$$\text{subject to } x_3 \leqslant 2, \qquad \mathbf{x} \geqslant \mathbf{0}$$

due to Wolfe (1972). Show that the objective function is convex but not $\mathbb{C}^2$ and that the solution is $\mathbf{x}^* = (0, 0, 2)^{\mathrm{T}}$. Solve the problem from $\mathbf{x}^{(1)} = (0, a, 0)^{\mathrm{T}}$ where $0 < a \leqslant \sqrt{2}/4$ using algorithm (11.2.2). Allow any point to be an acceptable solution in step (b) and use the steepest descent method in step (d). Show that $\mathbf{x}^{(2)} = \frac{1}{2}(a, 0, \sqrt{a})^{\mathrm{T}}$ and hence that, for $k \geqslant 2$,

$$\mathbf{x}^{(k)} = \begin{cases} (0, \alpha, \beta)^{\mathrm{T}} & \text{if } k \text{ is odd} \\ (\alpha, 0, \beta)^{\mathrm{T}} & \text{if } k \text{ is even} \end{cases}$$

where $\alpha = (\frac{1}{2})^{k-1}a$ and $\beta = \frac{1}{2}\sum_{j=0}^{k-2}(a/2^j)^{1/2}$. Hence show that $\mathbf{x}^{(k)} \to (0, 0, (1 + \frac{1}{2}\sqrt{2})\sqrt{a})^{\mathrm{T}}$ which is neither optimal nor a KT point.
7. After a constraint index q is removed from $\mathscr{A}$ in step (c) of the active set method, consider proving that the subsequent search direction $\mathbf{s}^{(k)}$ is downhill ($\mathbf{s}^{(k)\mathrm{T}}\mathbf{g}^{(k)} < 0$) and strictly feasible ($\mathbf{s}^{(k)\mathrm{T}}\mathbf{a}_q > 0$). The $\mathbf{Z}$ matrix is augmented

by a column vector $\mathbf{z}$ as $\mathbf{Z} := [\mathbf{Z} : \mathbf{z}]$. Show that the required conditions are obtained if $\mathbf{Z}^{\mathrm{T}}\mathbf{G}^{(k)}\mathbf{Z}$ is positive definite for the new $\mathbf{Z}$ matrix. If $\mathbf{Z}^{\mathrm{T}}\mathbf{G}^{(k)}\mathbf{Z}$ is positive definite for the old $\mathbf{Z}$ matrix only, show that the choice $\mathbf{s}^{(k)} = \mathbf{T}\mathbf{e}_q$ described in Section 10.4 has the required properties. In both cases assume that a second order multiplier estimate (11.1.19) is used.

Consider proving the same result for a quasi-Newton method when update (11.2.3) is used, when a first order estimate (11.1.18) for $\boldsymbol{\lambda}^{(k)}$ is used, and when the orthogonal factorization method defines $\mathbf{S}$ and $\mathbf{Z}$. Show that the column $\mathbf{z}$ which augments $\mathbf{Z}$ is $\mathbf{z} = \mathbf{a}_q^+$ which is the column of $\mathbf{A}^{+\mathrm{T}}$ (= $\mathbf{S}$) corresponding to constraint q. Hence show that $\mathbf{s}^{(k)\mathrm{T}}\mathbf{g}^{(k)} < 0$ and $\mathbf{s}^{(k)\mathrm{T}}\mathbf{a}_q = 1$, which satisfies the required conditions.

Chapter 12

Nonlinear Programming

12.1 Penalty and Barrier Functions

Nonlinear programming is the general case of (7.1.1) in which both the objective and constraint functions may be non-linear, and is the most difficult of the smooth optimization problems. Indeed there is no general agreement on the best approach and much research is still to be done. Historically the earliest developments were the *sequential minimization methods* described in Sections 12.1 and 12.2. These methods suffer from some computational disadvantages and are not entirely efficient. Nonetheless, especially for no-derivative problems in the absence of alternative software, the methods of Section 12.2 can still be recommended. Also the simplicity of the methods in this section (especially the *shortcut method* – see later) will continue to attract the unsophisticated user. Thus sequential techniques still have a useful part to play and are described in some detail.

When first (and possibly second) derivatives are available, then the *Lagrangian methods* of Section 12.3 have proved to be considerably more efficient, and software is becoming available. To some extent robustness is introduced by using a line search in which an exact penalty function is minimized. However this approach can fail, and a theoretically viable approach to the problem of inducing global convergence whilst retaining a rapid rate of convergence is described in Sections 14.3 and 14.5. This involves the use of a *non-differentiable exact penalty function* in conjunction with a restricted step method. Current research interest in these areas is strong, and further developments can be expected. No-derivative methods, possibly using finite differencing to obtain derivative estimates, can be expected to follow once the best approach has been determined. However it is likely that the effort involved in evaluating expressions for first derivatives will pay off in terms of the efficiency and reliability of the resulting software. Another idea which has attracted a lot of attention is that of a *feasible direction method* described in Section 12.4. It is shown that this is essentially equivalent to a non-linear generalized elimination of variables. There are inherent difficulties, however, in determining a fully reliable method. Some other interesting but not currently favoured approaches to the solution of nonlinear programming problems are reviewed in Section 12.5. To simplify the presentation methods are discussed either in terms of the equality constraint problem

$$\begin{aligned}&\underset{\mathbf{x}}{\text{minimize}}\ f(\mathbf{x})\\&\text{subject to } \mathbf{c}(\mathbf{x})=\mathbf{0}\end{aligned}\tag{12.1.1}$$

where $\mathbf{c}(\mathbf{x})$ is $\mathbb{R}^n \to \mathbb{R}^m$, or the inequality constraint problem

$$\begin{aligned}&\underset{\mathbf{x}}{\text{minimize}}\ f(\mathbf{x})\\&\text{subject to } c_i(\mathbf{x})\geqslant 0, \qquad i=1,2,\ldots,m.\end{aligned}\tag{12.1.2}$$

Usually the generalization to solve the mixed problem (7.1.1) is straightforward.

When solving a general nonlinear programming problem in which the constraints cannot easily be eliminated, it is necessary to balance the aims of reducing the objective function and staying inside or close to the feasible region, in order to induce global convergence (that is convergence to a local solution from any initial approximation). This inevitably leads to the idea of a *penalty function* which is some combination of f and $\mathbf{c}$ which enables f to be minimized whilst controlling constraint violations (or near constraint violations) by penalizing them. Early penalty functions were smooth so as to enable efficient techniques for smooth unconstrained optimization to be used: currently non-differentiable penalty functions are also receiving attention. For the equality problem the earliest penalty function (Courant, 1943) is

$$\begin{aligned}\phi(\mathbf{x},\sigma)&=f(\mathbf{x})+\tfrac{1}{2}\sigma\sum_i (c_i(\mathbf{x}))^2\\&=f(\mathbf{x})+\tfrac{1}{2}\sigma\mathbf{c}(\mathbf{x})^{\mathrm{T}}\mathbf{c}(\mathbf{x}).\end{aligned}\tag{12.1.3}$$

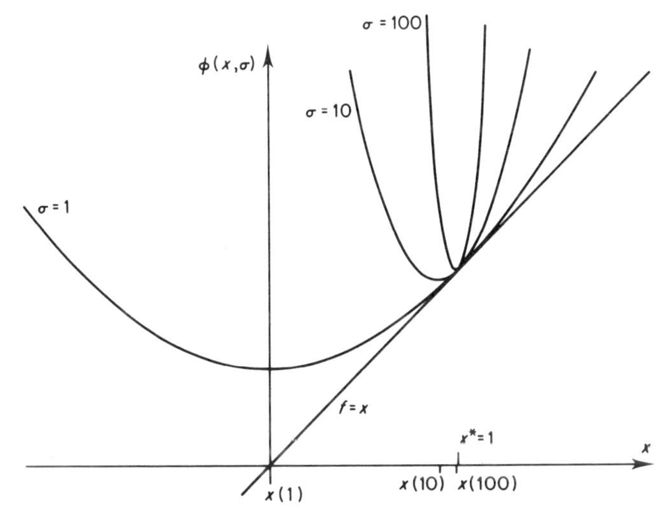

Figure 12.1.1 Convergence of the Courant penalty function

The penalty is formed from a sum of squares of constraint violations and the parameter σ determines the amount of the penalty. Some graphs of $\phi(\mathbf{x}, \sigma)$ are given in Figure 12.1.1 for the trivial problem: min x subject to $c(x) \triangleq x - 1 = 0$ for which $\phi = x + \frac{1}{2}\sigma(x-1)^2$. If the solution $x^* = 1$ is compared with the points which minimize $\phi(x, \sigma)$, it is clear that x^* is a limit point of the latter as $\sigma \to \infty$. Thus the technique of solving a *sequence* of minimization problems is suggested. This is traditionally implemented as follows.

(i) Choose a fixed sequence $\{\sigma^{(k)}\} \to \infty$, typically $\{1, 10, 10^2, 10^3, \ldots\}$.
(ii) For each $\sigma^{(k)}$ find a local minimizer, $\mathbf{x}(\sigma^{(k)})$ say, to $\min_{\mathbf{x}} \phi(\mathbf{x}, \sigma^{(k)})$. (12.1.4)
(iii) Terminate when $\mathbf{c}(\mathbf{x}(\sigma^{(k)}))$ is sufficiently small.

The effect of this iteration on the problem

$$\begin{aligned} &\text{minimize } -x_1 - x_2 \\ &\text{subject to } 1 - x_1^2 - x_2^2 = 0 \end{aligned} \qquad (12.1.5)$$

for which the solution and optimum Lagrange multiplier is $x_1^* = x_2^* = \lambda^* = 1/\sqrt{2}$, is shown in Table 12.1.1. It can be seen that $\mathbf{x}(\sigma^{(k)}) \to \mathbf{x}^*$ and that linear convergence is obtained with one extra decimal place being obtained at each iteration. This behaviour can in fact be justified for all problems, and this is done in theorem 12.1.2, equation (12.1.13), below. It must be emphasized that in practice step (ii) in (12.1.4) is done *numerically*, that is by the application of an unconstrained minimization method. The choice of this method will depend on whether or not derivatives are available and on the size of the problem (see Volume 1). Often $\mathbf{x}(\sigma^{(k)})$ is used as an initial approximation when minimizing $\phi(\mathbf{x}, \sigma^{(k+1)})$, and other information such as inverse Hessian approximations can also be passed forward from one iteration to the next. In fact algorithm (12.1.4) is idealized in that step (ii) cannot be solved exactly in a finite number of operations. It is assumed that $\mathbf{x}(\sigma^{(k)})$ is obtained as accurately as possible: in fact bounds on the accuracy can be given (equation (12.1.21) below) which still guarantee convergence. It has been assumed that the local minimizer $\mathbf{x}(\sigma^{(k)})$ exists. This may not be so, not only when the nonlinear programming problem is unbounded, but also in cases when local solutions exist. In the latter case a remedy (not guaranteed to work) is to increase the initial value $\sigma^{(1)}$ and repeat.

A variety of results relating to the convergence of this sequential penalty function can be given. In doing this, quantities derived from $\sigma^{(k)}$ like $\mathbf{x}(\sigma^{(k)})$, $f(\mathbf{x}(\sigma^{(k)}))$, etc., are denoted by $\mathbf{x}^{(k)}$, $f^{(k)}$, etc. It is assumed for the first theorem that $f(\mathbf{x})$ is bounded below on the (non-empty) feasible region so that

$$f^* = \inf f(\mathbf{x}) \qquad \forall \{\mathbf{x} : \mathbf{c}(\mathbf{x}) = \mathbf{0}\} \qquad (12.1.6)$$

exists. If global solutions can be computed in (12.1.4), step (ii), then the most simple result is the following.

Table 12.1.1 Application of the Courant penalty function

k	1	2	3	4	5	6
$\sigma^{(k)}$	1	10	100	1000	10000	100000
$x_1^{(k)} = x_2^{(k)}$	0.8846462	0.7308931	0.7095936	0.7073566	0.7071318	0.7071093
$c^{(k)}$	−0.5651978	−0.0684094	−0.0070462	−0.0007067	−0.0000708	−0.0000071
$\lambda^{(k)}$	0.5651978	0.684094	0.70462	0.7067	0.708	0.71
$\phi^{(k)}$	−1.609568	−1.438387	−1.416705	−1.414463	−1.414239	−1.414216
$\phi^{(k)} - \frac{1}{2}\boldsymbol{\lambda}^{(k)\mathrm{T}}\mathbf{c}^{(k)}$	−1.449844	−1.414988	−1.414222	−1.414213	−1.414214	−1.414213

Theorem 12.1.1 (Penalty function convergence)

If $\{\sigma^{(k)}\} \uparrow \infty$ *then*

(i) $\{\phi(\mathbf{x}^{(k)}, \sigma^{(k)})\}$ *is non-decreasing,*
(ii) $\{\mathbf{c}^{(k)\mathrm{T}}\mathbf{c}^{(k)}\}$ *is non-increasing,*
(iii) $\{f^{(k)}\}$ *is non-decreasing.*

Also $\mathbf{c}^{(k)} \to \mathbf{0}$ *and any accumulation point* $\mathbf{x}^*$ *of* $\{\mathbf{x}^{(k)}\}$ *solves (12.1.1).*

Proof

Let $\sigma^{(k)} < \sigma^{(\ell)}$; then from the definition of $\mathbf{x}^{(k)}$ and (12.1.3),

$$\phi(\mathbf{x}^{(k)}, \sigma^{(k)}) \leqslant \phi(\mathbf{x}^{(\ell)}, \sigma^{(k)}) \leqslant \phi(\mathbf{x}^{(\ell)}, \sigma^{(\ell)}) \leqslant \phi(\mathbf{x}^{(k)}, \sigma^{(\ell)}).$$

The first two inequalities give case (i). Subtracting the inner and outer inequalities gives

$$(\sigma^{(\ell)} - \sigma^{(k)})(\mathbf{c}^{(k)\mathrm{T}}\mathbf{c}^{(k)} - \mathbf{c}^{(\ell)\mathrm{T}}\mathbf{c}^{(\ell)}) \geqslant 0$$

and hence case (ii). Substituting in the first inequality then gives case (iii). By definition of $\mathbf{x}^{(k)}$,

$$\phi(\mathbf{x}^{(k)}, \sigma^{(k)}) \leqslant \inf_{\mathbf{x}:\mathbf{c}(\mathbf{x})=\mathbf{0}} \phi(\mathbf{x}, \sigma^{(k)}) = f^* \tag{12.1.7}$$

by (12.1.6). Thus using case (iii) above in (12.1.3), $\sigma^{(k)} \uparrow \infty$ implies $\mathbf{c}^{(k)} \to \mathbf{0}$. If $\mathbf{x}^{(k)} \to \mathbf{x}^*$ it follows that $\mathbf{c}(\mathbf{x}^*) = \mathbf{0}$, so $f(\mathbf{x}^*) \geqslant f^*$ as defined in (12.1.6). But from (12.1.3) and (12.1.7), $f^{(k)} \leqslant f^*$ so it follows that $f(\mathbf{x}^*) = f^*$, that is $\mathbf{x}^*$ solves (12.1.1). $\square$

It is interesting to observe that this result is obtained in absence of differentiability or Kuhn–Tucker regularity assumptions.

It is also possible to prove similar results when local minima are computed (with different assumptions on the problem), and also to get asymptotic estimates of the rate of convergence. In doing this the vector

$$\boldsymbol{\lambda}^{(k)} = -\sigma^{(k)}\mathbf{c}^{(k)} \tag{12.1.8}$$

is defined, and can be regarded as a *Lagrange multiplier estimate* by virtue of (12.1.9). The notation $\mathbf{h}^{(k)} = \mathbf{x}^{(k)} - \mathbf{x}^*$, $\mathbf{a}_i = \nabla c_i$, and $\mathbf{g} = \nabla f$, etc., as in Chapter 9 is used.

Theorem 12.1.2

If $\sigma^{(k)} \to \infty$ *and* $\mathbf{x}^{(k)} \to \mathbf{x}^*$ *is any accumulation point, and if rank* $\mathbf{A}^* = m$, *then* $\mathbf{x}^*$ *is a KT point and it follows that*

$$\boldsymbol{\lambda}^{(k)} = \boldsymbol{\lambda}^* + o(1) \tag{12.1.9}$$

$$\mathbf{c}^{(k)} = -\boldsymbol{\lambda}^*/\sigma^{(k)} + o(1/\sigma) \tag{12.1.10}$$

$$\sigma^{(k)}\mathbf{c}^{(k)\mathrm{T}}\mathbf{c}^{(k)} = \boldsymbol{\lambda}^{*\mathrm{T}}\boldsymbol{\lambda}^*/\sigma^{(k)} + o(1/\sigma). \tag{12.1.11}$$

Furthermore if second order sufficient conditions (9.3.5) hold at $\mathbf{x}^*$, $\boldsymbol{\lambda}^*$ *then*

$$f^* = \phi^* = f^{(k)} + \tfrac{1}{2}\sigma^{(k)}\mathbf{c}^{(k)\mathrm{T}}\mathbf{c}^{(k)} + o(1/\sigma) \tag{12.1.12}$$

$$\mathbf{h}^{(k)} = -\mathbf{T}^*\boldsymbol{\lambda}^*/\sigma^{(k)} + o(1/\sigma) \tag{12.1.13}$$

where $\mathbf{T}^*$ *is defined by*

$$\begin{bmatrix} \mathbf{W}^* & -\mathbf{A}^* \\ -\mathbf{A}^{*\mathrm{T}} & \mathbf{0} \end{bmatrix}^{-1} = \begin{bmatrix} \mathbf{H}^* & -\mathbf{T}^* \\ -\mathbf{T}^{*\mathrm{T}} & \mathbf{U}^* \end{bmatrix}. \tag{12.1.14}$$

Proof

The fact that $\mathbf{x}^{(k)}$ minimizes (12.1.14) implies that

$$\nabla\phi(\mathbf{x}^{(k)}, \sigma^{(k)}) = \mathbf{g}^{(k)} + \sigma^{(k)}\mathbf{A}^{(k)}\mathbf{c}^{(k)} = \mathbf{0} \tag{12.1.15}$$

and hence from (12.1.8) that

$$\mathbf{g}^{(k)} = \mathbf{A}^{(k)}\boldsymbol{\lambda}^{(k)}. \tag{12.1.16}$$

Since rank $\mathbf{A}^* = m$ it follows that $\mathbf{A}^{(k)+}$ exists and is bounded for all k sufficiently large and hence that

$$\boldsymbol{\lambda}^{(k)} = \mathbf{A}^{(k)+}\mathbf{g}^{(k)} = \boldsymbol{\lambda}^* + o(1) \tag{12.1.17}$$

where $\boldsymbol{\lambda}^*$ is defined by $\boldsymbol{\lambda}^* = \mathbf{A}^{*+}\mathbf{g}^*$. It also follows from (12.1.16) by continuity that $\mathbf{g}^* = \mathbf{A}^*\boldsymbol{\lambda}^*$, and from (12.1.17) and (12.1.8) that $\mathbf{c}^{(k)} = -\boldsymbol{\lambda}^*/\sigma^{(k)} + o(1/\sigma)$ so that $\mathbf{c}^* = \mathbf{0}$ in the limit $\sigma^{(k)} \to \infty$. Thus $\mathbf{x}^*$ satisfies KT conditions (Section 9.1) and (12.1.9) and (12.1.10) are established. Equation (12.1.11) follows directly and shows that $\lim \phi^{(k)} \triangleq \phi^* = f^*$. Without assuming second order conditions it is also possible to show from a Taylor series for $f(\mathbf{x})$ about $\mathbf{x}^{(k)}$ and (12.1.16) that

$$\begin{aligned} f^* &= f^{(k)} - \mathbf{h}^{(k)\mathrm{T}}\mathbf{g}^{(k)} + o(\|\mathbf{h}^{(k)}\|) \\ &= f^{(k)} - \mathbf{h}^{(k)\mathrm{T}}\mathbf{A}^{(k)}\boldsymbol{\lambda}^{(k)} + o(\|\mathbf{h}^{(k)}\|) \\ &= f^{(k)} - \mathbf{c}^{(k)\mathrm{T}}\boldsymbol{\lambda}^{(k)} + o(\|\mathbf{h}^{(k)}\|) \end{aligned}$$

using a similar Taylor series for $\mathbf{c}(\mathbf{x})$. Thus

$$\phi^* = \phi^{(k)} + \tfrac{1}{2}\sigma^{(k)}\mathbf{c}^{(k)\mathrm{T}}\mathbf{c}^{(k)} + o(\|\mathbf{h}^{(k)}\|) \tag{12.1.18}$$

from (12.1.4) and (12.1.8). Second order sufficient conditions for the equality constraint problem (9.3.5) and rank $\mathbf{A}^* = m$ imply that the Lagrangian matrix is non-singular at $\mathbf{x}^*$, $\boldsymbol{\lambda}^*$ so that the inverse in (12.1.14) exists (see Question 12.4). An expansion of the Lagrangian function about $\mathbf{x}^*$, $\boldsymbol{\lambda}^*$, and using (9.1.7) and (12.1.16) gives

$$\begin{pmatrix} \mathbf{0} \\ \mathbf{c}^{(k)} \end{pmatrix} = \begin{bmatrix} \mathbf{W}^* & -\mathbf{A}^* \\ -\mathbf{A}^{*\mathrm{T}} & \mathbf{0} \end{bmatrix} \begin{pmatrix} \mathbf{h}^{(k)} \\ \boldsymbol{\lambda}^{(k)} - \boldsymbol{\lambda}^* \end{pmatrix} + o(\max(\|\mathbf{h}^{(k)}\|, \|\boldsymbol{\lambda}^{(k)} - \boldsymbol{\lambda}^*\|)).$$

It follows from (12.1.10) and the existence of (12.1.14) that $\mathbf{h}^{(k)} = O(1/\sigma)$ which extends (12.1.18) to give (12.1.12). Multiplying on the left by (12.1.14) and using (12.1.8) then gives (12.1.13). □

The convergence of (12.1.9) and (12.1.10) can be observed easily in Table 12.1.1 and also the convergence of $\phi^{(k)} \to f^* = \sqrt{2}$ which (12.1.11) implies. The other results of theorem 12.1.2 can be used in a more sophisticated algorithm. Equation (12.1.12) gives a $o(1/\sigma)$ estimate of f^* which is better than the $O(1/\sigma)$ estimate given by $\phi^{(k)}$ itself. The asymptotic form of $\mathbf{h}^{(k)}$ in (12.1.13) can be used in an extrapolation scheme to estimate $\mathbf{x}^*$ (as given by Fiacco and McCormick (1968) in the context of barrier functions). These estimates can be used to terminate the penalty function iteration and also to provide better initial approximations when minimizing $\phi(\mathbf{x}, \sigma^{(k)})$. It is also interesting to observe that the rank assumption on $\mathbf{A}^*$ cannot easily be relaxed. For example if (12.1.3) has no feasible point then $\mathbf{c}^* \neq \mathbf{0}$ must occur and so as $\sigma \to \infty$ it is necessary from (12.1.8) and (12.1.15) that $\mathbf{A}^{(k)}\mathbf{c}^{(k)} \to \mathbf{0}$, that is $\mathbf{A}^*$ has dependent columns.

This well-developed theoretical background may make it appear that, apart from the inefficiency of sequential minimization, the method is a robust one which can be used with confidence. In fact this is not true at all and there are severe numerical difficulties which arise when the method is used in practice. These are caused by the fact that as $\sigma^{(k)} \to \infty$, it is increasingly difficult to solve the minimization problem in (12.1.4), step (ii). To illustrate this behaviour, the contours of $\phi(\mathbf{x}, \sigma)$ in (12.1.3) for the problem (12.1.5) are shown in Figure 12.1.2 for increasing values of σ. For $\sigma = 100$, it can be seen that whilst the solution $\mathbf{x}(100)$ is well determined in a radial direction, this is not so tangential to the constraint boundary, so that the exact location of $\mathbf{x}(100)$ is very difficult to determine numerically. This is expressed mathematically by the fact that (for $0 < m < n$) the Hessian matrix $\nabla^2\phi(\mathbf{x}^{(k)}, \sigma^{(k)})$ becomes increasingly ill conditioned as $\sigma^{(k)} \to \infty$. This result follows by virtue of

$$\nabla^2\phi(\mathbf{x}^{(k)}, \sigma^{(k)}) = \mathbf{W}^{(k)} + \sigma^{(k)}\mathbf{A}^{(k)}\mathbf{A}^{(k)\mathrm{T}} \tag{12.1.19}$$

where (12.1.8) is used in the definition

$$\mathbf{W}^{(k)} = \nabla_x^2\mathscr{L}(\mathbf{x}^{(k)}, \boldsymbol{\lambda}^{(k)}). \tag{12.1.20}$$

The $\sigma\mathbf{A}\mathbf{A}^{\mathrm{T}}$ term in (12.1.19) has rank m and so there are m eigenvalues of $\nabla^2\phi$ which approach ∞ as $\sigma^{(k)} \to \infty$. That the remaining eigenvalues are bounded is a consequence (see Lancaster, 1969) of the Courant–Fisher theorem. Thus the condition number of $\nabla^2\phi$ approaches ∞. In practice this shows up in that large values of $\nabla\phi$ are obtained whilst the minimization routine is unable to make progress in reducing ϕ.

These remarks have implications for the choice of the sequence $\{\sigma^{(k)}\}$. Choosing a very large $\sigma^{(1)}$, or increasing σ very rapidly in the sequence, gives minimization problems which are very difficult to solve accurately. The alternative of choosing $\sigma^{(1)}$ small and increasing σ slowly, keeps $\mathbf{x}^{(k)}$ close to the minimizer of $\phi(\mathbf{x}, \sigma^{(k+1)})$ and makes it easier to get accurate solutions, but is very inefficient. The typical sequence in (12.1.4), step (i), is a trade-off between these two effects. Probably

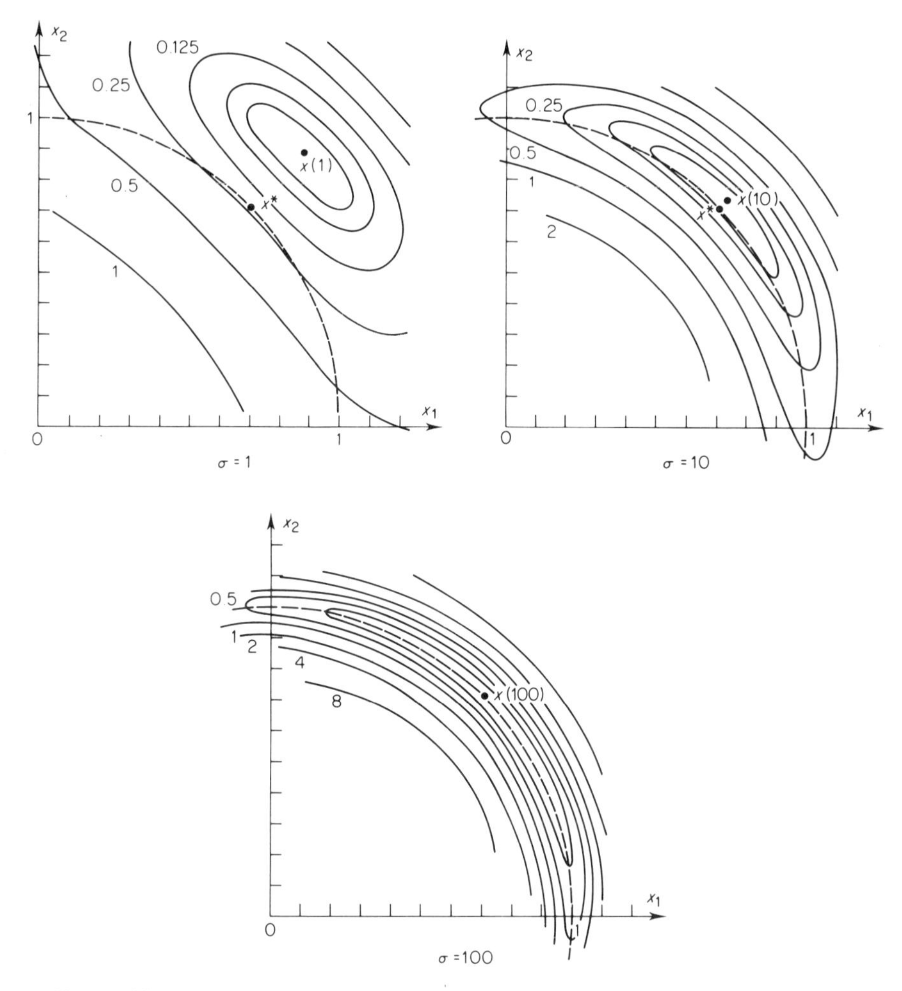

Figure 12.1.2 Increasing ill-conditioning in penalty functions (contours of $\phi - \phi_{\min}$ in powers of 2)

$\sigma^{(1)}$ should be chosen to balance the f and $\frac{1}{2}\sigma\mathbf{c}^T\mathbf{c}$ terms in (12.1.3) or to minimize the magnitude of $\nabla\phi$. This discussion also highlights the fact that it may not be possible to make good use of an accurate estimate of $\mathbf{x}^*$ in algorithm (12.1.4) since $\mathbf{x}^{(1)}$ will be remote from $\mathbf{x}^*$. In fact many users do not carry out the sequential technique at all, but carry out a *shortcut method* in which they minimize (12.1.3) for a single largish value of σ. In doing this they must be prepared to accept errors in the extent to which first order conditions are satisfied. Such users would be well advised to observe the constraint errors, and to estimate both the error in objective function using (12.1.12) and the error in the KT condition $\mathbf{g}^{(k)} = \mathbf{A}^{(k)}\boldsymbol{\lambda}^{(k)}$, in order to decide whether these errors are acceptable. If not, then it is quite easy to go on to use a multiplier penalty function as in Section 12.2. However, if derivatives and software permit, a Lagrangian method (Section 12.3) is likely to prove to be more efficient in the long run.

As mentioned earlier, algorithm (12.1.4) and theorem 12.1.1 are idealized in that they assume exact minimization of the penalty function. In fact it is straightforward to give the same results as in theorem 12.1.2 when approximate minimization is allowed. To do this let the test for terminating the minimization of $\phi(\mathbf{x}, \sigma^{(k)})$ with an approximate minimizer $\mathbf{x}^{(k)}$ be

$$\|\nabla \phi(\mathbf{x}^{(k)}, \sigma^{(k)})\| \leqslant \nu \|\mathbf{c}^{(k)}\|, \tag{12.1.21}$$

where $\nu > 0$ is pre-set and where $\mathbf{c}^{(k)} = \mathbf{c}(\mathbf{x}^{(k)})$, etc. With the same assumptions as in theorem 12.1.2, and using the fact that there exists a constant $\alpha > 0$ such that

$$\|\mathbf{A}^{(k)}\mathbf{c}^{(k)}\| \geqslant \alpha \|\mathbf{c}^{(k)}\| \tag{12.1.22}$$

for k sufficiently large (see Question 12.5), it follows from (12.1.21) and (12.1.15) that

$$\nu \|\mathbf{c}^{(k)}\| \geqslant \|\mathbf{g}^{(k)} + \sigma^{(k)}\mathbf{A}^{(k)}\mathbf{c}^{(k)}\| \geqslant \sigma^{(k)}\alpha \|\mathbf{c}^{(k)}\| - \mathbf{g}^{(k)}.$$

This shows that $\mathbf{c}^{(k)} \to \mathbf{0}$. Then (12.1.21) yields

$$\mathbf{g}^{(k)} = \mathbf{A}^{(k)}\boldsymbol{\lambda}^{(k)} + o(1)$$

and the rest of the results in theorem 12.1.2 follow as before. It is interesting to relate (12.1.21) to the practical difficulties caused by ill-conditioning, in that (12.1.21) definitely limits the extent to which large gradients can be accepted at an approximate minimizer of $\phi(\mathbf{x}, \sigma^{(k)})$.

A penalty function for the inequality constraint problem (12.1.2) can be given in an analogous way to (12.1.3) by

$$\phi(\mathbf{x}, \sigma) = f(\mathbf{x}) + \sigma \sum_i [\min(c_i(\mathbf{x}), 0)]^2. \tag{12.1.23}$$

An illustration is provided by the trivial problem: $\min x$ subject to $x \geqslant 1$. Then Figure 12.1.1 also illustrates the graph of ϕ, except that ϕ and f are equal for $x \geqslant 1$. Likewise if the constraint in (12.1.5) is replaced by $1 - x_1^2 - x_2^2 \geqslant 0$ then Figure 12.1.2 illustrates contours of ϕ, except that inside the unit circle the linear contours of $-x_1 - x_2$ must be drawn. These figures illustrate the jump discontinuity in the second derivative of (12.1.23) when $\mathbf{c} = \mathbf{0}$ (for example at $\mathbf{x}^*$). They also illustrate that $\mathbf{x}^{(k)}$ approaches $\mathbf{x}^*$ from the infeasible side of the inequality constraints, which leads to the term *exterior penalty function* for (12.1.23). Exactly similar results to those in theorems 12.1.1 and 12.1.2 follow on replacing $c_i^{(k)}$ by $\min(c_i^{(k)}, 0)$.

Another class of sequential minimization techniques is available to solve the inequality constraint problem (12.1.2), known as *barrier function methods*. These are characterized by their property of preserving strict constraint feasibility at all times, by using a barrier term which is infinite on the constraint boundaries. This can be advantageous if the objective function is not defined when the constraints are violated. The sequence of minimizers is also feasible, therefore, and hence the techniques are sometimes referred to as *interior point methods*. The two most important cases are the inverse barrier function

$$\phi(\mathbf{x}, r) = f(\mathbf{x}) + r \sum_i [c_i(\mathbf{x})]^{-1} \tag{12.1.24}$$

(Carroll, 1961) and the logarithmic barrier function

$$\phi(\mathbf{x}, r) = f(\mathbf{x}) - r \sum_i \log(c_i(\mathbf{x})) \tag{12.1.25}$$

(Frisch, 1955). As with σ in (12.1.3) the coefficient r is used to control the barrier function iteration. In this case a sequence $\{r^{(k)}\} \to 0$ is chosen, which ensures that the barrier term becomes more and more negligible except close to the boundary. Also $\mathbf{x}(r^{(k)})$ is defined as the minimizer of $\phi(\mathbf{x}, r^{(k)})$. Otherwise the procedure is the same as that in (12.1.4). Typical graphs of (12.1.24) for a sequence of values of $r^{(k)}$ are illustrated in Figure 12.1.3, and it can be seen that $\mathbf{x}(r^{(k)}) \to \mathbf{x}^*$ as $r^{(k)} \to 0$. This behaviour can be established rigorously in a similar way to theorem 12.1.1 (Osborne, 1972). Other features such as estimates of Lagrange multipliers and the asymptotic behaviour of $\mathbf{h}^{(k)}$ also follow, which can be used to determine a suitable sequence $\{r^{(k)}\}$ and to allow the use of extrapolation techniques (see Questions 12.6 and 12.7).

Unfortunately barrier function algorithms suffer from the same numerical difficulties in the limit as the penalty function algorithm does, in particular the badly determined nature of $\mathbf{x}(r^{(k)})$ tangential to the constraint surface, and the difficulty of locating the minimizer due to ill-conditioning and large gradients. Moreover there are additional problems which arise. The barrier function is undefined for infeasible points, and the simple expedient of setting it to infinity can make the line search

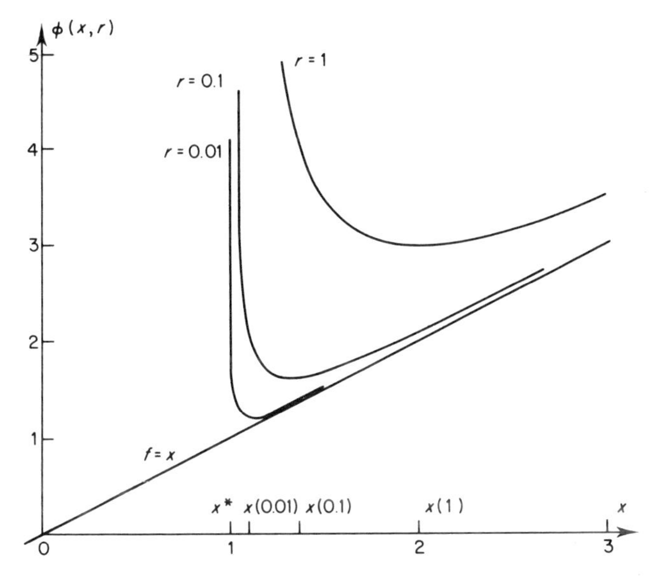

Figure 12.1.3 Increasing ill-conditioning in the barrier function (12.1.24)

inefficient. Also the singularity makes conventional quadratic or cubic interpolations in the line search work less efficiently. Thus special purpose line searches are required (Fletcher and McCann, 1969), and the aim of a simple-to-use algorithm is lost. Another difficulty is that an initial interior feasible point is required, and this in itself is a non-trivial problem involving the strict solution of a set of inequalities. In view of these difficulties and the general inefficiency of sequential techniques, barrier functions currently attract little interest.

12.2 Multiplier Penalty Functions

The way in which the penalty function (12.1.3) is used to solve (12.1.1) can be envisaged as an attempt to create a local minimizer at $\mathbf{x}^*$ in the limit $\sigma^{(k)} \to \infty$ (see Figures 12.1.1 and 12.1.2). However $\mathbf{x}^*$ can be made to minimize ϕ for finite σ by changing the origin of the penalty term. This suggests using the function

$$\begin{aligned}\phi(\mathbf{x}, \boldsymbol{\theta}, \boldsymbol{\sigma}) &= f(\mathbf{x}) + \tfrac{1}{2}\Sigma\sigma_i(c_i(\mathbf{x}) - \theta_i)^2 \\ &= f(\mathbf{x}) + \tfrac{1}{2}(\mathbf{c}(\mathbf{x}) - \boldsymbol{\theta})^T\mathbf{S}(\mathbf{c}(\mathbf{x}) - \boldsymbol{\theta}) \qquad (12.2.1)\end{aligned}$$

(Powell, 1969), where $\boldsymbol{\theta}, \boldsymbol{\sigma} \in \mathbb{R}^m$ and $\mathbf{S} = \text{diag}\, \sigma_i$. The parameters θ_i correspond to shifts of origin and the $\sigma_i \geqslant 0$ control the size of the penalty, like σ in (12.1.3). For the trivial problem: min x subject to $x = 1$ again, if the correct shift θ is chosen (which depends on σ), then it can be observed in Figure 12.2.1 that $\mathbf{x}^*$ minimizes $\phi(\mathbf{x}, \boldsymbol{\theta}, \boldsymbol{\sigma})$. This suggests an algorithm which attempts to locate the optimum value of $\boldsymbol{\theta}$ whilst keeping $\boldsymbol{\sigma}$ finite and so *avoids the ill-conditioning in the limit* $\boldsymbol{\sigma} \to \infty$. This is also illustrated for problem (12.1.5) with $\sigma = 1$, in which the optimum shift is $\theta = 1/\sqrt{2}$. The resulting contours are shown in Figure 12.2.2 and the contrast with Figure 12.1.2 as $\sigma^{(k)} \to \infty$ can be seen.

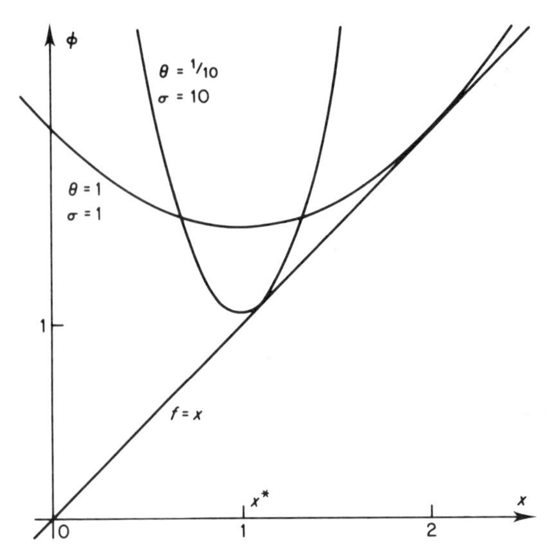

Figure 12.2.1 Multiplier penalty functions

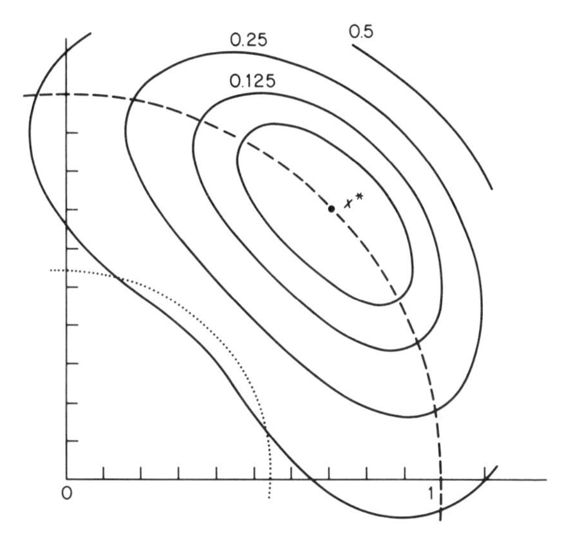

Figure 12.2.2 A multiplier penalty function ($\sigma = 1$, $\theta = 1/\sqrt{2}$) (contours of $\phi - \phi_{\min}$ in powers of 2)

In fact it is more convenient to define

$$\lambda_i = \theta_i \sigma_i, \qquad i = 1, 2, \ldots, m \tag{12.2.2}$$

and ignore the term $\frac{1}{2}\Sigma\sigma_i\theta_i^2$ (independent of $\mathbf{x}$) giving

$$\phi(\mathbf{x}, \boldsymbol{\lambda}, \boldsymbol{\sigma}) = f(\mathbf{x}) - \boldsymbol{\lambda}^{\mathrm{T}}\mathbf{c}(\mathbf{x}) + \tfrac{1}{2}\mathbf{c}(\mathbf{x})^{\mathrm{T}}\mathbf{S}\mathbf{c}(\mathbf{x}) \tag{12.2.3}$$

(Hestenes, 1969). There exists a corresponding optimum value of $\boldsymbol{\lambda}$ for which $\mathbf{x}^*$ minimizes $\phi(\mathbf{x}, \boldsymbol{\lambda}, \boldsymbol{\sigma})$, which in fact is the Lagrange multiplier $\boldsymbol{\lambda}^*$ at the solution to (12.1.1). This result is now true independent of $\boldsymbol{\sigma}$, so it is usually convenient to ignore the dependence on $\boldsymbol{\sigma}$ and write $\phi(\mathbf{x}, \boldsymbol{\lambda})$ in (12.2.3), and to use $\boldsymbol{\lambda}$ as the control parameter in a sequential minimization algorithm as follows.

(i) Determine a sequence $\{\boldsymbol{\lambda}^{(k)}\} \to \boldsymbol{\lambda}^*$.
(ii) For each $\boldsymbol{\lambda}^{(k)}$ find a local minimizer, $\mathbf{x}(\boldsymbol{\lambda}^{(k)})$ say, to $\min_{\mathbf{x}} \phi(\mathbf{x}, \boldsymbol{\lambda})$. (12.2.4)
(iii) Terminate when $\mathbf{c}(\mathbf{x}(\boldsymbol{\lambda}^{(k)}))$ is sufficiently small.

The main difference between this algorithm and (12.1.4) is that $\boldsymbol{\lambda}^*$ is not known in advance so the sequence in step (i) cannot be predetermined. However it is shown below how such a sequence can be constructed. Because (12.2.3) is obtained from (12.1.3) by adding a multiplier term $-\boldsymbol{\lambda}^{\mathrm{T}}\mathbf{c}$, (12.2.3) is often referred to as a *multiplier penalty function*. Alternatively (12.2.3) is the Lagrangian function (9.1.6) in which the objective f is augmented by the term $\frac{1}{2}\mathbf{c}^{\mathrm{T}}\mathbf{S}\mathbf{c}$. Hence the term *augmented Lagrangian function* is also used to describe (12.2.3).

The result that $\boldsymbol{\lambda}^*$ is the optimum choice of the control parameter vector in (12.2.3) is expressed in the following.

Theorem 12.2.1

If second order sufficient conditions hold at $\mathbf{x}^*$, $\boldsymbol{\lambda}^*$ *then there exists* $\boldsymbol{\sigma}' \geqslant \mathbf{0}$ *such that for any* $\boldsymbol{\sigma} > \boldsymbol{\sigma}'$, $\mathbf{x}^*$ *is an isolated local minimizer of* $\phi(\mathbf{x}, \boldsymbol{\lambda}^*, \boldsymbol{\sigma})$, *that is* $\mathbf{x}^* = \mathbf{x}(\boldsymbol{\lambda}^*)$.

Proof

Differentiating (12.2.3) gives

$$\nabla\phi(\mathbf{x}, \boldsymbol{\lambda}^*, \boldsymbol{\sigma}) = \mathbf{g} - \mathbf{A}\boldsymbol{\lambda}^* + \mathbf{ASc}. \tag{12.2.5}$$

The second order conditions require $\mathbf{x}^*$, $\boldsymbol{\lambda}^*$ to be a KT point, that is $\mathbf{g}^* = \mathbf{A}^*\boldsymbol{\lambda}^*$ and $\mathbf{c}^* = \mathbf{0}$, so it follows that $\nabla\phi(\mathbf{x}^*, \boldsymbol{\lambda}^*, \boldsymbol{\sigma}) = \mathbf{0}$. Differentiating (12.2.5) gives

$$\nabla^2\phi(\mathbf{x}, \boldsymbol{\lambda}, \boldsymbol{\sigma}) = \mathbf{W} + \mathbf{ASA}^{\mathrm{T}} \triangleq \mathbf{W}_\sigma \tag{12.2.6}$$

say, where $\mathbf{W} = \nabla^2 f - \Sigma_i(\lambda_i - \sigma_i c_i)\nabla^2 c_i$, and hence

$$\mathbf{W}_\sigma^* \triangleq \nabla^2\phi(\mathbf{x}^*, \boldsymbol{\lambda}^*, \boldsymbol{\sigma}) = \mathbf{W}^* + \mathbf{A}^*\mathbf{SA}^{*\mathrm{T}}. \tag{12.2.7}$$

Let rank $\mathbf{A}^* = r \leqslant m$ and let $\mathbf{B}$ $(n \times r)$ be an orthonormal basis matrix $(\mathbf{B}^{\mathrm{T}}\mathbf{B} = \mathbf{I})$ for $\mathbf{A}^*$, so that $\mathbf{A}^* = \mathbf{BC}$ where $\mathbf{C}$ $(= \mathbf{B}^{\mathrm{T}}\mathbf{A}^*)$ has rank r. Consider any vector $\mathbf{u} \neq \mathbf{0}$ and let $\mathbf{u} = \mathbf{v} + \mathbf{Bw}$ where $\mathbf{B}^{\mathrm{T}}\mathbf{v} = \mathbf{0} = \mathbf{A}^{*\mathrm{T}}\mathbf{v}$. Then

$$\mathbf{u}^{\mathrm{T}}\mathbf{W}_\sigma^*\mathbf{u} = \mathbf{v}^{\mathrm{T}}\mathbf{W}^*\mathbf{v} + 2\mathbf{v}^{\mathrm{T}}\mathbf{W}^*\mathbf{Bw} + \mathbf{w}^{\mathrm{T}}\mathbf{B}^{\mathrm{T}}\mathbf{W}^*\mathbf{Bw} + \mathbf{w}^{\mathrm{T}}\mathbf{CSC}^{\mathrm{T}}\mathbf{w}.$$

From (9.3.5) and (9.3.4) there exists a constant $a > 0$ such that $\mathbf{v}^{\mathrm{T}}\mathbf{W}^*\mathbf{v} \geqslant a\|\mathbf{v}\|_2^2$. Let b be the greatest singular value of $\mathbf{W}^*\mathbf{B}$ and let $d = \|\mathbf{B}^{\mathrm{T}}\mathbf{W}^*\mathbf{B}\|_2$. Let $\sigma = \min_i \sigma_i \geqslant 0$ and let $\mu > 0$ be the smallest eigenvalue of $\mathbf{CC}^{\mathrm{T}}$. Then

$$\mathbf{u}^{\mathrm{T}}\mathbf{W}_\sigma^*\mathbf{u} \geqslant a\|\mathbf{v}\|_2^2 - 2b\|\mathbf{v}\|_2\|\mathbf{w}\|_2 + (\sigma\mu - d)\|\mathbf{w}\|_2^2.$$

Since $\|\mathbf{v}\| = \|\mathbf{w}\| = 0$ cannot hold, if $\sigma > \sigma'$ where $\sigma' = (d + b^2/a)/\mu$ then it follows that $\mathbf{u}^{\mathrm{T}}\mathbf{W}_\sigma^*\mathbf{u} > 0$. Thus both $\nabla\phi(\mathbf{x}^*, \boldsymbol{\lambda}^*, \boldsymbol{\sigma}) = \mathbf{0}$ and $\nabla^2\phi(\mathbf{x}^*, \boldsymbol{\lambda}^*, \boldsymbol{\sigma})$ is positive definite so $\mathbf{x}^*$ is an isolated local minimizer of $\phi(\mathbf{x}, \boldsymbol{\lambda}^*, \boldsymbol{\sigma})$ for $\boldsymbol{\sigma}$ sufficiently large. □

The assumption of second order conditions is important and not easily relaxed. For example if $f = x_1^4 + x_1x_2$ and $c = x_2$ then $\mathbf{x}^* = \mathbf{0}$ solves (12.1.1) with the unique multiplier $\lambda^* = 0$. However second order conditions are not satisfied, and in fact $\mathbf{x}^* = \mathbf{0}$ does not minimize $\phi(\mathbf{x}, \mathbf{0}, \boldsymbol{\sigma}) = x_1^4 + x_1x_2 + \frac{1}{2}\sigma x_2^2$ for any value of σ. Henceforth it is assumed that second order sufficient conditions hold and $\boldsymbol{\sigma}$ is sufficiently large.

The minimizer $\mathbf{x}(\boldsymbol{\lambda})$ of $\phi(\mathbf{x}, \boldsymbol{\lambda})$ can also be regarded as having been determined by solving the non-linear equations

$$\nabla\phi(\mathbf{x}, \boldsymbol{\lambda}) = \mathbf{0}. \tag{12.2.8}$$

Because $\nabla^2\phi(\mathbf{x}^*, \boldsymbol{\lambda}^*)$ is positive definite and is the Jacobian matrix of this system, it follows from the implicit function theorem (Apostol, 1957) that there exist open neighbourhoods $\Omega_\lambda \subset \mathbb{R}^m$ about $\boldsymbol{\lambda}^*$ and $\Omega_x \subset \mathbb{R}^n$ about $\mathbf{x}^*$ and a $\mathbb{C}^1$ function

$\mathbf{x}(\boldsymbol{\lambda})$ $(\Omega_\lambda \to \Omega_x)$ such that $\nabla\phi(\mathbf{x}(\boldsymbol{\lambda}), \boldsymbol{\lambda}) = \mathbf{0}$. Also $\nabla^2\phi(\mathbf{x}, \boldsymbol{\lambda})$ is positive definite for all $\mathbf{x} \in \Omega_x$ and $\boldsymbol{\lambda} \in \Omega_\lambda$, so $\mathbf{x}(\boldsymbol{\lambda})$ is the minimizer of $\phi(\mathbf{x}, \boldsymbol{\lambda})$. It may be that $\phi(\mathbf{x}, \boldsymbol{\lambda})$ has local minima so that various solutions to (12.2.8) exist: it is assumed that a consistent choice of $\mathbf{x}(\boldsymbol{\lambda})$ is made by the minimization routine as that solution which exists in Ω_x. The vector $\mathbf{x}(\boldsymbol{\lambda})$ can also be interpreted in yet another way as the solution to a neighbouring problem to (12.1.1) (see Question 12.14) and this can sometimes be convenient.

It is important to examine the function

$$\psi(\boldsymbol{\lambda}) \triangleq \phi(\mathbf{x}(\boldsymbol{\lambda}), \boldsymbol{\lambda}) \tag{12.2.9}$$

which is derived from $\boldsymbol{\lambda}$ by first finding $\mathbf{x}(\boldsymbol{\lambda})$. By the local optimality of $\mathbf{x}(\boldsymbol{\lambda})$ on Ω_x, for any $\boldsymbol{\lambda}$ in Ω_λ it follows that

$$\psi(\boldsymbol{\lambda}) = \phi(\mathbf{x}(\boldsymbol{\lambda}), \boldsymbol{\lambda}) \leqslant \phi(\mathbf{x}^*, \boldsymbol{\lambda}) = \phi(\mathbf{x}^*, \boldsymbol{\lambda}^*) = \psi(\boldsymbol{\lambda}^*) \tag{12.2.10}$$

(using $\mathbf{c}^* = \mathbf{0}$). Thus $\boldsymbol{\lambda}^*$ *is a local unconstrained maximizer of* $\psi(\boldsymbol{\lambda})$. This result is also true globally if $\mathbf{x}(\boldsymbol{\lambda})$ is a global maximizer of $\phi(\mathbf{x}, \boldsymbol{\lambda})$. Thus methods for generating a sequence $\boldsymbol{\lambda}^{(k)} \to \boldsymbol{\lambda}^*$ can be derived by applying unconstrained minimization methods to $-\psi(\boldsymbol{\lambda})$. To do this requires formulae for the derivatives $\nabla\psi$ and $\nabla^2\psi$ of ψ with respect to $\boldsymbol{\lambda}$. Using the matrix notation $[\partial\mathbf{x}/\partial\boldsymbol{\lambda}]_{ij}$ to denote $\partial x_i/\partial\lambda_j$, it follows from the chain rule that

$$[\mathrm{d}\psi/\mathrm{d}\boldsymbol{\lambda}] = [\partial\phi/\partial\mathbf{x}]\,[\partial\mathbf{x}/\partial\boldsymbol{\lambda}] + [\partial\phi/\partial\boldsymbol{\lambda}],$$

using the total derivative to denote variations through both $\mathbf{x}(\boldsymbol{\lambda})$ and $\boldsymbol{\lambda}$. Since $[\partial\phi/\partial\mathbf{x}] = \mathbf{0}$ from (12.2.8) and $\partial\phi/\partial\lambda_i = -c_i$ from (12.2.3) it follows that

$$\nabla\psi(\boldsymbol{\lambda}) = -\mathbf{c}(\mathbf{x}(\boldsymbol{\lambda})). \tag{12.2.11}$$

Also by the chain rule

$$[\mathrm{d}\mathbf{c}/\mathrm{d}\boldsymbol{\lambda}] = [\partial\mathbf{c}/\partial\mathbf{x}]\,[\partial\mathbf{x}/\partial\boldsymbol{\lambda}] = \mathbf{A}^{\mathrm{T}}[\partial\mathbf{x}/\partial\boldsymbol{\lambda}].$$

Operating on (12.2.8) by $[\mathrm{d}/\mathrm{d}\boldsymbol{\lambda}]$ gives

$$[\mathrm{d}\nabla\phi(\mathbf{x}(\boldsymbol{\lambda}), \boldsymbol{\lambda})/\mathrm{d}\boldsymbol{\lambda}] = [\partial\nabla\phi/\partial\mathbf{x}]\,[\partial\mathbf{x}/\partial\boldsymbol{\lambda}] + [\partial\nabla\phi/\partial\boldsymbol{\lambda}] = \mathbf{0}.$$

But $[\partial\nabla\phi/\partial\mathbf{x}] = \nabla^2\phi(\mathbf{x}(\boldsymbol{\lambda}), \boldsymbol{\lambda}) = \mathbf{W}_\sigma$ and $[\partial\nabla\psi/\partial\boldsymbol{\lambda}] = -\mathbf{A}$ so it follows that

$$\nabla^2\psi(\boldsymbol{\lambda}) = -[\mathrm{d}\mathbf{c}/\mathrm{d}\boldsymbol{\lambda}] = -\mathbf{A}^{\mathrm{T}}\mathbf{W}_\sigma^{-1}\mathbf{A}\Big|_{\mathbf{x}(\lambda)}. \tag{12.2.12}$$

Since $\mathbf{c}(\mathbf{x}(\boldsymbol{\lambda}^*)) = \mathbf{c}(\mathbf{x}^*) = \mathbf{0}$ and $\mathbf{W}_\sigma^*$ is positive definite it follows that $\nabla\psi(\boldsymbol{\lambda}^*) = \mathbf{0}$ and (when rank $\mathbf{A}^* = m$) that $\nabla^2\psi(\boldsymbol{\lambda}^*)$ is negative definite, which reinforces the maximization result in (12.1.10).

The most obvious sequence $\{\boldsymbol{\lambda}^{(k)}\}$ to be used in step (i) of algorithm (12.2.4) is obtained by using Newton's method from an initial estimate $\boldsymbol{\lambda}^{(1)}$, giving the iteration

$$\boldsymbol{\lambda}^{(k+1)} = \boldsymbol{\lambda}^{(k)} - (\mathbf{A}^{\mathrm{T}}\mathbf{W}_\sigma^{-1}\mathbf{A})^{-1}\mathbf{c}\Big|_{\mathbf{x}(\lambda^{(k)})}. \tag{12.2.13}$$

This method requires $\mathbf{W}_\sigma$, and hence explicit formulae for second derivatives, which is disadvantageous. However when only first derivatives are available and a quasi-Newton method is used to find $\mathbf{x}(\boldsymbol{\lambda}^{(k)})$, then the resulting $\mathbf{H}$ matrix (see Section 3.2) is a very good approximation to $\mathbf{W}_\sigma^{-1}$. Using this matrix in (12.2.3) (Fletcher, 1975) enables the advantages of Newton's method to be obtained whilst only requiring first derivatives. A different method is suggested by Powell (1969) and by Hestenes (1969), and is best motivated by the fact that for large $\boldsymbol{\sigma}$,

$$(\mathbf{A}^T\mathbf{W}_\sigma^{-1}\mathbf{A})^{-1} \approx \mathbf{S} \tag{12.2.14}$$

(see Question 12.12). When this approximation is used in (12.2.13) the iteration

$$\boldsymbol{\lambda}^{(k+1)} = \boldsymbol{\lambda}^{(k)} - \mathbf{S}\mathbf{c}^{(k)} \tag{12.2.15}$$

is obtained. No derivatives are required by this formula so it is particularly convenient when the routine which minimizes $\phi(\mathbf{x}, \boldsymbol{\lambda})$ does not calculate or estimate derivatives. Furthermore by making $\mathbf{S}$ sufficiently large, an arbitrarily fast rate of linear convergence of $\boldsymbol{\lambda}^{(k)}$ to $\boldsymbol{\lambda}^*$ can be obtained (see Question 12.13).

An illustration of the methods based on using (12.2.13) or (12.2.15), applied to solve the problem (12.1.15), is given in Table 12.2.1, starting from $\lambda^{(1)} = 0$. It can be observed that $\lambda^{(k)} \to \lambda^*$ and $c^{(k)} \to 0$ in all cases. The Powell–Hestenes method exhibits linear convergence at a rate 0.26 when $\sigma = 1$ and 0.034 when $\sigma = 10$. This illustrates the fact that increasing σ_i by 10 will asymptotically reduce the rate of convergence of c_i by one-tenth. Newton's method is seen to converge at an approximately quadratic rate and is a little better than the Powell–Hestenes method with $\sigma = 10$.

Although these methods have good local convergence properties, they must be supplemented in a general algorithm to ensure global convergence. This can be done by an algorithm due to Powell (1969).

$$\begin{array}{rl}
\text{(i)} & \text{Initially set } \boldsymbol{\lambda} = \boldsymbol{\lambda}^{(1)},\ \boldsymbol{\sigma} = \boldsymbol{\sigma}^{(1)},\ k = 0,\ \|\mathbf{c}^{(0)}\|_\infty = \infty. \\
\text{(ii)} & \text{Find the minimizer } \mathbf{x}(\boldsymbol{\lambda}, \boldsymbol{\sigma}) \text{ of } \phi(\mathbf{x}, \boldsymbol{\lambda}, \boldsymbol{\sigma}) \text{ and denote } \mathbf{c} = \mathbf{c}(\mathbf{x}(\boldsymbol{\lambda}, \boldsymbol{\sigma})). \\
\text{(iii)} & \text{If } \|\mathbf{c}\|_\infty > \tfrac{1}{4}\|\mathbf{c}^{(k-1)}\|_\infty \text{ set } \sigma_i = 10\sigma_i\ \forall\ i : |c_i| > \tfrac{1}{4}\|\mathbf{c}^{(k-1)}\|_\infty \\
& \text{and go to (ii).} \\
\text{(iv)} & \text{Set } k = k+1,\ \boldsymbol{\lambda}^{(k)} = \boldsymbol{\lambda},\ \boldsymbol{\sigma}^{(k)} = \boldsymbol{\sigma},\ \mathbf{c}^{(k)} = \mathbf{c}. \\
\text{(v)} & \text{Set } \boldsymbol{\lambda} = \boldsymbol{\lambda}^{(k)} - \mathbf{S}^{(k)}\mathbf{c}^{(k)} \text{ and go to (ii).}
\end{array} \tag{12.2.16}$$

The aim of the algorithm is to achieve linear convergence at a rate 0.25 or better. If any component c_i is not reduced at this rate, the corresponding penalty σ_i is increased tenfold which induces more rapid convergence (see Question 12.12). A simple proof that $\mathbf{c}^{(k)} \to \mathbf{0}$ can be given along the following lines. Clearly this happens unless the inner iteration ((iii) $\to$ (ii)) fails to terminate. In this iteration $\boldsymbol{\lambda}$ is fixed, and if for any i, $|c_i| > \frac{1}{4}\|\mathbf{c}^{(k-1)}\|_\infty$ occurs infinitely often, then $\sigma_i \to \infty$. As in theorem 12.1.1 this implies $c_i \to 0$, which contradicts the infinitely often assumption and proves termination. This convergence result is true whatever formula is used in step (v) of (12.2.16). It follows as in theorem 12.1.2 that any limit point is a KT point of (12.1.1) with $\mathbf{x}^{(k)} \to \mathbf{x}^*$ and $\boldsymbol{\lambda}^{(k)} - \mathbf{S}^{(k)}\mathbf{c}^{(k)} \to \boldsymbol{\lambda}^*$. For formulae (12.2.13) and (12.2.15), the required rate of convergence is obtained when $\boldsymbol{\sigma}$ is sufficiently large, and then the basic iteration takes over in which $\boldsymbol{\sigma}$ stays constant and only the $\boldsymbol{\lambda}$ parameters are changed.

Table 12.2.1 Different formulae for changing $\lambda^{(k)}$

	Newton method (12.2.13), $\sigma = 1$		Powell–Hestenes method (12.2.15) $\sigma = 1$		$\sigma = 10$	
Iteration	$\lambda^{(k)}$	$c^{(k)}$	$\lambda^{(k)}$	$c^{(k)}$	$\lambda^{(k)}$	$c^{(k)}$
1	0	−0.5651977	0	−0.5651977	0	−0.0684095
2	0.6672450	−0.0296174	0.5651977	−0.1068981	0.6840946	−0.0022228
3	0.7068853	−0.0001637	0.6720958	−0.0259956	0.7063222	−0.0000758
4	0.7071068	$-0.149_{10}{-7}$	0.6980914	−0.0066692	0.7070801	−0.0000026
5			0.7047606	−0.0017339	0.7071058	$-0.894_{10}{-7}$
6			0.7064945	−0.0004524		
7			0.7069469	−0.0001181		
8			0.7070650	−0.0000308		

In practice this proof is not as powerful as it might seem. Unfortunately increasing $\boldsymbol{\sigma}$ can lead to difficulties caused by ill-conditioning as described in Section 12.1, in which case accuracy in the solution is lost. Furthermore when no feasible point exists, this situation is not detected, and $\boldsymbol{\sigma}$ is increased without bound which is an unsatisfactory situation. I think there is scope for other algorithms to be determined which keep $\boldsymbol{\sigma}$ fixed at all times and induce convergence by ensuring that $\psi(\boldsymbol{\lambda})$ is increased sufficiently at each iteration, for example by using a restricted step modification of Newton's method (12.2.13), as described in Chapter 5, Volume 1.

There is no difficulty in modifying these penalty functions to handle inequality constraints. For the inequality constraint problem (12.1.2) a suitable modification of (12.2.3) (Fletcher, 1975) is

$$\phi(\mathbf{x}, \boldsymbol{\theta}, \boldsymbol{\sigma}) = f(\mathbf{x}) + \tfrac{1}{2} \sum_i \sigma_i (c_i(\mathbf{x}) - \theta_i)_-^2 \tag{12.2.17}$$

where $a_- = \min(a, 0)$. The effect of this on the problem: $\min x$ subject to $x \geqslant 1$ can also be seen in Figure 12.2.1, the only difference being that for $c \geqslant \theta$ (that is $x \geqslant 1 + \theta$) the graph of ϕ is identical with that of f. Thus although $\phi(\mathbf{x}, \boldsymbol{\theta}, \boldsymbol{\sigma})$ has second derivative jump discontinuities at $c_i(\mathbf{x}) = \theta_i$, these are usually remote from the solution and in practice do not appear to affect the performance of the unconstrained minimization routine adversely. Another example of this is the inequality constraint version of problem (12.1.5). The surface $c(\mathbf{x}) = \theta = 1/\sqrt{2}$ on which this discontinuity occurs is illustrated by the dotted circle in Figure 12.2.2, which is remote from $\mathbf{x}^*$. The contours of (12.2.17) differ from those given only within this dotted circle, where they become the unpenalized linear contours of $f(\mathbf{x})$. As with (12.2.1) it is possible to rearrange the function, using (12.2.2) and omitting terms independent of $\mathbf{x}$, to give the multiplier penalty function

$$\phi(\mathbf{x}, \boldsymbol{\lambda}, \boldsymbol{\sigma}) = f(\mathbf{x}) + \sum_i \begin{cases} -\lambda_i c_i + \tfrac{1}{2}\sigma_i c_i^2 & \text{if } c_i \leqslant \lambda_i/\sigma_i \\ -\tfrac{1}{2}\lambda_i^2/\sigma_i & \text{if } c_i \geqslant \lambda_i/\sigma_i \end{cases} \tag{12.2.18}$$

where $c_i = c_i(\mathbf{x})$. Rockafeller (1974) first suggested this type of function, and as with (12.2.3) it is the most convenient form for developing the theory.

Most of the theoretical results can be extended to the inequality case. If strict complementarity holds then the extension of theorem 12.2.1 is immediate, although the result can also be proved in the absence of this condition (Fletcher, 1975). The dual function $\psi(\boldsymbol{\lambda})$ is again defined by (12.2.9) and an analogous global result to (12.2.10) is

$$\begin{aligned} \psi(\boldsymbol{\lambda}) &= \phi(\mathbf{x}(\boldsymbol{\lambda}), \boldsymbol{\lambda}) \leqslant \phi(\mathbf{x}^*, \boldsymbol{\lambda}) \\ &= f^* + \sum_i \begin{cases} -c_i^*\lambda_i + \tfrac{1}{2}\sigma_i c_i^{*2} & c_i^* \leqslant \lambda_i/\sigma_i \\ -\tfrac{1}{2}\lambda_i^2/\sigma_i & c_i^* \geqslant \lambda_i/\sigma_i \end{cases} \\ &\leqslant f^* + \sum_i \begin{cases} -\tfrac{1}{2}\sigma_i c_i^{*2} \\ -\tfrac{1}{2}\lambda_i^2/\sigma_i \end{cases} \\ &\leqslant f^* = \phi(\mathbf{x}^*, \boldsymbol{\lambda}^*) = \psi(\boldsymbol{\lambda}^*). \end{aligned} \tag{12.2.19}$$

This result is also true locally if strict complementarity holds and can probably be extended in the absence of this condition, again by following Fletcher (1975).

Derivative expressions analogous to (12.2.11) and (12.2.12) are easily obtained as

$$\mathrm{d}\psi(\boldsymbol{\lambda})/\mathrm{d}\lambda_i = -\min(c_i, \theta_i), \qquad i = 1, 2, \ldots, m \tag{12.2.20}$$

where $c_i = c_i(\mathbf{x}(\boldsymbol{\lambda}))$ and $\theta_i = \lambda_i/\sigma_i$, and

$$\nabla^2\psi(\boldsymbol{\lambda}) = \begin{bmatrix} -\mathbf{A}^{\mathrm{T}}\mathbf{W}_\sigma^{-1}\mathbf{A} & \mathbf{0} \\ \mathbf{0} & -\mathbf{S}^{-1} \end{bmatrix}_{\mathbf{x}(\lambda)} \tag{12.2.21}$$

where the columns of $\mathbf{A}$ correspond to indices $i : c_i < \theta_i$ and those of $\mathbf{S}^{-1}$ to indices $i : c_i \geqslant \theta_i$. Algorithms for determining the sequence $\{\boldsymbol{\lambda}^{(k)}\}$ in step (i) of algorithm (12.2.4) can be determined using these derivative expressions. An equivalent form of Newton's method (12.2.3) is possible, although in view of the implicit inequalities $\boldsymbol{\lambda} \geqslant \mathbf{0}$ it is probably preferable to choose $\boldsymbol{\lambda}^{(k+1)}$ to solve a subproblem

$$\begin{aligned} &\underset{\boldsymbol{\lambda}}{\text{maximize}}\ q^{(k)}(\boldsymbol{\lambda}) \\ &\text{subject to } \boldsymbol{\lambda} \geqslant \mathbf{0} \end{aligned} \tag{12.2.22}$$

where $q^{(k)}(\boldsymbol{\lambda})$ is obtained by truncating a Taylor series for $\psi(\boldsymbol{\lambda})$ about $\boldsymbol{\lambda}^{(k)}$ after the quadratic term. In fact the simple structure of (12.2.21) enables a more simple problem to be solved in terms of just the indices $i : c_i < \theta_i$. The result in (12.2.14) can also be used to determine an extension of (12.2.15), that is

$$\lambda_i^{(k+1)} = \lambda_i^{(k)} - \min(\sigma_i c_i^{(k)}, \lambda_i^{(k)}), \qquad i = 1, 2, \ldots, m. \tag{12.2.23}$$

These formulae can be incorporated in a globally convergent algorithm like (12.2.16) by using $\|\nabla\psi\|_\infty$ in place of $\|\mathbf{c}\|_\infty$ to monitor the rate of convergence.

A selection of numerical experiments is given by Fletcher (1975), which seems to indicate that whilst both the Newton-like formulae or the Powell–Hestenes formulae for updating $\boldsymbol{\lambda}^{(k)}$ are effective, the Newton-like method is somewhat more efficient. Local convergence is rapid and high accuracy can usually be achieved in about four to six minimizations. When this occurs with modest values of $\boldsymbol{\sigma}$ then no difficulties due to ill-conditioning and loss of accuracy are observed. Furthermore the Hessian matrix can be carried forward, and updated if σ_i is increased, so the computational effort for the successive minimizations goes down rapidly. Since $\phi(\mathbf{x}, \boldsymbol{\lambda}, \boldsymbol{\sigma})$ is always well defined, there is no difficulty in coping with infeasible points, and it is easy to program the method using an existing quasi-Newton subroutine. The main disadvantage is that the sequential nature of the method is less efficient than the more direct approach of Section 12.3. Also the global convergence result based on increasing $\boldsymbol{\sigma}$, whilst powerful in theory, does not always work well in practice, and there are practical applications in which it has caused ill-conditioning and low accuracy.

Two final points are worthy of note. Firstly if the problem is a mixture of linear and non-linear constraints, then it may be worth incorporating only the non-linear constraints into the penalty function, and the linear constraints can be included when $\phi(\mathbf{x}, \boldsymbol{\lambda}, \boldsymbol{\sigma})$ is minimized, for example at step (ii) in algorithm (12.1.4). This is especially true for bounds on the variables, since minimization with bounds is not a significant complication on an unconstrained minimization routine. Another point is that approximate minimization of the penalty function (see (12.1.21) for example) can also be considered, and a review of recent work in this area is given by Coope and Fletcher (1979). In this case the algorithms start to become more like the direct methods of Section 12.3 and Coope and Fletcher give an algorithm which incorporates the Lagrangian correction defined in (12.3.5).

12.3 The Lagrange–Newton (SOLVER) Method

A penalty function method is a somewhat indirect way of attempting to solve non-linear constraint problems. A more direct and efficient approach is to iterate on the basis of certain approximations to the problem functions $f(\mathbf{x})$ and $\mathbf{c}(\mathbf{x})$, in particular by using linear approximations to the constraint functions $\mathbf{c}(\mathbf{x})$. This has to be done with some care to ensure rapid convergence properties close to the solution, and one particular method is seen to be fundamental in this respect. This method is most simply explained as being Newton's method applied to find the stationary point of the Lagrangian function (9.1.6), and hence might be referred to as the *Lagrange–Newton* method. The Lagrangian function is defined in terms of variables $\mathbf{x}$ and $\boldsymbol{\lambda}$, so a feature of the resulting methods is that a sequence of approximations $\mathbf{x}^{(k)}$, $\boldsymbol{\lambda}^{(k)}$ to both the solution vector $\mathbf{x}^*$ and the vector of optimum Lagrange multipliers $\boldsymbol{\lambda}^*$ is generated.

In the first instance, consider applying the method to the equality constraint problem (12.1.1) and define $\boldsymbol{\nabla}$ as in (9.1.8) so that equation (9.1.7) is the stationary point (KT) condition at $\mathbf{x}^*$, $\boldsymbol{\lambda}^*$. As usual a Taylor series for $\boldsymbol{\nabla}\mathscr{L}$ about $\mathbf{x}^{(k)}$, $\boldsymbol{\lambda}^{(k)}$ gives

$$\boldsymbol{\nabla}\mathscr{L}(\mathbf{x}^{(k)}+\boldsymbol{\delta}\mathbf{x},\ \boldsymbol{\lambda}^{(k)}+\boldsymbol{\delta\lambda}) = \boldsymbol{\nabla}\mathscr{L}^{(k)} + [\boldsymbol{\nabla}^2\mathscr{L}^{(k)}]\begin{pmatrix}\boldsymbol{\delta}\mathbf{x}\\ \boldsymbol{\delta\lambda}\end{pmatrix} + \cdots \tag{12.3.1}$$

where $\boldsymbol{\nabla}\mathscr{L}^{(k)} = \boldsymbol{\nabla}\mathscr{L}(\mathbf{x}^{(k)}, \boldsymbol{\lambda}^{(k)})$, etc. Neglecting higher order terms and setting the left hand side to zero by virtue of (9.1.7) gives the iteration

$$[\boldsymbol{\nabla}^2\mathscr{L}^{(k)}]\begin{pmatrix}\boldsymbol{\delta}\mathbf{x}\\ \boldsymbol{\delta\lambda}\end{pmatrix} = -\boldsymbol{\nabla}\mathscr{L}^{(k)}. \tag{12.3.2}$$

This is solved to give corrections $\boldsymbol{\delta}\mathbf{x}$ and $\boldsymbol{\delta\lambda}$ and is of course Newton's method for the stationary point problem. Formulae for $\boldsymbol{\nabla}\mathscr{L}$ and $\boldsymbol{\nabla}^2\mathscr{L}$ are readily obtained from (9.1.6), giving the system

$$\begin{bmatrix}\mathbf{W}^{(k)} & -\mathbf{A}^{(k)}\\ -\mathbf{A}^{(k)\mathrm{T}} & \mathbf{0}\end{bmatrix}\begin{pmatrix}\boldsymbol{\delta}\mathbf{x}\\ \boldsymbol{\delta\lambda}\end{pmatrix} = \begin{pmatrix}-\mathbf{g}^{(k)} + \mathbf{A}^{(k)}\boldsymbol{\lambda}^{(k)}\\ \mathbf{c}^{(k)}\end{pmatrix}. \tag{12.3.3}$$

$\mathbf{A}^{(k)}$ is the Jacobian matrix of constraint normals evaluated at $\mathbf{x}^{(k)}$ and

$$\mathbf{W}^{(k)} = \boldsymbol{\nabla}^2 f(\mathbf{x}^{(k)}) - \sum_i \lambda_i^{(k)}\boldsymbol{\nabla}^2 c_i(\mathbf{x}^{(k)}) \tag{12.3.4}$$

is the Hessian matrix $\boldsymbol{\nabla}_x^2\mathscr{L}(\mathbf{x}^{(k)}, \boldsymbol{\lambda}^{(k)})$. In fact it is more convenient to write $\boldsymbol{\lambda}^{(k+1)} = \boldsymbol{\lambda}^{(k)} + \boldsymbol{\delta\lambda}$ and $\boldsymbol{\delta}^{(k)} = \boldsymbol{\delta}\mathbf{x}$, and to solve the equivalent system

$$\begin{bmatrix}\mathbf{W}^{(k)} & -\mathbf{A}^{(k)}\\ -\mathbf{A}^{(k)\mathrm{T}} & \mathbf{0}\end{bmatrix}\begin{pmatrix}\boldsymbol{\delta}\\ \boldsymbol{\lambda}\end{pmatrix} = \begin{pmatrix}-\mathbf{g}^{(k)}\\ \mathbf{c}^{(k)}\end{pmatrix} \tag{12.3.5}$$

to determine $\boldsymbol{\delta}^{(k)}$ and $\boldsymbol{\lambda}^{(k+1)}$. Then $\mathbf{x}^{(k+1)}$ is given by

$$\mathbf{x}^{(k+1)} = \mathbf{x}^{(k)} + \boldsymbol{\delta}^{(k)}. \tag{12.3.6}$$

The method requires initial approximations $\mathbf{x}^{(1)}$ and $\boldsymbol{\lambda}^{(1)}$, and uses (12.3.5) and (12.3.6) to generate the iterative sequence $\{\mathbf{x}^{(k)}, \boldsymbol{\lambda}^{(k)}\}$.

An alternative way of deriving this method is to observe that the second order sufficient conditions at $\mathbf{x}^*$, $\boldsymbol{\lambda}^*$ imply that $\mathbf{x}^*$ (that is $\boldsymbol{\delta} = \mathbf{0}$) solves the problem

$$\begin{aligned} &\underset{\boldsymbol{\delta}}{\text{minimize}}\ \mathscr{L}(\mathbf{x}^* + \boldsymbol{\delta}, \boldsymbol{\lambda}^*) \\ &\text{subject to } \mathbf{A}^{*\mathrm{T}}\boldsymbol{\delta} = \mathbf{0} \end{aligned} \tag{12.3.7}$$

because (9.3.5) and (9.3.4) ensure strictly positive curvature in the feasible region. Adding $0 = \boldsymbol{\lambda}^{*\mathrm{T}}\mathbf{A}^{*\mathrm{T}}\boldsymbol{\delta} = \mathbf{g}^{*\mathrm{T}}\boldsymbol{\delta}$ into the objective yields an equivalent problem

$$\begin{aligned} &\underset{\boldsymbol{\delta}}{\text{minimize}}\ \tfrac{1}{2}\boldsymbol{\delta}^{\mathrm{T}}\mathbf{W}^*\boldsymbol{\delta} + \mathbf{g}^{*\mathrm{T}}\boldsymbol{\delta} + f^* \\ &\text{subject to } \mathbf{A}^{*\mathrm{T}}\boldsymbol{\delta} = \mathbf{0} \end{aligned} \tag{12.3.8}$$

which is also solved by $\boldsymbol{\delta} = \mathbf{0}$, but which has $\boldsymbol{\lambda}^*$ as the Lagrange multiplier vector of the constraints. The constraints in (12.3.8) are linear approximations at $\mathbf{x}^*$ to $\mathbf{c}(\mathbf{x}) = \mathbf{0}$, so by analogy, if $\mathbf{x}^{(k)}$, $\boldsymbol{\lambda}^{(k)}$ are approximations to $\mathbf{x}^*$, $\boldsymbol{\lambda}^*$, then solution of the subproblem

$$\begin{aligned} &\underset{\boldsymbol{\delta}}{\text{minimize}}\ q^{(k)}(\boldsymbol{\delta}) \\ &\text{subject to } \boldsymbol{\ell}^{(k)}(\boldsymbol{\delta}) = \mathbf{0} \end{aligned} \tag{12.3.9}$$

is suggested, where

$$q^{(k)}(\boldsymbol{\delta}) \triangleq \tfrac{1}{2}\boldsymbol{\delta}^{\mathrm{T}}\mathbf{W}^{(k)}\boldsymbol{\delta} + \mathbf{g}^{(k)\mathrm{T}}\boldsymbol{\delta} + f^{(k)} \tag{12.3.10}$$

and

$$\boldsymbol{\ell}^{(k)}(\boldsymbol{\delta}) \triangleq \mathbf{A}^{(k)\mathrm{T}}\boldsymbol{\delta} + \mathbf{c}^{(k)}. \tag{12.3.11}$$

The second order sufficient conditions at $\mathbf{x}^*$, $\boldsymbol{\lambda}^*$ ensure that (12.3.9) has a well-determined solution when $\mathbf{x}^{(k)}$, $\boldsymbol{\lambda}^{(k)}$ are close to $\mathbf{x}^*$, $\boldsymbol{\lambda}^*$ (see Question 12.7). Thus the following iterative method is suggested, given initial estimates $\mathbf{x}^{(1)}$ and $\boldsymbol{\lambda}^{(1)}$.

For $k = 1, 2, \ldots$:
(i) Solve (12.3.9) (or (12.3.13)) to determine $\boldsymbol{\delta}^{(k)}$ and let $\boldsymbol{\lambda}^{(k+1)}$ be the Lagrange multipliers of the linear constraints. (12.3.12)
(ii) Set $\mathbf{x}^{(k+1)} = \mathbf{x}^{(k)} + \boldsymbol{\delta}^{(k)}$.

This method clearly indicates that for the inequality non-linear constraint problem (12.1.2) a suitable generalization is to solve the subproblem

$$\begin{aligned} &\underset{\boldsymbol{\delta}}{\text{minimize}}\ q^{(k)}(\boldsymbol{\delta}) \\ &\text{subject to } \boldsymbol{\ell}^{(k)}(\boldsymbol{\delta}) \geqslant \mathbf{0}. \end{aligned} \tag{12.3.13}$$

This is the basis of the SOLVER method (Wilson, 1963) as interpreted by Beale (1967) and it is convenient to refer to it by this name. Both (12.3.9) and (12.3.13) are *quadratic programming subproblems* and are readily solved by the methods of Chapter 10.

In fact equations (12.3.5) are the KT conditions implied by the solution of (12.3.9) so the two are clearly equivalent when a unique solution to (12.3.9) exists. However (12.3.9) is more fundamental as it requires the correct curvature conditions on $\mathbf{W}^{(k)}$ for a solution to exist, which (12.3.5) does not. The situation is analogous to that in which Newton's method (see Section 3.1) is best interpreted as minimizing certain quadratic approximations $q^{(k)}(\boldsymbol{\delta})$. The subproblem (12.3.9) obtained on the kth iteration can be interpreted as being derived from (12.1.1) by replacing the constraints $\mathbf{c}(\mathbf{x}) = \mathbf{0}$ by their linear Taylor series approximation $\boldsymbol{\ell}^{(k)}(\boldsymbol{\delta}) = \mathbf{0}$ at $\mathbf{x}^{(k)}$ given by (12.3.11) and by replacing the objective function $f(\mathbf{x})$ by the quadratic approximation $q^{(k)}(\boldsymbol{\delta})$ in (12.3.10). This is a quadratic Taylor series approximation at $\mathbf{x}^{(k)}$, but with the addition of constraint curvature terms in the Hessian. Including the second order constraint terms in the quadratic programming subproblem is important in that otherwise a second order rate of convergence for non-linear constraints would not be obtained. This is well illustrated by problem (12.1.5) in which the objective function is linear so that it is the curvature of the constraint which causes a solution to exist. In this case the sequence of quadratic programming subproblems only becomes well defined if the constraint curvature terms are included.

A numerical example which illustrates many of the features of the method is given by the inequality constraint problem (9.1.15) from $\mathbf{x}^{(1)} = (\frac{1}{2}, 1)^T$ and $\boldsymbol{\lambda}^{(1)} = \mathbf{0}$ (see Table 12.3.1). Since $\boldsymbol{\lambda}^{(1)} = \mathbf{0}$ no constraint curvature terms occur in $\mathbf{W}^{(1)}$, which is therefore the zero matrix since $f(\mathbf{x})$ is linear. Thus the initial subproblem is a linear programming problem and $\mathbf{x}^{(2)}$ is the vertex of the constraints linearized about $\mathbf{x}^{(1)}$. In fact, even though the constraint $c_1(\mathbf{x}) \geqslant 0$ is not active at the solution, the presence of its linearization is necessary to permit the first subproblem to be solvable. Moreover there exist different $\mathbf{x}^{(1)}$ for which the initial LP is unbounded, all of which illustrates that the well-behaved nature of (12.3.9) only holds necessarily in a neighbourhood of $\mathbf{x}^*$, $\boldsymbol{\lambda}^*$. In this case however the solution to the LP is well defined, and $\boldsymbol{\lambda}^{(2)} = (\frac{1}{3}, \frac{2}{3})^T$ is the multiplier vector at its solution, indicating that both linearized constraints are active. Thus for the second iteration

Table 12.3.1 The SOLVER method applied to (9.1.15)

k	$x_1^{(k)}$	$x_2^{(k)}$	$\lambda_1^{(k)}$	$\lambda_2^{(k)}$	$c_1^{(k)}$	$c_2^{(k)}$
1	$\frac{1}{2}$	1	0	0	$\frac{3}{4}$	$-\frac{1}{4}$
2	$\frac{11}{12}$	$\frac{2}{3}$	$\frac{1}{3}$	$\frac{2}{3}$	−0.173611	−0.284722
3	0.747120	0.686252	0	0.730415	0.128064	−0.029130
4	0.708762	0.706789	0	0.706737	0.204445	−0.001893
5	0.707107	0.707108	0	0.707105	0.207108	$-0.28_{10}{-5}$

$$\mathbf{W}^{(2)} = \mathbf{0} - \tfrac{1}{3}\begin{bmatrix} -2 & \\ & 0 \end{bmatrix} - \tfrac{2}{3}\begin{bmatrix} -2 & \\ & -2 \end{bmatrix} = \begin{bmatrix} 2 & \\ & \tfrac{4}{3} \end{bmatrix}$$

which is nicely positive definite. Solving the resulting QP problem causes $\ell_1^{(2)}(\boldsymbol{\delta}) \geqslant 0$ to become inactive so that $\lambda_1^{(3)} = 0$. Thus the correct active set is established, and the rapid convergence associated with Newton's method is observed on subsequent iterations. In comparing this with Table 12.2.1 say, it should be remembered that in the latter, each iteration requires an unconstrained minimization calculation and therefore many evaluations of the problem functions and their derivatives. In contrast, the SOLVER method only requires one evaluation of the problem function and derivatives to determine the coefficients of the finite quadratic programming subproblem. Thus the SOLVER method, if it works, is generally considerably superior in terms of the number of function and derivative evaluations which are required.

An important feature of the method which Table 12.3.1 illustrates is that ultimately the rate of convergence is second order. If second order sufficient conditions for the equality constraint problem (12.1.1) hold at $\mathbf{x}^*$, $\boldsymbol{\lambda}^*$, and if rank $\mathbf{A}^* = m$, then the Lagrangian matrix

$$\nabla^2\mathscr{L}^* = \begin{bmatrix} \mathbf{W}^* & -\mathbf{A}^* \\ -\mathbf{A}^{*\mathrm{T}} & \mathbf{0} \end{bmatrix} \tag{12.3.14}$$

is non-singular (see Question 12.4). The second order convergence of iteration (12.3.5) and (12.3.6) then follows by virtue of theorem 6.2.1 applied to the system of $n + m$ equations $\nabla\mathscr{L}(\mathbf{x}, \boldsymbol{\lambda}) = \mathbf{0}$. This requires both $\mathbf{x}^{(k)}$ and $\boldsymbol{\lambda}^{(k)}$ to be sufficiently close to $\mathbf{x}^*$ and $\boldsymbol{\lambda}^*$ for some k. In fact the multiplier estimates $\boldsymbol{\lambda}^{(k)}$ play a relatively minor role, in that they only arise in the second order term involving $\mathbf{W}^{(k)}$, and this can be exploited to give a stronger result.

Theorem 12.3.1

If $\mathbf{x}^{(1)}$ is sufficiently close to $\mathbf{x}^*$, if the Lagrangian matrix $\begin{bmatrix} \mathbf{W}^{(1)} & -\mathbf{A}^{(1)} \\ -\mathbf{A}^{(1)\mathrm{T}} & \mathbf{0} \end{bmatrix}$ is non-singular, and if second order sufficient conditions hold at $\mathbf{x}^*$, $\boldsymbol{\lambda}^*$ with rank $\mathbf{A}^* = m$, then the Lagrange–Newton iteration (12.3.5) and (12.3.6) converges and the rate is second order.

Proof

Define errors $\mathbf{h}^{(k)} = \mathbf{x}^{(k)} - \mathbf{x}^*$ and $\boldsymbol{\Delta}^{(k)} = \boldsymbol{\lambda}^{(k)} - \boldsymbol{\lambda}^*$, and assume that f and the c_i are $\mathbb{C}^2$ and the elements of their Hessian matrices satisfy Lipschitz conditions, so that the Taylor series about $\mathbf{x}^{(k)}$

$$\mathbf{c}^* = \mathbf{c}^{(k)} - \mathbf{A}^{(k)\mathrm{T}}\mathbf{h}^{(k)} + O(\|\mathbf{h}^{(k)}\|^2)$$

$$\mathbf{g}^* = \mathbf{g}^{(k)} - \nabla^2 f^{(k)}\mathbf{h}^{(k)} + O(\|\mathbf{h}^{(k)}\|^2)$$

$$\mathbf{a}_i^* = \mathbf{a}_i^{(k)} - \nabla^2 c_i^{(k)}\mathbf{h}^{(k)} + O(\|\mathbf{h}^{(k)}\|^2), \qquad i = 1, 2, \ldots, m,$$

are valid. It follows from (12.3.5) that $\mathbf{h}^{(k+1)}$, $\boldsymbol{\Delta}^{(k+1)}$ satisfy the equations

$$\begin{bmatrix} \mathbf{W}^{(k)} & -\mathbf{A}^{(k)} \\ -\mathbf{A}^{(k)\mathrm{T}} & \mathbf{0} \end{bmatrix} \begin{pmatrix} \mathbf{h}^{(k+1)} \\ \boldsymbol{\Delta}^{(k+1)} \end{pmatrix} = \begin{pmatrix} -\Sigma_i \Delta_i^{(k)} \nabla^2 c_i^{(k)} \mathbf{h}^{(k)} + O(\|\mathbf{h}^{(k)}\|^2) \\ O(\|\mathbf{h}^{(k)}\|^2) \end{pmatrix}$$

$$= \begin{pmatrix} O(\|\mathbf{h}^{(k)}\|^2) + O(\|\mathbf{h}^{(k)}\| \, \|\boldsymbol{\Delta}^{(k)}\|) \\ O(\|\mathbf{h}^{(k)}\|^2) \end{pmatrix}. \tag{12.3.15}$$

At $\mathbf{x}^*$, $\boldsymbol{\lambda}^*$ the Lagrangian matrix is non-singular (see Question 12.4) so for $\mathbf{x}^{(k)}$, $\boldsymbol{\lambda}^{(k)}$ in some neighbourhood of $\mathbf{x}^*$, $\boldsymbol{\lambda}^*$,

$$\begin{pmatrix} \mathbf{h}^{(k+1)} \\ \boldsymbol{\Delta}^{(k+1)} \end{pmatrix} = O(\|\mathbf{h}^{(k)}\|^2) + O(\|\mathbf{h}^{(k)}\| \, \|\boldsymbol{\Delta}^{(k)}\|)$$

and so there exists a constant $c > 0$ such that

$$\max(\|\mathbf{h}^{(k+1)}\|, \|\boldsymbol{\Delta}^{(k+1)}\|) \leqslant c\|\mathbf{h}^{(k)}\| \max(\|\mathbf{h}^{(k)}\|, \|\boldsymbol{\Delta}^{(k)}\|). \tag{12.3.16}$$

Thus, in a smaller neighbourhood, if $1 > c\max(\|\mathbf{h}^{(k)}\|, \|\boldsymbol{\Delta}^{(k)}\|) = \alpha$, say, then

$$\max(\|\mathbf{h}^{(k+1)}\|, \|\boldsymbol{\Delta}^{(k+1)}\|) \leqslant \alpha\|\mathbf{h}^{(k)}\| \leqslant \alpha \max(\|\mathbf{h}^{(k)}\|, \|\boldsymbol{\Delta}^{(k)}\|),$$

so the iteration converges and the rate is seen to be quadratic from (12.3.16). Now let only $\mathbf{x}^{(1)}$ be in a neighbourhood of $\mathbf{x}^*$, so that $\mathbf{A}^{(1)}$ has full rank, and let $\boldsymbol{\lambda}^{(1)}$ be such that the Lagrangian matrix is non-singular. Then $\|\boldsymbol{\Delta}^{(1)}\| \geqslant \|\mathbf{h}^{(1)}\|$ and so as above there exists a constant, d say, such that

$$\max(\|\mathbf{h}^{(2)}\|, \|\boldsymbol{\Delta}^{(2)}\|) \leqslant d\|\mathbf{h}^{(1)}\| \, \|\boldsymbol{\Delta}^{(1)}\|.$$

If $\mathbf{x}^{(1)}$ is sufficiently close to $\mathbf{x}^*$ in that $\|\mathbf{h}^{(1)}\| < 1/(cd\|\boldsymbol{\Delta}^{(1)}\|)$ then $\max(\|\mathbf{h}^{(2)}\|, \|\boldsymbol{\Delta}^{(2)}\|) < 1/c$ and so $\mathbf{x}^{(2)}$, $\boldsymbol{\lambda}^{(2)}$ is in the neighbourhood for which convergence occurs. □

Essentially therefore the minor role played by $\boldsymbol{\lambda}^{(k)}$ is illustrated by the fact that $\|\boldsymbol{\Delta}^{(k)}\|$ only occurs linearly on the right hand side of (12.3.15), and it is this fact that is exploited in the theorem. For example, if $\mathbf{x}^{(1)} = \mathbf{x}^*$, then $\mathbf{x}^{(2)} = \mathbf{x}^*$ and $\boldsymbol{\lambda}^{(2)} = \boldsymbol{\lambda}^*$ irrespective of errors in $\boldsymbol{\lambda}^{(1)}$. This suggests that when using the method, it is more important to have $\mathbf{x}^{(1)}$ accurate than $\boldsymbol{\lambda}^{(1)}$. This is in contrast to the multiplier penalty function method of Section 12.2 where an inaccurate value of $\boldsymbol{\lambda}^{(1)}$ definitely limits the extent to which $\mathbf{x}(\boldsymbol{\lambda}^{(1)})$ agrees with $\mathbf{x}^*$, assuming $\boldsymbol{\sigma}$ is fixed. The result of theorem 12.3.1 however concerns (12.3.5) and not (12.3.9), and if large errors in $\boldsymbol{\lambda}^{(1)}$ are made then the curvature of $\mathbf{W}^{(k)}$ can be such that no solution exists to (12.3.9). Thus some care must be taken in interpreting the theorem. There is no difficulty in extending the theorem to handle inequality constraints and the details are sketched out in Question 12.18.

These results show that the local properties of the SOLVER method are very satisfactory, so that the main difficulty which exists is the fact that the iteration may fail to converge remote from the solution, and that the solution to the subproblem (12.3.9) or (12.3.13) may not even exist (it may either be unbounded or infeasible). To induce global convergence requires some measure of goodness of $\mathbf{x}^{(k)}$ and $\boldsymbol{\lambda}^{(k)}$ to be available which is minimized locally at the solution. One way to do this is to define the error in the KT conditions

$$\phi(\mathbf{x}, \boldsymbol{\lambda}) = \| \mathbf{c} \|_2^2 + \| \mathbf{g} - \mathbf{A}\boldsymbol{\lambda} \|_2^2$$

where $\mathbf{c} = \mathbf{c}(\mathbf{x})$, etc., but this is not entirely satisfactory in that it does not give any bias to minimizing the objective function and can therefore cause convergence to any stationary point to occur. A more suitable possibility is to use an *exact penalty function*, that is any function $\phi(\mathbf{x})$, defined in terms of $f(\mathbf{x})$, $\mathbf{c}(\mathbf{x})$, and possibly derivatives thereof, which is minimized locally by $\mathbf{x}^*$. An important exact penalty function which can be used is the non-differentiable function $\phi = \nu f + \| \mathbf{c} \|_1$ which is described in more detail in Section 14.3. A smooth exact penalty function is given by the augmented Lagrangian function $\phi(\mathbf{x}, \boldsymbol{\lambda}^*, \boldsymbol{\sigma})$ in (12.2.3) (see theorem 12.2.1) but this requires a knowledge of $\boldsymbol{\lambda}^*$. Yet another exact penalty function is described briefly in Section 12.5. One possibility for stabilizing the SOLVER iteration therefore is to use the resulting correction $\boldsymbol{\delta}$ to define a search direction, along which some exact penalty function is minimized. This possibility is explored later in the section. However it is important to realize that this idea alone can still fail, just as a line search can fail to induce convergence in the Newton–Raphson method for solving equations, which is a special case of the SOLVER method (Fletcher, 1980b).

The most successful way of inducing global convergence for unconstrained minimization is arguably the restricted step or trust region approach or the use of a Levenberg–Marquardt parameter (see Volume 1, Chapter 5). Both these ideas have been suggested in the context of SOLVER-like methods (in fact a step restriction does occur in Beale's (1967) description of the SOLVER method). Modifying $\mathbf{W}^{(k)}$ by adding a Marquardt–Levenberg term $\nu\mathbf{I}$ may help to ensure that (12.3.9) is not unbounded, but no longer has the effect of giving a bias towards steepest descent. In fact for a non-linear equations problem ($m = n$) the correction $\boldsymbol{\delta}$ is unchanged, so the modification has no effect and therefore does not provide a guarantee of convergence. If a step length restriction $\| \boldsymbol{\delta} \| \leqslant h^{(k)}$ is added to (12.3.9) or (12.3.13) then the possibility of an unbounded correction is entirely removed. In this case the difficulty is that if $\mathbf{x}^{(k)}$ is infeasible and $h^{(k)}$ is sufficiently small then the resulting subproblem has no feasible point. Thus the idea in Section 5.1 of forcing convergence by reducing $h^{(k)}$ if necessary so as to give a descent step, is no longer valid. Therefore the straightforward generalization of the SOLVER method in this way is unsatisfactory. However it is possible to incorporate the restricted step approach with a method which makes an approximation to the exact L_1 penalty function in such a way as to guarantee global convergence and yet retain many of the features of the SOLVER method. This approach is described in Sections 14.3 and 14.5 and is very promising, although the results regarding rate of convergence are somewhat weakened (see again Section 14.5).

Another practical disadvantage of the SOLVER method is the need to compute second derivatives. Variations of the method have been suggested in which updating formulae, analogous to those in quasi-Newton methods, are used to revise a matrix

$\mathbf{B}^{(k)}$ which approximates $\mathbf{W}^{(k)}$. Han (1976) suggests using the DFP formula (see Volume 1, Section 3.2) but with $\boldsymbol{\gamma}^{(k)}$ being defined by

$$\boldsymbol{\gamma}^{(k)} = \nabla\mathscr{L}(\mathbf{x}^{(k+1)}, \boldsymbol{\lambda}^{(k+1)}) - \nabla\mathscr{L}(\mathbf{x}^{(k)}, \boldsymbol{\lambda}^{(k+1)}) \qquad (12.3.17)$$

and shows that the resulting algorithm is superlinearly convergent. Powell (1978a) prefers to keep the matrix $\mathbf{B}^{(k)}$ positive definite so that the solution to the subproblem is always well defined. He does this by defining the vector

$$\boldsymbol{\eta}^{(k)} = \theta\boldsymbol{\gamma}^{(k)} + (1-\theta)\mathbf{B}^{(k)}\boldsymbol{\delta}^{(k)}, \qquad 0 \leqslant \theta \leqslant 1 \qquad (12.3.18)$$

($\boldsymbol{\gamma}^{(k)}$ as in (12.3.17)) that is closest to $\boldsymbol{\gamma}^{(k)}$ subject to the condition

$$\boldsymbol{\delta}^{(k)\mathrm{T}}\boldsymbol{\eta}^{(k)} \geqslant 0.2\,\boldsymbol{\delta}^{(k)\mathrm{T}}\mathbf{B}^{(k)}\boldsymbol{\delta}^{(k)}.$$

$\boldsymbol{\eta}^{(k)}$ is used in place of $\boldsymbol{\gamma}^{(k)}$ in the updating formula and Powell prefers the BFGS formula on account of its success in solving unconstrained minimization problems. It might be thought that this device is somewhat artificial since the matrix $\mathbf{W}^*$ may not be positive definite, so that $\mathbf{B}^{(k)}$ may never be a close approximation to $\mathbf{W}^*$. However the projections of $\mathbf{B}^{(k)}$ and $\mathbf{W}^{(k)}$ into the tangent hyperplanes of the active constraints are likely to be close, and this is the important part of $\mathbf{B}^{(k)}$ insofar as the solution of the subproblem is concerned. Powell (1978b) is able to exploit this observation to prove superlinear convergence even when $\mathbf{W}^*$ is indefinite. Han (1977) also suggests a line search based on using the L_1 exact penalty function. He gives a global convergence result assuming firstly that the matrices $\mathbf{B}^{(k)}$ are positive definite and that $\mathbf{B}^{(k)}$ and $\mathbf{B}^{(k)^{-1}}$ are bounded, and secondly that the multipliers $\boldsymbol{\lambda}^{(k)}$ are bounded. He also makes another assumption which excludes an inconsistent linearization which can occur when the vectors $\mathbf{a}_i^\infty$ $i \in \mathscr{A}^\infty$ are dependent. Using assumptions like these is unsatisfactory and should not obscure the need to study more certain ways of inducing global convergence. Powell (1978a) also uses a line search using an L_1 exact penalty function but for computational reasons changes the penalty parameters μ_i (see (14.3.17)). A strong global convergence result for Powell's algorithm also does not hold, and failure has been observed (Chamberlain, 1979). However good practical results have been observed with these algorithms on a number of standard test problems, which substantiates to some extent the claim that they provide a robust modification of the SOLVER method. Table 12.3.2 compares the number of function and gradient evaluations required to solve Colville's (1968) first three problems and gives some idea of the relative improvement that is obtained. Even allowing for the fact that Powell's method solves a QP problem on each iteration, the improvement is still substantial. However Powell (1977) reports some difficulty in solving the Dembo 7 test problem (Dembo, 1976) (as also do Coope and Fletcher (1979) with an augmented Lagrangian method), whilst it has been possible to solve this problem with an algorithm of the type described in Section 14.5 (Fletcher, 1980b). Fletcher also gives two other problems which are solved by the latter but not by a SOLVER-like method and this is an indication of the more robust nature of the type of algorithm given in Section 14.5.

Table 12.3.2 Comparison of nonlinear programming techniques

Problem	Extrapolated barrier function	Multiplier penalty function	Powell (1978a)
TP1	177	47	6
TP2	245	172	17
TP3	123	73	3

12.4 Non-linear Elimination and Feasible Direction Methods

An apparently attractive approach to the nonlinear programming problem is to try to produce direct methods which generalize the ideas in Chapters 10 and 11. These methods attempt to maintain feasibility by searching from one feasible point to another along feasible arcs (lines in Chapters 10 and 11) and hence are referred to as *feasible direction methods.* They date back at least as far as Zoutendijk (1960) and Rosen (1960, 1961). Inequality constraints are handled by an active constraint strategy as in Sections 10.3 and 11.2 but attention is initially directed towards methods suitable for solving the equality problem (12.1.1). The presence of non-linear constraints does not allow a line search to maintain feasibility so a simplistic explanation of what is done is that at any feasible point $\mathbf{x}^{(k)}$ a search direction $\mathbf{s}^{(k)}$ in the tangent plane is calculated and a feasible arc is then obtained by projecting any point on the resulting line into a corresponding point in the feasible region (see Figure 12.4.1). A search along the resulting feasible arc is then made to reduce $f(\mathbf{x})$. Many methods of this type have been suggested. Another possible approach to nonlinear programming is to use variables to eliminate constraints, if necessary by solving a non-linear system of equations at each iteration. It is shown below that this is a special case of a *non-linear generalized elimination method.* This turns out to be equivalent to the idea of a feasible direction method and a number of common methods are shown to be special cases of elimination. The projection step in the feasible direction method is then equivalent to the solution of the non-linear system of equations in the elimination method.

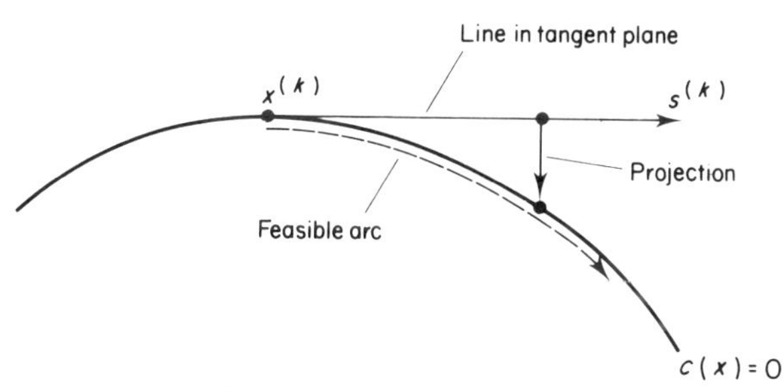

Figure 12.4.1 Feasible direction search

It is shown in Section 10.1 that many methods for linear constraints are generalized elimination methods and correspond to different ways of choosing $\mathbf{V}$ in (10.1.20). Another way of regarding this construction is that a linear transformation to new variables $\mathbf{y}$ is made, defined by

$$\mathbf{y} = \begin{pmatrix} \mathbf{y}_1 \\ \mathbf{y}_2 \end{pmatrix} = [\mathbf{A} : \mathbf{V}]^{\mathrm{T}}\mathbf{x} - \begin{pmatrix} \mathbf{b} \\ \mathbf{0} \end{pmatrix}$$

where the partitions are $\mathbf{y}_1 \in \mathbb{R}^m$ and $\mathbf{y}_2 \in \mathbb{R}^{n-m}$. Since $[\mathbf{A} : \mathbf{V}]$ is non-singular the transformation is one to one. Then the method is derived by keeping $\mathbf{y}_1 = \mathbf{0}$, thus satisfying the constraints, and minimizing with respect to the remaining variables $\mathbf{y}_2$ (corresponding to $\mathbf{y}$ in (10.1.10)). A similar approach is possible when solving non-linear constraint problems. It is assumed that $\mathbf{x}^{(k)}$ is a feasible point at which $\mathbf{A}^{(k)}$ has full rank. It is then appropriate to consider an analogous non-linear transformation $\mathbf{x} \leftrightarrow \mathbf{y}$ defined by

$$\mathbf{y} = \begin{pmatrix} \mathbf{y}_1 \\ \mathbf{y}_2 \end{pmatrix} = \begin{pmatrix} \mathbf{c}(\mathbf{x}^{(k)} + \boldsymbol{\delta}) \\ \mathbf{V}^{\mathrm{T}}\boldsymbol{\delta} \end{pmatrix}. \tag{12.4.1}$$

$\mathbf{V}$ is such that $[\mathbf{A}^{(k)} : \mathbf{V}]$ is non-singular in which case the transformation is one-to-one in some neighbourhood of $\mathbf{x}^{(k)}$; in fact as (12.4.1) indicates, new $\mathbf{y}$ variables are defined on each iteration to help keep the transformation well defined. If $\mathbf{y}_2$ is any value of the free variables, the corresponding $\boldsymbol{\delta}$ is required which solves (12.4.1) with $\mathbf{y}_1 = \mathbf{0}$. This can be calculated by the Newton–Raphson iteration (Section 6.2) in an inner (r) iteration. The Jacobian matrix of the transformation (12.4.1) is $[\mathbf{A} : \mathbf{V}]$ and a suitable initial approximation is

$$\boldsymbol{\delta}^{(1)} = \mathbf{Z}\mathbf{y}_2. \tag{12.4.2}$$

The sequence of iterates is defined for $r \geqslant 1$ by

$$\begin{aligned} \boldsymbol{\delta}^{(r+1)} &= \boldsymbol{\delta}^{(r)} - [\mathbf{A} : \mathbf{V}]^{-\mathrm{T}}\begin{pmatrix} \mathbf{c}^{(r)} \\ \mathbf{0} \end{pmatrix} \\ &= \boldsymbol{\delta}^{(r)} - \mathbf{S}\mathbf{c}^{(r)} \end{aligned} \tag{12.4.3}$$

where $\mathbf{A}$ (and hence $\mathbf{S}$) and $\mathbf{c}^{(r)}$ are evaluated at $\mathbf{x}^{(k)} + \boldsymbol{\delta}^{(r)}$. When the r-iteration is deemed to have converged, $\boldsymbol{\delta}^{(r+1)}$ becomes the required vector $\boldsymbol{\delta}$. Both $\boldsymbol{\delta}$ and $\mathbf{x}$ $(= \mathbf{x}^{(k)} + \boldsymbol{\delta})$ can be regarded as functions of the free variables $\mathbf{y}_2$. In practice it is more efficient in (12.4.3) to evaluate $\mathbf{A}$ and hence $\mathbf{S}$ at $\mathbf{x}^{(k)}$, although the r-iteration no longer has a second order rate of convergence. Convergence of the r-iteration can be proved if (12.4.2) is a sufficiently good initial approximation which is true for small enough $\mathbf{y}_2$. In terms of Figure 12.4.1, $\boldsymbol{\delta}^{(1)}$ in (12.4.2) can be regarded as a step in the tangent plane and the iteration (12.4.3) is the projection of $\mathbf{x}^{(k)} + \boldsymbol{\delta}^{(1)}$ so that $\mathbf{x}^{(k)} + \boldsymbol{\delta}$ satisfies the non-linear constraints. The direction of the projection is seen from (12.4.3) to lie in the column space of $\mathbf{S}$ and will therefore differ depending on how $\mathbf{S}$ is defined in (10.1.20).

The idea can be modified to give a search method by defining $\mathbf{p}^{(k)} \in \mathbb{R}^{n-m}$ as a search direction in the space of the free variables and taking $\mathbf{y}_2 = \alpha\mathbf{p}^{(k)}$ as a step in this direction. Then $\boldsymbol{\delta}$ $(= \boldsymbol{\delta}(\alpha))$ is calculated from (12.4.2) and (12.4.3) and defines a feasible arc $\mathbf{x}(\alpha) = \mathbf{x}^{(k)} + \boldsymbol{\delta}(\alpha)$. The direction of the arc at $\mathbf{x}^{(k)}$ is the feasible direction $\mathbf{s}^{(k)} = \mathbf{Z}\mathbf{p}^{(k)}$ (see Figure 12.4.1). It is then possible to choose $\alpha^{(k)}$ to optimize $f(\mathbf{x}^{(k)} + \boldsymbol{\delta})$ by a line search along this feasible arc, assuming the arc exists and can be calculated for sufficiently large α, which may not always be the case.

One method of this type is the *GRG method* of Abadie and Carpentier (1969) which is equivalent to the direct elimination of variables. Thus $\mathbf{x}$ is partitioned into $\mathbf{x}_1 \in \mathbb{R}^m$ and $\mathbf{x}_2 \in \mathbb{R}^{n-m}$ and the variables $\mathbf{x}_2$ become the free variables $\mathbf{y}_2$. The matrix $\mathbf{S}$ is $[\mathbf{A}_1^{-1} : \mathbf{0}]^{\mathrm{T}}$ (see (10.1.22)) and $\mathbf{Z}$ is $[-\mathbf{A}_2\mathbf{A}_1^{-1} : \mathbf{I}]^{\mathrm{T}}$. How $\mathbf{p}^{(k)}$ is chosen is described below; otherwise the method follows the above scheme. One way of writing the projection iteration (12.4.2) and (12.4.3) is to have $\{\mathbf{x}^{(k+1,r)}\}$ as a sequence of estimates of $\mathbf{x}^{(k+1)}$ defined by

$$\mathbf{x}^{(k+1,1)} = \mathbf{x}^{(k)} + \alpha^{(k)}\mathbf{Z}\mathbf{p}^{(k)}$$

$$\mathbf{x}_1^{(k+1,r+1)} = \mathbf{x}_1^{(k+1,r)} - \mathbf{A}_1^{-\mathrm{T}}\mathbf{c}^{(k+1,r)}, \qquad r = 1, 2, \ldots, \tag{12.4.4}$$

whilst $\mathbf{x}_2^{(k+1)} = \mathbf{x}_2^{(k+1,1)}$ stays fixed, which shows essentially that the $\mathbf{x}_1$ variables are being chosen so as to eliminate the constraints. Another feasible direction method is the *gradient projection method* of Rosen (1961) in which the feasible direction $\mathbf{s}^{(k)}$ is the projection of the negative gradient vector into the tangent plane. Rosen achieves this by representing the projection matrix directly, but a more stable approach is to use the orthogonal projection method of Section 10.1 to define $\mathbf{S}$ and $\mathbf{Z}$, and then to proceed as described in this section. A feasible direction method which generalizes the Wolfe (1963a) reduced gradient method is also possible in a similar way.

The next step is to consider how the direction $\mathbf{s}^{(k)}$ $(= \mathbf{Z}\mathbf{p}^{(k)})$ of the feasible arc might be calculated. The objective function can be regarded as defining a function $f_y(\mathbf{y}) = f_x(\mathbf{x})$ where $\mathbf{x}$ and $\mathbf{y}$ are related by (12.4.1). As in Volume 1, Section 3.3, the chain rule shows that first derivatives are related by

$$\mathbf{g}_x = [\mathbf{A} : \mathbf{V}]\begin{pmatrix}\mathbf{g}_{y_1}\\ \mathbf{g}_{y_2}\end{pmatrix} \tag{12.4.5}$$

so that using (10.1.20), $\mathbf{g}_{y_2} = \mathbf{Z}^{\mathrm{T}}\mathbf{g}_x$ can again be interpreted as a reduced gradient vector. Thus early methods (Abadie and Carpentier, 1969; Rosen, 1961) choose $\mathbf{y}_2 = \alpha\mathbf{p}^{(k)}$ where $\mathbf{p}^{(k)} = -\mathbf{Z}^{\mathrm{T}}\mathbf{g}^{(k)}$ is the reduced steepest descent direction, and the direction of the feasible arc is $\mathbf{s}^{(k)} = -\mathbf{Z}\mathbf{Z}^{\mathrm{T}}\mathbf{g}^{(k)}$. It is possible to improve these methods with little complication by using conjugate gradients (see Section 4.1) and the resulting methods have been used effectively on some problems. Nonetheless ideas of curvature do not figure strongly in these methods and better algorithms can be expected. The analogue of Newton's method (Section 3.1) requires second derivatives of f with respect to $\mathbf{x}$ or $\mathbf{y}$. Differentiating in (12.4.5) yields

$$\mathbf{G}_x = \sum_{i=1}^{m} \nabla^2 c_i\, \partial f/\partial y_i + [\mathbf{A} : \mathbf{V}]\mathbf{G}_y[\mathbf{A} : \mathbf{V}]^{\mathrm{T}} \tag{12.4.6}$$

so that the reduced Hessian at $\mathbf{x}^{(k)}$ (that is $\mathbf{y}_2 = \mathbf{0}$) is defined by

$$\nabla_{y_2}^2 f_y(\mathbf{0}) = \mathbf{Z}^{\mathrm{T}}\mathbf{W}^{(k)}\mathbf{Z} \tag{12.4.7}$$

where $\mathbf{W}^{(k)} = \mathbf{G}^{(k)} - \Sigma \lambda_i^{(k)} \nabla^2 c_i^{(k)}$ with $\boldsymbol{\lambda}^{(k)} = \mathbf{S}^T \mathbf{g}^{(k)}$. Thus the Hessian of the Lagrangian function duly occurs along with what can be regarded a first order estimate of Lagrange multipliers (see (11.1.18)). The basic Newton's method therefore defines a correction $\mathbf{y}_2^{(k)}$ by solving the system

$$[\mathbf{Z}^T \mathbf{W}^{(k)} \mathbf{Z}] \mathbf{y}_2 = -\mathbf{Z}^T \mathbf{g}^{(k)} \tag{12.4.8}$$

which is then used with (12.4.2) and (12.4.3). Use of this formula to define a search direction, or its use in a modified Newton method (see Volume 1, Section 3.1) are all possible extensions of this technique. It should be noted that the matrices $\mathbf{Z}$ and $\mathbf{S}$ which arise above should strictly be written as $\mathbf{Z}^{(k)}$ and $\mathbf{S}^{(k)}$ since they are determined by $\mathbf{A}^{(k)}$, which is no longer constant as it is in Chapter 10.

An illustration of this Newton method to solve problem (12.1.5) from the feasible point $\mathbf{x}^{(1)} = (0.8, 0.6)^T$ is given in Table 12.4.1. Orthogonal factorizations $\mathbf{Z}^T\mathbf{A} = \mathbf{0}$, $\mathbf{Z}^T\mathbf{Z} = \mathbf{I}$, and $\mathbf{S} = \mathbf{A}^+$ are used in the method. The tabulated values are related to (12.4.3) by $\mathbf{x}^{(k+1,r)} = \mathbf{x}^{(k)} + \boldsymbol{\delta}^{(r)}$. The convergence of the inner iteration (12.4.3) at a very rapid linear rate can be seen, although it is not quadratic since $\mathbf{S}$ is evaluated at $\mathbf{x}^{(k)}$ and not $\mathbf{x}^{(k+1,r)}$. The rapid convergence of the outer iteration which uses (12.4.8) can also be observed. An extension to quasi-Newton methods is also possible, in which the matrix $\mathbf{Z}^{(k)T}\mathbf{W}^{(k)}\mathbf{Z}^{(k)}$ is updated using the BFGS formula say, involving differences in reduced gradient vectors.

$$\gamma^{(k)} = \mathbf{Z}^{(k+1)T}\mathbf{g}^{(k+1)} - \mathbf{Z}^{(k)T}\mathbf{g}^{(k)} \tag{12.4.9}$$

and differences in reduced coordinates $\boldsymbol{\delta}^{(k)} = \mathbf{y}_2^{(k)} - \mathbf{0} = \mathbf{y}_2^{(k)}$.

Another second order feasible direction method is summarized by Sargent (1974). This is based on using a non-linear version of (12.3.5) in which the constraint linearization $\mathbf{c}^{(k)} + \mathbf{A}^{(k)T}\boldsymbol{\delta} = \mathbf{0}$ is replaced by $\mathbf{c}(\mathbf{x}^{(k)} + \boldsymbol{\delta}) = \mathbf{0}$ as in (12.4.1). To allow the possibility of changing the length of the correction, a parameter α is introduced, giving the non-linear transformation

$$\begin{aligned} \mathbf{W}^{(k)}\boldsymbol{\delta} &= \alpha(\mathbf{A}^{(k)}\boldsymbol{\lambda} - \mathbf{g}^{(k)}) \\ \mathbf{c}(\mathbf{x}^{(k)} + \boldsymbol{\delta}) &= \mathbf{0} \end{aligned} \tag{12.4.10}$$

Table 12.4.1 Newton's method for nonlinear programming

k	r	$x_1^{(k,r)}$	$x_2^{(k,r)}$	$\lambda^{(k)}$	$y_2^{(k)}$	$c^{(r)}$
1	–	0.8	0.6	0.7	−0.142857	
2	1	0.714286	0.714286			−0.020408
	2	0.706122	0.708163			−0.000104
	3	0.706081	0.708132			−0.000001
	4	0.706080	0.708132			$-1_{10^{-8}}$
2	–			0.707106	0.001451	
3	1	0.707108	0.707108			−0.000002
	2	0.707107	0.707107			$-4_{10^{-13}}$
3	–			0.707107	0	

which is solved to give $\boldsymbol{\delta}^{(k)}$ and $\boldsymbol{\lambda}^{(k+1)}$. This system can also be solved in an inner (r) iteration by the Newton–Raphson method in the form

$$\begin{aligned} \mathbf{W}^{(k)}\boldsymbol{\delta}^{(k,r+1)} &= \alpha^{(k)}(\mathbf{A}^{(k)}\boldsymbol{\lambda}^{(k+1,r+1)} - \mathbf{g}^{(k)}) \\ \mathbf{c}(\mathbf{x}^{(k)} + \boldsymbol{\delta}^{(k,r)}) &+ \mathbf{A}^{\mathrm{T}}(\boldsymbol{\delta}^{(k,r+1)} - \boldsymbol{\delta}^{(k,r)}) = \mathbf{0}. \end{aligned} \tag{12.4.11}$$

Setting $\mathbf{A} = \mathbf{A}^{(k)}$ at the expense of the second order rate of convergence enables the linear system (12.4.11) to be solved for $\boldsymbol{\delta}^{(k,r+1)}$ and $\boldsymbol{\lambda}^{(k+1,r+1)}$ by any of the usual techniques described in Sections 10.1 and 10.2. Initial values $\boldsymbol{\delta}^{(k,1)}$ and $\boldsymbol{\lambda}^{(k+1,1)}$ for the r-iteration are obtained by solving the linearization of (12.4.10) in a similar way. When the r-iteration is deemed to have converged, $\mathbf{x}^{(k+1)}$ is set to $\mathbf{x}^{(k)} + \boldsymbol{\delta}^{(k,r+1)}$ and $\boldsymbol{\lambda}^{(k+1)}$ to $\boldsymbol{\lambda}^{(k+1,r+1)}$. To make the algorithm more robust, a line search is included, along the arc defined by solving (12.4.10) for different values of α, in order to reduce $f(\mathbf{x}^{(k)} + \boldsymbol{\delta})$ sufficiently. To ensure convergence, some sort of bias towards the (reduced) steepest descent direction must be included. This can be done for example by adding a Levenberg–Marquardt term $\nu\mathbf{I}$ to $\mathbf{W}^{(k)}$ Alternatively, a quasi-Newton approach is possible, and use of Powell's modification of the BFGS formula (see (12.3.18)) would be a good way of ensuring that $\mathbf{W}^{(k)}$ stays positive definite.

Convergence proofs for feasible direction methods have been given, but there are some hidden pitfalls, as Sargent (1974) points out. He considers the possibility of adding a penalty term, but then the distinction between direct methods and penalty function methods using low accuracy minimization becomes indistinct. As I see it, the main difficulty of feasible direction methods is the requirement to converge the inner r-iteration, because the basic Newton–Raphson method is not guaranteed to converge, and more elaborate alternatives would unduly complicate the overall method. Although the basic Newton–Raphson method may converge for small enough α, this may place an undue restriction on the length of step. Also there is a correlation between the convergence of the Newton–Raphson method and the rank of the Jacobian matrix, and it is not attractive to have to make implicit assumptions about the latter.

Additional complications arise when attention is switched from equality problems to consider the inequality constraint problem (12.1.2). The most obvious approach is to use an active set strategy as in Section 11.2. The general idea of keeping $c_i(\mathbf{x}) = 0$, $i \in \mathscr{A}$, and of systematically adding and dropping constraints remains the same, and estimates of Lagrange multipliers and precautions against zigzagging need also to be made in a similar way. There is however an additional difficulty which arises when the correction $\mathbf{Z}\mathbf{y}_2$ in (12.3.2) causes $\mathbf{x}^{(k+1,1)}$ to violate constraints not in the active set. Ideally the analogue of the linear constraint algorithm (11.2.2), step (e), would be to iterate the search along the feasible arc with respect to changes in α in an attempt to find the largest value of $\alpha = \bar{\alpha}^{(k)}$ which gives a feasible point. The constraint which is then limiting an increase in α can then be added into the active set, if appropriate. This process was originally used by Rosen (1961) but it adds yet another level of iteration into the method, which can therefore become very inefficient. Furthermore it is often possible to judge directly which constraint will become active, and this constraint can be zeroed with the other active constraints by extending the dimension of the r-iteration (12.4.3). However there are then difficulties in writing down a strategy which covers all possibilities. For example $\mathbf{x}^{(k+1,1)}$ may violate so many constraints that it is not clear which ones to include in the r-iteration. Also constraints violated by $\mathbf{x}^{(k+1,1)}$ might not be active at all on the feasible arc. It is possible to introduce additional ad hoc rules to cover these cases but the resulting algorithm becomes cumbersome and unappealing.

In some ways it might be better to disregard the active set strategy and think in terms of a more complicated non-linear transformation to replace (12.4.1) or (12.4.10). Sargent (1974) suggests

$$\begin{aligned} \mathbf{W}^{(k)}\boldsymbol{\delta} &= \alpha(\mathbf{A}^{(k)}\boldsymbol{\lambda} - \mathbf{g}^{(k)}) \\ \mathbf{c}(\mathbf{x}^{(k)} + \boldsymbol{\delta}) &\geqslant 0, \qquad \boldsymbol{\lambda} \geqslant \mathbf{0} \\ \boldsymbol{\lambda}^{\mathrm{T}}\mathbf{c}(\mathbf{x}^{(k)} + \boldsymbol{\delta}) &= 0 \end{aligned}$$

as the appropriate analogue of (12.4.10) for an inequality constraint problem. This can be solved by an inner iteration involving a sequence of quadratic programming subproblems. This is a very elegant solution to the difficulties caused by inequality constraints, although it does not help in solving the fundamental difficulty which is to force convergence of the inner iteration.

12.5 Other Methods

Many other methods have been suggested for nonlinear programming, and a few of these are reviewed in this section. Although these methods are perhaps currently not as popular as those described previously, they often exhibit considerable ingenuity and contain interesting ideas which are worthy of note. An interesting penalty function for the equality problem (12.1.1) arises from the observation that the function

$$\phi(\mathbf{x}, f) = \left\| \begin{matrix} f(\mathbf{x}) - f \\ c_1(\mathbf{x}) \\ c_2(\mathbf{x}) \\ \vdots \\ c_m(\mathbf{x}) \end{matrix} \right\|_p \tag{12.5.1}$$

is minimized by $\mathbf{x}^*$ if the control parameter $f = f^*$. Thus a sequential penalty function method can be envisaged in which a sequence of estimates $\{f^{(k)}\} \to f^*$ is generated so that the minimizers $\mathbf{x}(f^{(k)}) \to \mathbf{x}^*$. Morrison (1968) suggests a method with $p = 2$ in which the sum of squares function $[\phi(\mathbf{x}, f)]^2$ is minimized sequentially, but there is a potential difficulty that if $m < n$ then $\nabla^2\phi(\mathbf{x}^*, f)$ tends to a singular matrix as $f \to f^*$. This can slow down the rate of convergence of conventional algorithms. Gill and Murray (1976a) introduce two new features, one of which is a special purpose method for solving ill-posed least squares problems. They also give a new formula for updating the control parameter $f^{(k)}$ which has a second order rate of convergence and controls cancellation errors.

Another penalty function which relates the ideas in (12.5.1) to those contained in multiplier penalty functions (Section 12.2) and L_∞ exact penalty functions (Section 14.3) is given by Bandler and Charalambous (1972). The function

$$\begin{aligned} \phi(\mathbf{x}, p, \boldsymbol{\alpha}, f) &= (\sum_i f_i^p)^{1/p}, \qquad 0 \leqslant i \leqslant m \\ f_0 &= f(\mathbf{x}) - f \\ f_i &= f(\mathbf{x}) - f - \alpha_i c_i(\mathbf{x}), \qquad i \geqslant 1 \end{aligned} \tag{12.5.2}$$

is defined, where $f < f^*$ and $f_i > 0 \;\forall\, i \geqslant 0$. There are many ways to force the minimizer $\mathbf{x}(p, \boldsymbol{\alpha}, f)$ of this function to converge to $\mathbf{x}^*$. In a neighbourhood of $\mathbf{x}^*$, and in the limit $p \to \infty$, the function becomes equivalent to the exact L_∞ penalty function. Thus one possible mode of application is a Polya-type algorithm in which f and $\boldsymbol{\alpha}$ are fixed, and a control sequence $\{p^{(k)}\} \to \infty$ is used. Another possibility is to fix p and $\boldsymbol{\alpha}$ and to choose a sequence of control parameters $f^{(k)} \to f^*$ as in (12.5.1). Finally p and f can remain fixed, in which case a sequence $\boldsymbol{\alpha}^{(k)} \to \boldsymbol{\lambda}^*/(m+1)$ of scaled Lagrange multiplier estimates can be used as control parameters. Charalambous (1977) shows how a suitable updating formula for the $\boldsymbol{\alpha}^{(k)}$ parameters can be determined.

Another attractive idea is to attempt to define an *exact penalty function* $\phi(\mathbf{x})$ which is minimized locally by the solution $\mathbf{x}^*$ to (12.1.1) or (12.1.2). A simple way to do this is described in Section 14.3 and gives rise to a non-differentiable function. This requires special purpose techniques to solve the unconstrained minimization problem, but is a promising approach, and one particular algorithm is described in Section 14.5. It is also possible to determine a *smooth* exact penalty function. The key idea here (Fletcher, 1973) is to use an augmented Lagrangian function in which $\boldsymbol{\lambda}$ is not an independent vector but is a function $\boldsymbol{\lambda}(\mathbf{x})$ determined by a finite calculation. In particular the choice $\boldsymbol{\lambda}(\mathbf{x}) = \mathbf{A}^{+\mathrm{T}}\mathbf{g}|_\mathbf{x}$, obtained by solving the over-determined least squares problem $\mathbf{A}\boldsymbol{\lambda} = \mathbf{g}$, gives rise to the penalty function

$$\phi(\mathbf{x}) = f(\mathbf{x}) - \boldsymbol{\lambda}(\mathbf{x})^\mathrm{T}\mathbf{c}(\mathbf{x}) + \mathbf{c}(\mathbf{x})^\mathrm{T}\mathbf{S}\mathbf{c}(\mathbf{x}). \tag{12.5.3}$$

An obvious choice for $\mathbf{S}$ is $\frac{1}{2}\sigma\mathbf{I}$, in which case the penalty function is exact for any value of σ above a certain threshold. In fact the choice $\mathbf{S} = \sigma\mathbf{A}^+\mathbf{A}^{+\mathrm{T}}$ turns out to be more significant, since it is then possible to rearrange $\phi(\mathbf{x})$ to give

$$\phi(\mathbf{x}) = f(\mathbf{x}) - \boldsymbol{\pi}(\mathbf{x})^\mathrm{T}\mathbf{c}(\mathbf{x}) \tag{12.5.4}$$

where

$$\boldsymbol{\pi}(\mathbf{x}) = \mathbf{A}^+(\mathbf{g} - \sigma\mathbf{A}^{+\mathrm{T}}\mathbf{c})|_\mathbf{x}$$

is the Lagrange multiplier vector of the subproblem

$$\begin{aligned} &\underset{\boldsymbol{\delta}}{\text{minimize}} \;\; \tfrac{1}{2}\sigma\boldsymbol{\delta}^\mathrm{T}\boldsymbol{\delta} + \mathbf{g}^\mathrm{T}\boldsymbol{\delta} \\ &\text{subject to } \mathbf{A}^\mathrm{T}\boldsymbol{\delta} + \mathbf{c} = \mathbf{0} \end{aligned} \tag{12.5.5}$$

where $\mathbf{g}$, $\mathbf{A}$, and $\mathbf{c}$ are evaluated at $\mathbf{x}$. In fact the solution $\boldsymbol{\delta}(\mathbf{x})$ of this system is discarded, but the multipliers $\boldsymbol{\pi}(\mathbf{x})$ are substituted into (12.5.4) to define $\phi(\mathbf{x})$. The constraints in (12.5.5) are linearizations of the non-linear constraints about $\mathbf{x}$

(see (12.3.11)) so a generalization to solve inequality constraints is immediate. The resulting function $\phi(\mathbf{x})$ can be minimized by any suitable smooth unconstrained minimization technique. The main difficulty with this approach is that the calculation of $\boldsymbol{\lambda}(\mathbf{x})$ or $\boldsymbol{\pi}(\mathbf{x})$ requires first derivatives of f and $\mathbf{c}$. Thus $\nabla\phi$ requires second derivatives of f and $\mathbf{c}$, which are often not available. However, if these derivatives are available, then an $O(\mathbf{h})$ estimate of $\nabla^2\phi(\mathbf{x}^*)$ can be made, which ensures second order convergence, and hence a Newton-like minimization method is very practicable in these circumstances.

Another idea which occurs frequently is to make linearizations of the non-linear functions which arise in the problem, and then solve LP or QP subproblems. Some early methods such as the cutting plane method and MAP make linear Taylor series approximations like (7.1.3) about the current iterate $\mathbf{x}^{(k)}$ to both the objective and constraint functions, and the resulting LP (possibly including some step-length restriction) is solved to determine $\mathbf{x}^{(k+1)}$. This may be satisfactory when the solution is at a vertex of the feasible region, but in general does not model the effect of curvature. Thus the convergence properties suffer and there can be numerical difficulties. The techniques of Section 12.3 are therefore recommended, since they avoid all these difficulties albeit at the expense of solving a QP subproblem. Another technique which uses linear approximations is *separable programming* (see Beale (1970) for instance) in which the objective and constraint functions are separable. A separable function is one which can be written as

$$f(\mathbf{x}) = \sum_i f_i(z_i) \tag{12.5.6}$$

where the scalar quantities z_i are either linear functions of $\mathbf{x}$ or other separable functions. Each separable function is replaced by a piecewise linear approximation whose knots are at fixed values of the variable z_i. It is then possible to reduce the problem to something like a linear programming problem, but with differences in the rules for changing the basis. The accuracy of the solution clearly depends on that of the piecewise linear approximations. In fact separable programming is most useful when a rough estimate of the solution is satisfactory and when the problem is substantially linear and sparse, such as in business or economic models.

Another type of algorithm which solves QP subproblems is given by Murray (1969). This is motivated by trying to avoid the ill-conditioning associated with $\sigma\to\infty$ in the penalty function (12.1.3). A simple interpretation of the method arises from the fact that the minimizer $\mathbf{x}(\sigma)$ of (12.1.3) equivalently solves the constrained problem

$$\begin{aligned} &\underset{\mathbf{x}}{\text{minimize}}\ f(\mathbf{x}) \\ &\text{subject to } \mathbf{c}(\mathbf{x}) = \mathbf{c}(\mathbf{x}(\sigma)). \end{aligned} \tag{12.5.7}$$

A sequence of values $\sigma^{(k)}\to\infty$ is chosen although this is done as the algorithm proceeds. Murray (1969) does not give details but suggests that the rate of increase should depend on how well the solution is approximated. The quadratic approximation (12.3.10) to $f(\mathbf{x})$ and the linear approximation (12.3.11) to $\mathbf{c}(\mathbf{x})$ are made, and using the asymptotic result (12.1.10) the subproblem

$$\begin{aligned} &\underset{\boldsymbol{\delta}}{\text{minimize}}\ q^{(k)}(\boldsymbol{\delta}) \\ &\text{subject to } \boldsymbol{\ell}^{(k)}(\boldsymbol{\delta}) = \frac{\mathbf{c}^{(k)}\sigma^{(k)}}{\sigma^{(k+1)}} \end{aligned} \tag{12.5.8}$$

is defined. This is solved to determine a search direction along which $\phi(\mathbf{x}, \sigma^{(k+1)})$ is minimized. The line search offsets the apparent disadvantage that the 'best' choice $\sigma^{(k+1)} = \infty$ is not usually made. The method can be improved by introducing estimates of Lagrange multipliers and such a method is described by Biggs (1975). This method has been used widely in practice and has performed well, but it turns out to be similar to the methods of Section 12.3 and a detailed discussion of the method will not be given here. A convergence proof for the method is given by Biggs (1978) but seems to rely strongly on the requirement of uniform independence of the columns of the Jacobian matrix which is an unrealistic assumption.

Questions for Chapter 12

1. Show that if the penalty function (12.1.3) is applied to a quadratic programming problem with equality constraints, then $\phi(\mathbf{x}, \sigma)$ is a quadratic function of $\mathbf{x}$. State the resulting Hessian matrix $\nabla_x^2\phi$.
2. Consider using the penalty function (12.1.3) to solve the problem: $\min -x_1x_2x_3$ subject to $72 - x_1 - 2x_2 - 2x_3 = 0$. Verify that the explicit expression for $\mathbf{x}(\sigma)$ given by $x_2 = x_3 = 24/(1 + \sqrt{(1 - 8/\sigma)})$, $x_1 = 2x_2$, satisfies $\nabla\phi(\mathbf{x}(\sigma), \sigma) = \mathbf{0}$. Verify also that $\mathbf{x}(\sigma) \to \mathbf{x}^*$ as $\sigma \to \infty$. Find $\mathbf{x}(\sigma)$ when $\sigma = 9$ and verify that $\nabla^2\phi(\mathbf{x}(9), 9)$ is positive definite.
3. Consider the problem (12.1.5). For the penalty function (12.1.3) show that the elements of the minimizer $\mathbf{x}(\sigma)$ satisfy the equations $x_1 = x_2$ and $2x_1^3 - x_1 - 1/(2\sigma) = 0$. Show that as $\sigma \to \infty$, $x_1 = \frac{1}{2}\sqrt{2} + a/\sigma + O(1/\sigma^2)$, and find a. Consider now the problem: $\min -x_1 - x_2$ subject to $1 - x_1^2 - x_2^2 \geqslant 0$ and $x_2 - x_1^2 \geqslant 0$. Show how the penalty function may be modified to solve this problem. For what values of σ do the minimizers of the two penalty functions agree?
4. Show that if second order conditions (9.3.5) hold, and if rank $\mathbf{A}^* = m$, then the Lagrangian matrix $\begin{bmatrix} \mathbf{W}^* & -\mathbf{A}^* \\ -\mathbf{A}^{*\mathrm{T}} & \mathbf{0} \end{bmatrix}$ is non-singular. Let $\begin{bmatrix} \mathbf{W}^* & -\mathbf{A}^* \\ -\mathbf{A}^{*\mathrm{T}} & \mathbf{0} \end{bmatrix}\begin{pmatrix} \mathbf{s} \\ \mathbf{t} \end{pmatrix} = \mathbf{0}$. Show that rank $\mathbf{A}^* = m$ implies $\mathbf{t} = \mathbf{0}$. If $\mathbf{s} \neq \mathbf{0}$ show that s satisfies (9.3.4) and $\mathbf{s}^{\mathrm{T}}\mathbf{W}^*\mathbf{s} = \mathbf{0}$ which contradicts (9.3.5). Hence conclude that the Lagrangian matrix is non-singular.
5. Let $\mathbf{c}^{(k)} \to \mathbf{c}^*$, $\mathbf{A}^{(k)} \to \mathbf{A}^*$ ($\mathbf{A}$ is $n \times m$, $n > m$) and let rank $\mathbf{A}^* = m$. Use the Rayleigh quotient result ($\mathbf{u}^{\mathrm{T}}\mathbf{M}\mathbf{u} \geqslant \lambda_n\mathbf{u}^{\mathrm{T}}\mathbf{u}\ \forall\ \mathbf{u}$) and the definition of a singular value σ_i of $\mathbf{A}$ (square root of an eigenvalue λ_i of $\mathbf{A}^{\mathrm{T}}\mathbf{A}$) to show that if $0 < \beta < \sigma_n$ then $\|\mathbf{A}^{(k)}\mathbf{c}^{(k)}\|_2 \geqslant \beta\|\mathbf{c}^{(k)}\|_2$ for all k sufficiently large. Hence establish (12.1.22).
6. For the inverse barrier function (12.1.24) show that estimates of Lagrange

multipliers analogous to (12.1.8) are given by $\lambda_i^{(k)} = r^{(k)}/c_i^{(k)2}$. Hence show that $\lambda_i^{(k)} \to 0$ for an inactive constraint and that any limit point $\mathbf{x}^*$, $\boldsymbol{\lambda}^*$ is a KT point. Show that $\mathbf{h}^{(k)}$ has asymptotic behaviour $O(r^{(k)\,1/2})$ which can be used in an acceleration technique (see SUMT of Fiacco and McCormick, 1968). Show also that $o(r)$ estimates of f^* can be made in a similar way to (12.1.12).

7. For the logarithmic barrier function (12.1.25) show that estimates of Lagrange multipliers are given by $\lambda_i^{(k)} = r^{(k)}/c_i^{(k)}$, that $\mathbf{x}^*$, $\boldsymbol{\lambda}^*$ is a KT point, that the asymptotic behaviour of $\mathbf{h}^{(k)}$ is $O(r)$, and that $o(r)$ estimates of f^* can be made.
8. Investigate the Hessian matrices of the barrier functions (12.1.24) and (12.1.25) at $\mathbf{x}(r^{(k)})$ and demonstrate ill-conditioning as $r^{(k)} \to 0$ if the number of active constraints is between 1 and $n - 1$. What happens when $\lambda_i^* = 0$ for an active constraint?
9. When the penalty function (12.1.23) is applied to the problem

$$\begin{aligned}
&\text{minimize } -x_1 - x_2 + x_3 \\
&\text{subject to } 0 \leqslant x_3 \leqslant 1 \\
&\qquad x_1^3 + x_3 \leqslant 1 \\
&\qquad x_1^2 + x_2^2 + x_3^2 \leqslant 1
\end{aligned}$$

the following data are obtained. Use this to estimate the optimum solution and multipliers, together with the active constraints, and give some indication of the accuracy which is achieved.

k	$\sigma^{(k)}$	$x_1(\sigma^{(k)})$	$x_2(\sigma^{(k)})$	$x_3(\sigma^{(k)})$
1	1	0.834379	0.834379	−0.454846
2	10	0.728324	0.728324	−0.087920
3	100	0.709557	0.709557	−0.009864
4	1000	0.707356	0.707356	−0.001017

10. Consider the problem

$$\begin{aligned}
&\underset{\mathbf{x}}{\text{minimize}} \;\; \exp(x_1 x_2 x_3 x_4 x_5) \\
&\text{subject to } x_1^2 + x_2^2 + x_3^2 + x_4^2 + x_5^2 = 10 \\
&\qquad x_2 x_3 = 5 x_4 x_5 \\
&\qquad x_1^3 + x_2^3 = -1
\end{aligned}$$

given by Powell (1969). Investigate how accurately a local solution $\mathbf{x}^*$ and multipliers $\boldsymbol{\lambda}^*$ can be obtained, using the penalty function (12.1.3) and the sequence $\{\sigma^{(k)}\} = 0.01, 0.1, 1, 10, 100$. Use a quasi-Newton subroutine which requires no derivatives. A suitable initial approximation to $\mathbf{x}(\sigma^{(1)})$ is the vector $(-2, 2, 2, -1, -1)^{\mathrm{T}}$, and it is sufficient to obtain each element of $\mathbf{x}(\sigma^{(k)})$ correct to three decimal places.
11. Consider using the multiplier penalty function (12.2.3) to solve the problem

in Question 12.2. Assume that the controlling parameters $\lambda^{(1)} = 0$ and $\sigma = 9$ are chosen initially, in which case the minimizer $\mathbf{x}(0, 9)$ is the same as the minimizer $\mathbf{x}(9)$ which has been determined in Question 12.2. Write down the new parameter $\lambda^{(2)}$ which would be given by using formulae (12.2.13) and (12.2.15). Which formula gives the better estimate of λ^*?

12. Use the Sherman–Morrison formula (Volume 1, Question 3.13) to show that if $\mathbf{S}$ is non-singular, then

$$\begin{bmatrix} \mathbf{W} + \mathbf{ASA}^{\mathrm{T}} & -\mathbf{A} \\ -\mathbf{A}^{\mathrm{T}} & \mathbf{0} \end{bmatrix}^{-1} = \begin{bmatrix} \mathbf{W} & -\mathbf{A} \\ -\mathbf{A}^{\mathrm{T}} & \mathbf{0} \end{bmatrix} - \begin{bmatrix} \mathbf{0} & \mathbf{0} \\ \mathbf{0} & \mathbf{S} \end{bmatrix}.$$

Hence use (10.2.6) to show that, in the notation of (12.2.6),

$$(\mathbf{A}^{\mathrm{T}}\mathbf{W}_\sigma^{-1}\mathbf{A})^{-1} = (\mathbf{A}^{\mathrm{T}}\mathbf{W}^{-1}\mathbf{A})^{-1} + \mathbf{S}.$$

Let $\boldsymbol{\sigma}_1$ be fixed so that $\mathbf{W}_{\sigma_1}$ is positive definite. Deduce that

$$(\mathbf{A}^{\mathrm{T}}\mathbf{W}_\sigma^{-1}\mathbf{A})^{-1} = \mathbf{S} + (\mathbf{A}^{\mathrm{T}}\mathbf{W}_{\sigma_1}^{-1}\mathbf{A})^{-1} - \mathbf{S}_1 = \mathbf{S} + O(1)$$

which is (12.2.14).

13. Consider any iteration function

$$\boldsymbol{\lambda}^{(k+1)} = \boldsymbol{\phi}(\boldsymbol{\lambda}^{(k)}) \triangleq \boldsymbol{\lambda}^{(k)} - \mathbf{M}^{(k)}\mathbf{c}^{(k)}$$

analogous to (12.2.13) or (12.2.15) for use with a multiplier penalty function. Show that

$$\nabla\boldsymbol{\phi}^{\mathrm{T}}(\boldsymbol{\lambda}^*) = \mathbf{I} - (\mathbf{A}^{*\mathrm{T}}\mathbf{W}_\sigma^{*-1}\mathbf{A}^*)\mathbf{M}^{*\mathrm{T}}.$$

Hence use the result of Question 12.12 and theorem 6.2.2 (Volume 1) to show that the Powell–Hestenes formula (12.2.15) converges at a linear rate, which can be made arbitrarily rapid by making $\mathbf{S}$ sufficiently large.

14. Show that the vector $\mathbf{x}(\boldsymbol{\lambda})$ which minimizes (12.2.1) equivalently solves the problem

$$\underset{\mathbf{x}}{\text{minimize}}\ f(\mathbf{x})$$
$$\text{subject to } \mathbf{c}(\mathbf{x}) = \mathbf{c}(\mathbf{x}(\boldsymbol{\lambda})).$$

If $\mathbf{x}(\boldsymbol{\lambda})$ minimizes (12.2.18) show that the equivalent problem is

$$\underset{\mathbf{x}}{\text{minimize}}\ f(\mathbf{x})$$
$$\text{subject to } c_i(\mathbf{x}) \geqslant \min(c_i(\mathbf{x}(\boldsymbol{\lambda})), \theta_i), \qquad i = 1, 2, \ldots, m$$

(Fletcher, 1975). These problems are neighbouring problems to (12.1.1) and (12.1.2) respectively.

15. Consider using the penalty function

$$\phi(\mathbf{x}, \boldsymbol{\lambda}, \boldsymbol{\sigma}) = f(\mathbf{x}) + \sum_i \sigma_i \exp(-\lambda_i c_i(\mathbf{x})/\sigma_i)$$

where $\sigma_i > 0$, to solve problem (12.1.1). Assume that a local solution $\mathbf{x}^*$ of this problem satisfies the second order sufficient conditions (9.3.5) and (9.3.4).

Show that if control parameters $\boldsymbol{\lambda} = \boldsymbol{\lambda}^*$ and any $\boldsymbol{\sigma}$ are chosen, then $\mathbf{x}^*$ is a stationary point of $\phi(\mathbf{x}, \boldsymbol{\lambda}^*, \boldsymbol{\sigma})$. Show also that if, for all i, both $\lambda_i^* \neq 0$ and σ_i is sufficiently small, then $\mathbf{x}^*$ is a local minimizer of $\phi(\mathbf{x}, \boldsymbol{\lambda}^*, \boldsymbol{\sigma})$. (You may use without proof the lemma that if $\mathbf{A}$ is symmetric, if $\mathbf{D}$ is diagonal with sufficiently large positive elements, and if $\mathbf{v}^T\mathbf{A}\mathbf{v} > 0$ for all $\mathbf{v}$ ($\neq \mathbf{0}$) such that $\mathbf{B}^T\mathbf{v} = \mathbf{0}$, then $\mathbf{A} + \mathbf{B}\mathbf{D}\mathbf{B}^T$ is positive definite.)

What can be said about the case when some $\lambda_i^* = 0$? To what extent is this penalty function comparable to the Powell–Hestenes function?

If $\mathbf{x}(\boldsymbol{\lambda})$ is the unique global minimizer of $\phi(\mathbf{x}, \boldsymbol{\lambda}, \boldsymbol{\sigma})$ for all $\boldsymbol{\lambda}$, where $\boldsymbol{\sigma}$ is constant, and if $\mathbf{x}^* = \mathbf{x}(\boldsymbol{\lambda}^*)$, show that $\boldsymbol{\lambda}^*$ maximizes the function $\psi(\lambda) \triangleq \phi(\mathbf{x}(\boldsymbol{\lambda}), \boldsymbol{\lambda}, \boldsymbol{\sigma})$. What are the practical implications of this result?

16. Solve the problem in Question 12.19 by generalized elimination of variables using the Newton method as shown in Table 12.4.1. Compare the sequence of estimates and the amount of work required with that in Question 12.19.
17. Assume that second order sufficient conditions (9.3.5) and (9.3.4) hold at $\mathbf{x}^*, \boldsymbol{\lambda}^*$. Let $(\mathbf{x}^{(k)}, \boldsymbol{\lambda}^{(k)}) \to (\mathbf{x}^*, \boldsymbol{\lambda}^*)$ and let a vector $\mathbf{s}^{(k)}$ ($\|\mathbf{s}^{(k)}\|_2 = 1$) exist such that $\mathbf{s}^{(k)T}\mathbf{W}^{(k)}\mathbf{s}^{(k)} \leqslant 0 \ \forall\ k$. Show by continuity that (9.3.5) and (9.3.4) are contradicted and hence that the subproblem (12.3.9) is well determined for $\mathbf{x}^{(k)}, \boldsymbol{\lambda}^{(k)}$ sufficiently close to $\mathbf{x}^*, \boldsymbol{\lambda}^*$.
18. Consider generalizing theorem 12.3.1 to apply to the inequality constraint problem (12.1.2). If strict complementarity holds at $\mathbf{x}^*, \boldsymbol{\lambda}^*$, show by virtue of the implicit function theorem that a non-singular Lagrangian matrix (12.3.14) implies that the solution of (12.3.5) changes smoothly in some neighbourhood of $\mathbf{x}^*, \boldsymbol{\lambda}^*$. Hence show that the solution of the equality constraint problem also solves the inequality problem, so that constraints $c_i(\mathbf{x})$, $i \notin \mathscr{A}^*$, can be ignored. If strict complementarity does not hold, assume that the vectors $\mathbf{a}_i^*$, $i \in \mathscr{A}^*$, are independent and that the second order sufficient conditions of theorem 9.3.2 hold. Deduce that the Lagrangian matrix (12.3.14) is non-singular for any subset of active constraints for which $\mathscr{A}_+^* \subseteq \mathscr{A} \subseteq \mathscr{A}^*$. Show that there is a neighbourhood of $\mathbf{x}^*, \boldsymbol{\lambda}^*$ in which the active constraints obtained from solving (12.3.13) satisfy this condition, so that a result like (12.3.16) holds for any such $\mathscr{A}$, and hence for $\mathscr{A}^*$. The rest of the theorem then follows.
19. Consider solving the problem: $\min x_1 + x_2$ subject to $x_2 \geqslant x_1^2$ by the Lagrange–Newton method (12.3.12) from $\mathbf{x}^{(1)} = \mathbf{0}$. Why does the method fail if $\lambda^{(1)} = 0$? Verify that rapid convergence occurs if $\lambda^{(1)} = 1$ is chosen.

Chapter 13

Other Optimization Problems

13.1 Integer Programming

In this chapter two other types of problem are studied which have the feature that they can be reduced to the solution of standard problems described in other chapters. *Integer programming* is the study of optimization problems in which some of the variables are required to take integer values. The most obvious example of this is the number of men, machines, or components, etc., required in some system or schedule. It also extends to cover such things as transformer settings, steel stock sizes, etc., which occur in a fixed ordered set of discrete values, but which may be neither integers, nor equally spaced. Most combinatorial problems can also be formulated as integer programming problems; this often requires the introduction of a *zero–one variable* which takes either of the values 0 or 1 only. Zero–one variables are also required when the model is dependent on the outcome of some decision on whether or not to take a certain course of action. Certain more complicated kinds of condition can also be handled by the methods of integer programming through the introduction of *special ordered sets*; see Beale (1978) for instance. There are many special types of integer programming problems but this section covers the fairly general category of *mixed integer programming* (both integer and continuous variables allowed, and any continuous objective and constraint functions as in (7.1.1)). The special case is also considered of *mixed integer LP* in which some integer variables occur in what is otherwise an LP problem. The *branch and bound method* is described for solving mixed integer programming problems, together with additional special features which apply to mixed integer LP. There are many special cases of integer programming problems and it is the case that the branch and bound method is not the most suitable in every case. Nonetheless this method is the one method which has a claim to be of wide generality and yet reasonably efficient. It solves a sequence of subproblems which are continuous problems of type (7.1.1) and is therefore well related to other material in this book. Other methods of less general applicability but which can be useful in certain special cases are reviewed by Beale (1978).

Integer programming problems are usually much more complicated and expensive to solve than the corresponding continuous problem on account of the discrete nature of the variables and the combinatorial number of feasible solutions which thus exists. A tongue-in-cheek possibility is the *Dantzig two-phase method.* The first phase is try to convince the user that he or she does not wish to solve an integer programming problem at all! Otherwise the continuous problem is solved and the minimizer is rounded off to the nearest integer. This may seem flippant but in fact is how many integer programming problems are treated in practice. It is important to realize that this approach can fail badly. There is of course no guarantee that a good solution can be obtained in this way, even by examining all integer points in some neighbourhood of the continuous solution. Even worse is the likelihood that the rounded-off points will be infeasible with respect to the continuous constraints in the problem (for example problem (13.1.5) below), so that the solution has no value. Nonetheless the idea of rounding up or down does have an important part to play when used in a systematic manner, and indeed is one main feature of the branch and bound method.

To develop the branch and bound method therefore, the problem is to find the solution $\mathbf{x}^*$ of the problem

$$\begin{aligned} P_I: &\text{ minimize } f(\mathbf{x}) \\ &\text{ subject to } \mathbf{x} \in R, \qquad x_i \text{ integer } \forall\, i \in I \end{aligned} \tag{13.1.1}$$

where I is the set of integer variables and R is the (closed) feasible region of the continuous problem

$$P\text{: minimize } f(\mathbf{x}) \quad \text{subject to } \mathbf{x} \in R, \tag{13.1.2}$$

which could be (7.1.1) for example. Let the minimizer $\mathbf{x}'$ of P exist: if it is feasible in P_I then it solves P_I. If not, then there exists an $i \in I$ for which x_i' is not an integer. In this case two problems can be defined by *branching* on variable x_i in problem P, giving

$$\begin{aligned} P^-: &\text{ minimize } f(\mathbf{x}) \\ &\text{ subject to } \mathbf{x} \in R, \quad x_i \leqslant [x_i'] \end{aligned} \tag{13.1.3}$$

and

$$\begin{aligned} P^+: &\text{ minimize } f(\mathbf{x}) \\ &\text{ subject to } \mathbf{x} \in R, \quad x_i \geqslant [x_i'] + 1, \end{aligned} \tag{13.1.4}$$

where $[x]$ means the largest integer not greater than x. Also define P_I^- as problem (13.1.3) together with the integrality constraints x_i, $i \in I$, integer, and P_I^+ likewise. Two observations which follow are that $\mathbf{x}^*$ is feasible in either P^- or P^+ but not both, in which case it solves P_I^- or P_I^+. Also any feasible point in P_I^- or P_I^+ is feasible in P_I.

It is usually possible to repeat the branching process by branching on P^- and P^+, and again on the resulting problems, so as to generate a *tree* structure (Figure 13.1.1). Each *node* corresponds to a continuous optimization problem, the *root* is the problem P, and the nodes are connected by the *branches* defined above. There are two special cases where no branching is possible at any given node: one is where the solution to the corresponding problem is integer feasible (□ in tree) and the other where the problem has no feasible point (● in tree). Otherwise each node is a

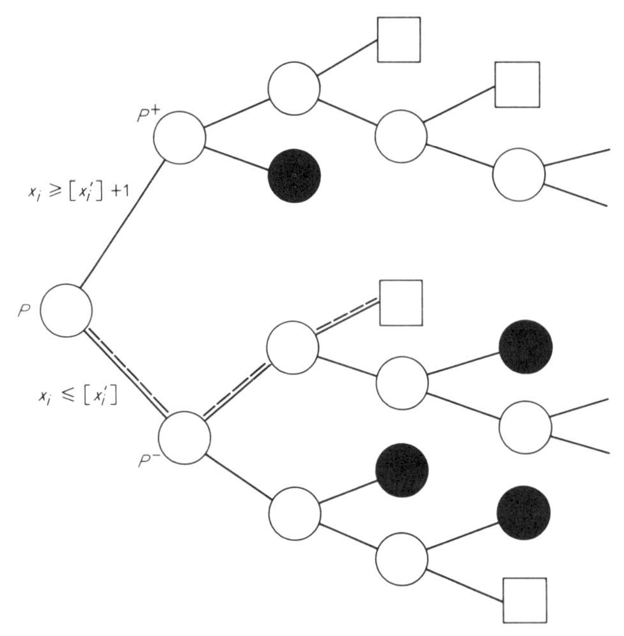

Figure 13.1.1 Tree structure in the branch and bound method

parent problem (○ in tree) and gives rise to two branched problems. If the feasible region is bounded then the tree is finite and each *path* through the tree ends in a □ or a ●. Assuming that the solution $\mathbf{x}^*$ of P_I exists, then it is feasible along just one path through the tree (the broken line, say, in Figure 13.1.1) and is feasible in, and therefore solves, the □ problem which must exist at the end of the path. The solution of every □ problem is feasible in P_I and so the required solution vector $\mathbf{x}^*$ is that solution of a □ problem which has least objective function value. Thus the aim of the branch and bound method is to seek out □ nodes by exploring the tree in order to find the one with the least value. Often in the tree there are ○ nodes whose solution violates more than one integrality constraint, so in fact the tree is not uniquely defined until the choice of branching variable is defined. One anomalous (although unlikely) possibility is that $\mathbf{x}^*$ may be well defined, but the continuous problem P may be unbounded ($f(\mathbf{x}) \to -\infty$). This possibility is easily excluded in most practical cases by adding bounds to the variables so as to make R bounded.

To find the type of each node in the tree requires the solution of a continuous optimization problem. Also the number of nodes grows exponentially with the number of variables and may not even be finite. Therefore to examine the whole tree is usually impossibly expensive and so the branch and bound method attempts to find the solution of P_I by making only a partial search of the tree. Assuming that a global solution of each problem can be calculated, adding the branching constraint makes the solution of the parent problem infeasible in the branched problem, and so the optimum objective function value of the parent problem is a lower bound for that of the branched problem. Assume that some but not all of the problems in the tree have been solved, and let f_i be the best optimum objective function value which has been obtained by solving the problem at a □ node i. The solution of this problem is feasible in P_I so $f^* \leqslant f_i$. Now if f_j is the value associated with any ○ node j, and $f_j \geqslant f_i$, then branches from node j cannot reduce f_j and so they cannot give a better value than f_i. Hence these branches need not be examined. This is the main principle which enables the solution to be obtained by making only a partial search of the tree.

Nonetheless because of the size of the tree, the branch and bound method only becomes effective if the partial search strategy is very carefully chosen. Two important decisions have to be made:

(a) which problem or node should be solved next, and
(b) on which variable should a branch be made.

The following would seem to be an up-to-date assessment of the state of the art (see Beale (1978), for example, for more detailed references). An early method favoured the use of a *stack* (a last-in first-out list or data structure). This algorithm keeps a stack of unsolved problems, each with a lower bound L on its minimum objective function value. This bound can be the value of the parent problem, although in linear programming problems a better bound can be obtained as described below. From those □ nodes which have been explored, the current best integer feasible solution $\hat{\mathbf{x}}$ with value $\hat{f}$ is stored. Initially $\hat{f} = \infty$, the continuous problem is put in the stack with $L = -\infty$, and the algorithm proceeds as follows.

(i) If no problem is in the stack, finish with $\hat{\mathbf{x}}, \hat{f}$ as $\mathbf{x}^*, f^*$; otherwise take the top problem from the stack.
(ii) If $L \geqslant \hat{f}$ reject the problem and go to (i).
(iii) Try to solve the problem: if no feasible point exists reject the problem and go to (i).
(iv) Let the solution be $\mathbf{x}'$ with value f': if $f' \geqslant \hat{f}$ reject the problem and go to (i).
(v) If $\mathbf{x}'$ is integer feasible then set $\hat{\mathbf{x}} = \mathbf{x}'$, $\hat{f} = f'$ and go to (i).
(vi) Select an integer variable i such that $[x_i'] < x_i'$, create two new problems by branching on x_i, place these on the stack with lower bound $L = f'$ (or a tighter lower bound derived from f'), and go to (i).

The rationale for these steps follows from the above discussion: in step (iv) any branched problems must have $f \geqslant f' \geqslant \hat{f} \geqslant f^*$ so cannot improve the optimum, so the node is rejected. Likewise in step (ii) where $f \geqslant L \geqslant \hat{f} \geqslant f^*$. Step (iii) corresponds to finding a ● node and step (iv) to a □ node. Step (vi) is the branching operation, and step (i) corresponds to having examined or excluded from consideration all nodes in the tree, in which case the solution at the best □ node gives $\mathbf{x}^*$. The effect of this stack method is to follow a single path deep into the tree to a □ node. Then the algorithm works back, either rejecting problems or creating further sub-trees by branching and perhaps updating the best integer feasible solution $\hat{\mathbf{x}}$ and $\hat{f}$. This is illustrated in Figure 13.1.2.

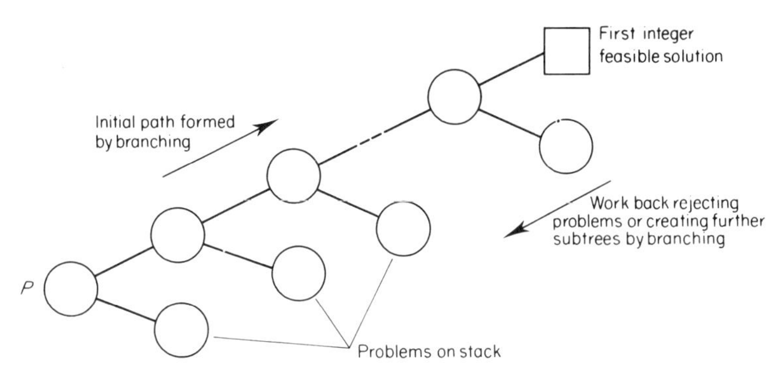

Figure 13.1.2 Progress of the stack method

So far the description of the algorithm is incomplete in that rules are required for choosing the branching variable if more than one possibility exists, and for deciding which branched problem goes on the stack first. To make this decision requires estimates e_i^+ and e_i^- (>0) of the increase in f when adding the branching constraints $x_i \geqslant [x_i'] + 1$ and $x_i \leqslant [x_i']$ respectively. These estimates can be made using information about first derivatives (and also second derivatives in other than LP problems) in a way which is described below. Then the rule is to choose i to solve $\max_i \max(e_i^+, e_i^-)$ and place the branched problem corresponding to $\max(e_i^+, e_i^-)$ on the stack first. Essentially the worst case is placed on the stack first, and so its solution is deferred until later. The motivation for this rule is hopefully that, being the worst case, this problem will have a high value of f and so can be rejected at a later stage with little or no branching. A numerical example of the stack method to the integer LP problem

$$\begin{aligned} &\text{minimize} && x_1 + 10x_2 \\ &\text{subject to} && 66x_1 + 14x_2 \geqslant 1430 \\ &&& -82x_1 + 28x_2 \geqslant 1306, \qquad x_1, x_2 \text{ integer} \end{aligned} \tag{13.1.5}$$

is illustrated in Figure 13.1.3. Some shading on the grid shows the effect of adding extra constraints to the feasible region, and numbered points indicate solutions to the subproblems. The solution tree is shown, and the typical progress described in Figure 13.1.2 can be observed. After branching on problem 7, the stack contains problems 9, 8, 6, 4, 2 in top to bottom order. Problem 9 gives rise to an integer feasible solution, which enables the remaining problems to be rejected in turn without further branching. Notice that it may of course be necessary to branch on the same variable more than once in the tree.

In practice it has become apparent that the stack method contains some good features but can also be inefficient. The idea of following a path deep into the tree in order to find an integer feasible solution works well in that it rapidly determines a good value of $\hat{f}$ which can be used to reject other branches. This feature also gives a feasible solution if time does not permit the search to be completed. Furthermore it is very efficient in the amount of storage required to keep details of unsolved pending problems in the tree. However the process of back-tracking from a □ node

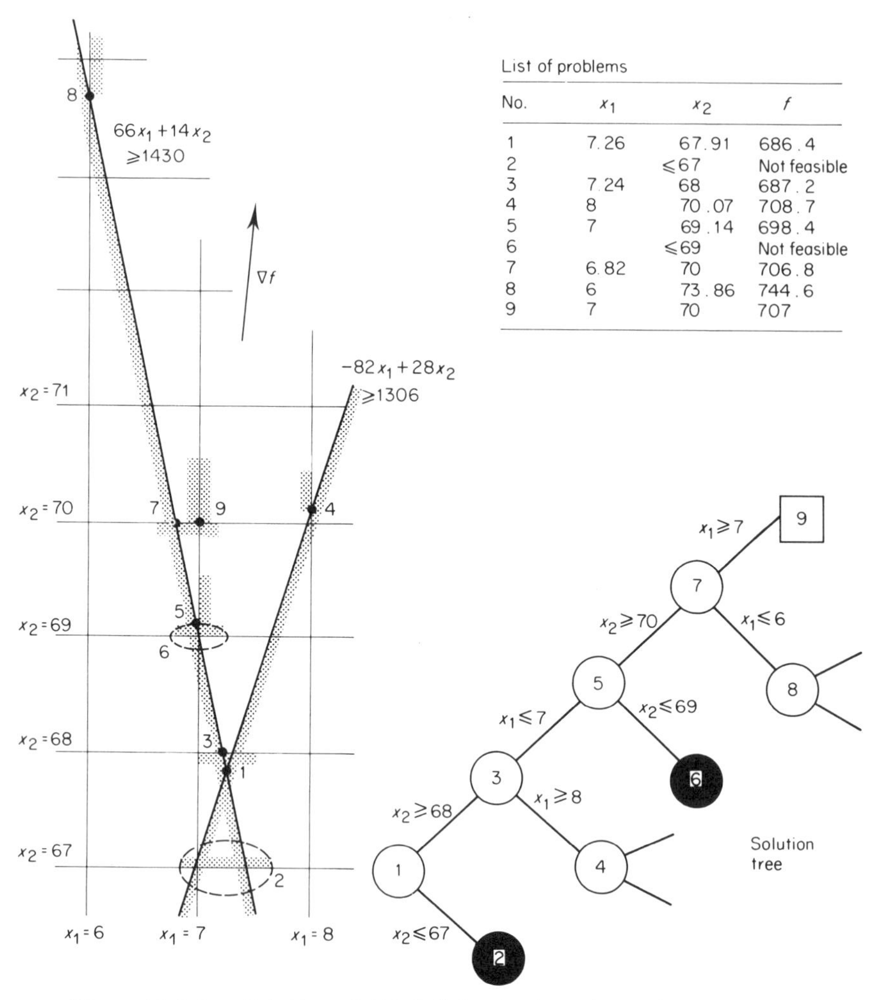

List of problems

No.	x_1	x_2	f
1	7.26	67.91	686.4
2		$\leqslant 67$	Not feasible
3	7.24	68	687.2
4	8	70.07	708.7
5	7	69.14	698.4
6		$\leqslant 69$	Not feasible
7	6.82	70	706.8
8	6	73.86	744.6
9	7	70	707

Figure 13.1.3 Solution of (13.1.5) by the branch and bound method

to neighbouring nodes in the tree can be unrewarding. Once a path can be followed no further it is preferable to select the best of all pending nodes to explore next, rather than that on top of the stack. To do this, it is convenient to associate an estimate E of its solution value with each unsolved problem which is created by branching. E is conveniently defined by $f' + e_i^+$ and $f' + e_i^-$ respectively, where f' is the value of the parent problem and e_i^+, e_i^- are the estimates referred to above. A path starting at the pending node with least E is then followed through the tree. Rules of this type are reviewed by Beale (1978) and Breu and Burdet (1974). Another useful idea is to have the user supply a cut-off value, above which solutions are of no interest. This can be used initially in place of $\hat{f} = \infty$ and may prevent deep exploration of unfavourable paths. A further disadvantage of the stack method is that the 'worst case' rule for choosing the branching variable is often inadequate. Present algorithms prefer to branch on the 'most important variable', that is the one which is likely to be most sensitive in determining the solution. This can be determined from user-assigned 'priorities' on the variables (often zero–one variables should be given high priority). To choose between variables of equal priority the variable which solves $\max_i \min(e_i^-, e_i^+)$ is selected, that is the one for which the least estimate of the increase in f is largest. This quantifies the definition of the most important variable. Other possible selection rules are discussed by Breu and Burdet (1974).

The branch and bound strategies depend critically on making good estimates e_i^+ and e_i^- of the increases in f caused by adding either of the branching constraints. Such estimates are readily made for an integer LP problem and are most easily described in terms of the active set method (Section 8.3). Let $\mathbf{x}'$ be the solution of a parent node, value f'. Then the directions of feasible edges at $\mathbf{x}'$ are the vectors $\mathbf{s}_q$ where $\mathbf{A}^{-\mathrm{T}} = [\mathbf{s}_1, \mathbf{s}_2, \ldots, \mathbf{s}_n]$. The slope of f along $\mathbf{s}_q$ is given by the multiplier λ_q ($\geqslant 0$). Consider finding e_i^+. A search along $\mathbf{s}_q$ increases x_i if $(\mathbf{s}_q)_i > 0$ at a rate $(\mathbf{s}_q)_i$, and increases f at a rate λ_q. Thus for small changes Δx_i, if a constraint $x_i \geqslant x_i' + \Delta x_i$ is added to the problem, then the increase Δf in the optimal solution value f' is given by $\Delta f = p_i^+ \Delta x_i$ where

$$p_i^+ = \min_{q : (\mathbf{s}_q)_i > 0} \frac{\lambda_q}{(\mathbf{s}_q)_i}. \tag{13.1.6}$$

For larger changes, previously inactive constraints may become active along the corresponding edge, and so further increases in f occur on restoring feasibility. Thus in general a lower bound $\Delta f \geqslant p_i^+ \Delta x_i$ on the increase in f is obtained. Thus if the fractional part of x_i' is ϕ_i $(= x_i' - [x_i'])$ then the required estimate of the change in f is

$$e_i^+ = p_i^+(1 - \phi_i). \tag{13.1.7}$$

It may be that no element $(\mathbf{s}_q)_i > 0$ exists, corresponding to the fact that no feasible direction exists which increases x_i. Thus the resulting branched problem is infeasible (a ● node). It can be convenient to define $e_i^+ = p_i^+ = \infty$ in this case. A similar development can be used when decreasing x_i. Consider a small change Δx_i (>0), and add the constraint $x_i \leqslant x_i' - \Delta x_i$ to the problem. Then $\Delta f = p_i^- \Delta x_i$ where

$$p_i^- = \min_{q : (\mathbf{s}_q)_i < 0} \frac{\lambda_q}{-(\mathbf{s}_q)_i}. \tag{13.1.8}$$

In general $\Delta f \geqslant p_i^- \Delta x_i$, and the required estimate is

$$e_i^- = p_i^- \phi_i. \tag{13.1.9}$$

The estimates (13.1.7) and (13.1.9) are used as described above in choosing the branching variable and estimating E for each unsolved problem created by branching. In fact since for integer LP these estimates give strict lower bounds on Δf, it follows that use of $L = E$ gives a tighter lower bound than the value of the parent problem $L = f'$. Other ways in which the structure of an LP problem can be used advantageously are given by Beale (1968). Often the dual formulation is recommended for solving the LP problem since constraints can be added whilst retaining feasibility. However it seems to me that the standard form given in Question 8.8 would also be very suitable, using a cost function like (8.4.5) to restore feasibility. When solving linear constraint problems in which the objective is non-linear and when the solution $\mathbf{x}'$ to any subproblem lies at a vertex, then similar estimates to the above can be made. These estimates are strict bounds if the function is convex. If the solution is not at a vertex then second order effects have to be introduced into (13.1.6) or (13.1.8) along the directions of zero slope which exist to avoid the trivial case $p_i^+ = p_i^- = 0$. I do not know whether this possibility has been investigated in practice although the information to do it is available. Similar estimates can also be made in integer problems with non-linear constraints.

An example in which these estimates are calculated is given by problem (13.1.5) whose solution is described in Figure 13.1.3. Consider possible branches at the solution of the root problem (1). Feasible edges are defined by

$$\mathbf{A}^{-1} = \frac{1}{2996}\begin{bmatrix} 28 & 82 \\ -14 & 66 \end{bmatrix} = \begin{bmatrix} \mathbf{s}_1^T \\ \mathbf{s}_2^T \end{bmatrix}$$

and

$$\boldsymbol{\lambda} = \mathbf{A}^{-1}\mathbf{c} = \frac{1}{2996}\begin{pmatrix} 848 \\ 646 \end{pmatrix}.$$

Consider branching on x_2: searching along either $\mathbf{s}_1$ or $\mathbf{s}_2$ increases x_2 so it can be predicted that the branch $x_2 \leqslant 67$ is infeasible and $p_2^- = e_2^- = \infty$. For increasing x_2,

$$p_2^+ = \min\left(\frac{848}{82}, \frac{646}{66}\right) = \frac{646}{66}$$

and $e_2^+ = 0.09 \times 646/66 \approx 0.9$ which agrees with the observed increase in value in going from problem 1 to 3, since no other constraints become active. For branching on x_1,

$$p_1^+ = 848/28 \approx 30, \qquad p_1^- = 646/14 \approx 46$$

$$e_1^+ = 0.74 \times 30 \approx 22, \qquad e_1^- \approx 0.26 \times 46 \approx 12$$

which again agrees with observed changes in value. These estimates also illustrate the different rules for choosing branching variables. The 'worst case' rule in the stack method is achieved by $e_2^- = \infty$ and gives rise to a branch on x_2. The 'most important variable' rule is solved by $e_1^- = 12$ and causes a branch on x_1. In fact this is a much better branch since it yields problems 4 and 5 directly, and this supports the current preference for this type of rule.

13.2 Geometric Programming

Another important type of problem which it is possible to transform to the solution of a more simple problem is known as the *geometric programming problem*. This is a nonlinear programming problem in which the problem functions are constructed from *terms* of the form

$$t_i = c_i \prod_{j=1}^{n} x_j^{a_{ij}}, \tag{13.2.1}$$

where the a_{ij} are real and may be negative or fractional for instance. The coefficients c_i and t_i however must be positive, and a sum of such terms is referred to as a *posynomial*. The problem

$$\begin{aligned} &\underset{\mathbf{x}}{\text{minimize}} \; g_0(\mathbf{x}) \\ &\text{subject to } g_k(\mathbf{x}) \leqslant 1, \qquad k = 1, 2, \ldots, m \\ &\qquad\qquad \mathbf{x} > \mathbf{0} \end{aligned} \tag{13.2.2}$$

in which the functions $g_k(\mathbf{x})$ are posynomials is a convenient standard form for a *geometric programming problem*. The terms which occur within $g_0, g_1, \ldots, g_m$ are numbered consecutively from 1 to p, say, and sets J_k which collect the indices of terms in each g_k are defined by

$$J_k = \{r_k, r_k + 1, \ldots, s_k\},$$

so that (13.2.3)

$$r_0 = 1, \qquad r_{k+1} = s_k + 1, \qquad \text{and} \qquad s_m = p.$$

Thus the posynomials $g_k(\mathbf{x})$ are defined by

$$g_k = \sum_{i \in J_k} t_i, \qquad k = 0, 1, \ldots, m. \tag{13.2.4}$$

Geometric programming has met with a number of applications in engineering (see, for instance, Duffin, Peterson, and Zener (1967), Dembo and Avriel (1978), and Bradley and Clyne (1976), amongst many). There could well be scope for even more applications (either directly or by making suitable posynomial approximations) if the technique were more widely understood and appreciated.

The geometric programming problem (13.2.2) has non-linear constraints but a transformation exists which causes considerable simplification. In some cases the solution can be found merely by solving a certain linear system, and in general it is possible to solve a more simple but equivalent minimization problem in which only linear constraints arise. In this problem the relative contribution of the various terms t_i is determined, and a subsequent calculation is then made to find the optimum variables $\mathbf{x}^*$. This transformation was originally developed as an application of the arithmetic–geometric mean inequality (see Duffin, Peterson, and Zener, 1967), hence the name geometric programming. However a more straightforward approach is to develop the transformation as an example of the duality theory of Section 9.5. First of all the transformation of variables

$$x_j = \exp(y_j), \qquad j = 1, 2, \ldots, n \tag{13.2.5}$$

is made, which removes the constraints $x_j > 0$ and allows (13.2.1) to be written

$$t_i = c_i \exp(\Sigma a_{ij} y_j). \tag{13.2.6}$$

In addition (13.2.2) is written equivalently as

$$\begin{aligned} &\underset{\mathbf{y}}{\text{minimize}} \ \log g_0(\mathbf{y}) \\ &\text{subject to} \log g_k(\mathbf{y}) \leqslant 0, \qquad k = 1, 2, \ldots, m. \end{aligned} \tag{13.2.7}$$

The main result then follows by an application of the duality transformation to (13.2.7).

So that this transformation is applicable it is first of all important to establish that each function $\log g_k(\mathbf{y})$ is a convex function on $\mathbb{R}^n$. To do this let $\mathbf{y}_0, \mathbf{y}_1 \in \mathbb{R}^n$ and $\theta \in (0, 1)$. If $u_i = [t_i(\mathbf{y}_0)]^{1-\theta}$ and $v_i = [t_i(\mathbf{y}_1)]^{\theta}$ are defined, then it follows from (13.2.6) that

$$u_i v_i = t_i(\mathbf{y}_\theta)$$

where $\mathbf{y}_\theta = (1 - \theta)\mathbf{y}_0 + \theta \mathbf{y}_1$. If $\mathbf{u}$ collects the elements u_i, $i \in J_k$, $\mathbf{v}$ similarly, then $\mathbf{u}^{\mathrm{T}}\mathbf{v} = g_k(\mathbf{y}_\theta)$. If $p = 1/(1 - \theta)$, $q = 1/\theta$ are defined, then Hölders inequality

$$\mathbf{u}^{\mathrm{T}}\mathbf{v} \leqslant \| \mathbf{u} \|_p \, \| \mathbf{v} \|_q \tag{13.2.8}$$

is valid (since $u_i > 0$, $v_i > 0$, and $p^{-1} + q^{-1} = 1$), where $\| \mathbf{u} \|_p \triangleq (\Sigma u_i^p)^{1/p}$. Taking logarithms of both sides in (13.2.8) establishes that

$$\log g_k(\mathbf{y}_\theta) \leqslant (1 - \theta) \log g_k(\mathbf{y}_0) + \theta \log g_k(\mathbf{y}_1)$$

and so $\log g_k(\mathbf{y})$ satisfies the definition (9.4.3) of a convex function.

It follows from this result that (13.2.7) is a convex programming problem (see (9.4.8)). Thus the dual problem (9.5.1) becomes

$$\begin{aligned} &\underset{\mathbf{y},\, \boldsymbol{\lambda}}{\text{maximize}} \ \mathscr{L}(\mathbf{y}, \boldsymbol{\lambda}) \triangleq \log g_0(\mathbf{y}) + \sum_{k=1}^{m} \lambda_k \log g_k(\mathbf{y}) \\ &\text{subject to} \ \nabla_y \mathscr{L}(\mathbf{y}, \boldsymbol{\lambda}) = \mathbf{0}, \qquad \boldsymbol{\lambda} \geqslant \mathbf{0}. \end{aligned} \tag{13.2.9}$$

The Lagrangian can be written more simply by defining $\lambda_0 = 1$ so that

$$\mathscr{L}(\mathbf{y}, \boldsymbol{\lambda}) = \sum_{k=0}^{m} \lambda_k \log g_k(\mathbf{y}). \tag{13.2.10}$$

As it stands, (13.2.9) has non-linear constraints so it is important to show that it can be simplified by a further change of variable. Write the equality constraints as

$$\frac{\partial \mathscr{L}}{\partial y_j} = 0 = \sum_{k=0}^{m} \lambda_k \sum_{i \in J_k} \frac{a_{ij} t_i}{g_k}, \qquad j = 1, 2, \ldots, n. \tag{13.2.11}$$

These equations suggest defining new non-negative variables δ_i by

$$\delta_i = \frac{\lambda_k t_i}{g_k}, \qquad \forall \ i \in J_k, \ \ k = 0, 1, \ldots, m, \tag{13.2.12}$$

which correspond to the terms (13.2.1), suitably weighted. Then (13.2.11) becomes

$$\sum_{i=1}^{p} \delta_i a_{ij} = 0, \qquad j = 1, 2, \ldots, n, \tag{13.2.13}$$

and these equations are known as *orthogonality* constraints. Thus the equality constraint in (13.2.9) is equivalent to a set of (under-determined) linear equations in terms of the δ_i variables. An equivalent form of (13.2.4) is also implied by (13.2.12), that is

$$\sum_{i \in J_k} \delta_i = \lambda_k, \qquad k = 0, 1, \ldots, m, \tag{13.2.14}$$

and is known as a *normalization* constraint. Furthermore it is possible to simplify the dual objective function (13.2.10), because (assuming 0 log 0 = 0)

$$\begin{aligned}
& \lambda_k \log g_k = \lambda_k \log \lambda_k + \lambda_k \log(g_k/\lambda_k) \\
\text{by (13.2.14)} \quad & = \lambda_k \log \lambda_k + \sum_{i \in J_k} \delta_i \log(g_k/\lambda_k) \\
\text{by (13.2.12)} \quad & = \lambda_k \log \lambda_k + \sum_{i \in J_k} \delta_i \log(t_i/\delta_i) \\
\text{by (13.2.6)} \quad & = \lambda_k \log \lambda_k + \sum_{i \in J_k} \delta_i \log(c_i/\delta_i) + \sum_{i \in J_k} \delta_i \sum_{j=1}^{n} a_{ij} y_j \\
\text{by (13.2.13)} \quad & = \lambda_k \log \lambda_k + \sum_{i \in J_k} \delta_i \log(c_i/\delta_i).
\end{aligned} \tag{13.2.15}$$

Thus the weighted terms δ_i and the Lagrange multipliers λ_k satisfy the following *dual geometric programming problem*

$$\begin{aligned}
\underset{\boldsymbol{\delta}, \boldsymbol{\lambda}}{\text{maximize}} \quad & v(\boldsymbol{\delta}, \boldsymbol{\lambda}) \triangleq \sum_{i=1}^{p} \delta_i \log(c_i/\delta_i) + \sum_{k=1}^{m} \lambda_k \log \lambda_k \\
\text{subject to} \quad & \sum_{i=1}^{p} \delta_i a_{ij} = 0 \qquad j = 1, 2, \ldots, n \\
& \sum_{i \in J_k} \delta_i = \lambda_k \qquad k = 0, 1, \ldots, m \\
& \delta_i \geqslant 0 \qquad i = 1, 2, \ldots, p \\
& \lambda_k \geqslant 0 \qquad k = 1, 2, \ldots, m
\end{aligned} \tag{13.2.16}$$

where $\lambda_0 = 1$.

This then is the resulting transformed problem which is more easily solved than (13.2.2). The problem has all linear constraints and a non-linear objective function so can be solved by the methods of Chapter 11. The number of degrees of freedom left when the constraints have been eliminated is known as the *degree of difficulty* of the problem, and is $p - n - 1$ if all the primal constraints are active. In some cases the degree of difficulty is known to be zero, in which case (13.2.16) reduces just to the solution of the orthogonality and normalization equations. Usually however it is necessary to solve (13.2.16) by a suitable numerical method. In fact (13.2.16) has other advantageous features. The dual objective function $v(\boldsymbol{\delta}, \boldsymbol{\lambda})$ is *separable* into one variable terms so that the Hessian matrix is diagonal and is readily calculated. If the λ_k variables are eliminated from (13.2.14) then the resulting function (of $\boldsymbol{\delta}$ only) is concave (see Duffin, Peterson, and Zener, 1967), so it follows from theorem 9.4.1 that any solution of (13.2.16) is a global solution. A more thorough study of how best to solve (13.2.16) by using a linearly constrained Newton method of the type described in Chapter 11 is given by Dembo (1979). Amongst the factors which he takes into account are the need to eliminate in a stable fashion as described in Chapter 10 here. He also considers how best to take sparsity of the exponent matrix $[a_{ij}]$ into account, and extends (13.2.2) to include lower and upper bound constraints

$$0 \leqslant \ell_i \leqslant x_i \leqslant u_i.$$

(These can be written as posynomial constraints $u_i^{-1}x_i \leqslant 1$ and $\ell_i x_i^{-1} \leqslant 1$ but it is more efficient to treat them separately.) A particularly interesting feature is that when a posynomial constraint $g_k \leqslant 1$ is inactive at the solution then a whole block of variables δ_i, $i \in J_k$, and λ_k are zero. Thus Dembo considers modifications to the active set strategy which allow the addition or removal of such blocks of indices to or from the active set.

Once the dual solution $\boldsymbol{\delta}^*$, $\boldsymbol{\lambda}^*$ has been found, then the primal function value is given by $\log g_0^* = v(\boldsymbol{\delta}^*, \boldsymbol{\lambda}^*)$ (see theorem 9.5.1). It is also important to be able to recover the optimum variables $\mathbf{y}^*$ and hence $\mathbf{x}^*$ in the primal problem. In principle it is possible to do this by solving the non-linear equations (13.2.4) and (13.2.6) for the active primal constraints for which $\lambda_k^* > 0$ and hence $g_k^* = 1$. However a more simple possibility is to determine the optimum multipliers w_j^*, $j = 1, 2, \ldots, n$, of the orthogonality constraints in (13.2.16). These multipliers are available directly if (13.2.16) is solved by an active set method such as in Chapter 11.2. Then

$$w_j^* = y_j^* = \log x_j^* \tag{13.2.17}$$

determines the primal solution. To justify (13.2.17), the first order conditions (9.1.3) applied to (13.2.16) give

$$\begin{aligned} \log(\delta_i/c_i) + 1 &= \mu_k + \sum_j a_{ij} w_j, \qquad i \in J_k \\ -\log \lambda_k - 1 &= -\mu_k \end{aligned} \tag{13.2.18}$$

where w_j and μ_k are the multipliers of the orthogonality and normalization constraints respectively and k indexes the active primal constraints. Eliminating μ_k gives

$$\log \frac{\delta_i}{c_i \lambda_k} = \sum_j a_{ij} w_j$$

and since $g_k^* = 1$, it follows from (13.2.12) and (13.2.6) that

$$\Sigma\, a_{ij} y_j = \Sigma\, a_{ij} w_j.$$

Since the exponent matrix $[a_{ij}]$ has more rows than columns and may be assumed to have full rank (see below), the result (13.2.17) follows. There is no loss of generality in assuming that $[a_{ij}]$ has full rank. If not, then from (13.2.6) it is possible to eliminate one or more y_j variable in favour of the remaining y_j variables, so as to give a matrix $[a_{ij}]$ of full rank.

Finally a simple example of the geometric programming transformation is given. Consider the geometric programming problem in two variables with one constraint

$$\begin{aligned} &\underset{\mathbf{x}}{\text{minimize}}\ g_0(\mathbf{x}) \triangleq \frac{2}{x_1 x_2^{1/2}} + x_1 x_2 \\ &\text{subject to } g_1(\mathbf{x}) \triangleq \tfrac{1}{2}x_1 + x_2 \leqslant 1, \qquad x_1, x_2 > 0. \end{aligned} \tag{13.2.19}$$

There are four terms, so the orthogonality constraints are

$$\begin{aligned} -\ \delta_1 + \delta_2 + \delta_3 \qquad &= 0 \\ -\tfrac{1}{2}\delta_1 + \delta_2 \qquad + \delta_4 &= 0 \end{aligned}$$

and the normalization constraints are

$$\begin{aligned} \delta_1 + \delta_2 \qquad\qquad &= 1 \\ \delta_3 + \delta_4 &= \lambda_1. \end{aligned}$$

Assuming that the primal constraint is active, the δ_i variables can be eliminated in terms of λ_1 (= λ) giving

$$\boldsymbol{\delta}(\lambda) = \frac{(2\lambda + 4, -2\lambda + 3, 3\lambda - 1, 4\lambda + 1)^{\mathrm{T}}}{7} \tag{13.2.20}$$

This illustrates that there is one degree of difficulty (= 4 − 2 − 1) and therefore a maximization problem in one variable (λ here) must be solved. Maximizing the dual objective function $v(\boldsymbol{\delta}(\lambda), \lambda)$ by a line search in λ gives $\lambda^* = 1.0247$ and $v^* = 1.10764512$. Also from (13.2.20) it follows that

$$\boldsymbol{\delta}^* = (0.86420, 0.13580, 0.72841, 0.29630)^{\mathrm{T}}.$$

Thus the dual geometric programming problem has been solved, and immediately $g_0^* = \exp v^* = 3.02722126$. Substituting in (13.2.18) for $k = 1$ and $J_1 = \{3, 4\}$ gives $\mu_1^* = 1.02441$ and hence

$$\begin{pmatrix} 0.35184 \\ -1.24078 \end{pmatrix} = \begin{bmatrix} 1 & 0 \\ 0 & 1 \end{bmatrix} \begin{pmatrix} w_1^* \\ w_2^* \end{pmatrix}.$$

Hence by (13.2.17) the vector of variables which solves the primal is $\mathbf{x}^* = (1.4217, 0.28915)^{\mathrm{T}}$. In fact the primal problem is fairly simple and it is also possible to obtain a solution to this directly by eliminating x_2 and carrying out a line search. This calculation confirms the solution given by the geometric programming transformation.

Questions for Chapter 13

1. Solve graphically the LP

$$\begin{aligned} \text{minimize} \quad & -x_1 - 2x_2 \\ \text{subject to} \quad & -2x_1 + 2x_2 \leqslant 3 \\ & 2x_1 + 2x_2 \leqslant 9, \end{aligned}$$

for the cases (i) no integer restriction, (ii) x_1 integer, (iii) x_1, x_2 integer. Show how case (iii) would be solved by the branch and bound method, determining problem lower bounds by $L = f' + e_i^+$ or $L = f' + e_i^-$. Give the part of the solution tree which is explored. Show also that although the method solves the problem finitely, the tree contains two infinite paths. Show that if the nearest integer is used as a criterion for which problem to explore first after branching, then it is possible that one of these infinite paths may be followed, depending on how ties are broken, so that the algorithm does not converge. Notice however that if the best pending node is selected after *every* branch then the algorithm does converge. If bounds $\mathbf{x} \geqslant \mathbf{0}$ are added to the problem, enumerate the whole tree: there are ten ○ nodes, six ● nodes and five □ nodes in all.

2. Solve the integer programming problem

$$\begin{aligned} \text{minimize} \quad & x_1^4 + x_2^4 + 16(x_1x_2 + (4 + x_2)^2) \\ \text{subject to} \quad & x_1, x_2 \text{ integer} \end{aligned}$$

by the branch and bound method. In doing this, problems in which $x_i \leqslant n_i$ or $x_i \geqslant n_i + 1$ arise. Treat these as equations, and solve the resulting problem by a line search in the remaining variable (or by using a quasi-Newton subroutine with the ith row and column of $\mathbf{H}^{(1)}$ zeroed). Calculate Lagrange multipliers at the resulting solution and verify that the inequality problem is solved. Use the distance to the nearest integer as the criterion for which branch to select. Give the part of the solution tree which is explored.

3. Consider the geometric programming problem

$$\begin{aligned} \text{minimize} \quad & x_1^{-1}x_2^{-1}x_3^{-1} \\ \text{subject to} \quad & x_1 + 2x_2 + 2x_3 \leqslant 72, \qquad \mathbf{x} > \mathbf{0} \end{aligned}$$

derived from Rosenbrock's parcel problem (see Question 9.3). Show that the problem has zero degree of difficulty and hence solve the equations (13.2.13) and (13.2.14) to solve (13.2.16). Show immediately that the solution value is 1/3456. Find the multipliers w_j^*, $j = 1, 2, 3$, of the orthogonality constraints and hence determine the optimum variables from (13.2.17).

4. Consider the geometric programming problem

$$\begin{aligned} \text{minimize} \quad & 40x_1x_2 + 20x_2x_3 \\ \text{subject to} \quad & \tfrac{1}{5}x_1^{-1}x_2^{-1/2} + \tfrac{3}{5}x_2^{-1}x_3^{-2/3} \leqslant 1, \qquad \mathbf{x} > \mathbf{0}. \end{aligned}$$

Show that the problem has zero degree of difficulty and can be solved in a similar way to the previous problem.

5. Consider the geometric programming problem

$$\text{minimize} \quad 40x_1^{-1}x_2^{-1/2}x_3^{-1} + 20x_1x_3 + 40x_1x_2x_3$$

$$\text{subject to} \quad \tfrac{1}{3}x_1^{-2}x_2^{-2} + \tfrac{4}{3}x_2^{1/2}x_3^{-1} \leqslant 1, \qquad \mathbf{x} > \mathbf{0}.$$

Show that the problem has one degree of difficulty. Hence express $\boldsymbol{\delta}$ in terms of the parameter λ ($= \lambda_1$) as in (13.2.20), and solve the problem numerically by maximizing the function $v(\boldsymbol{\delta}(\lambda), \lambda)$ by a line search in λ. Then solve for the variables using (13.2.17). This and the previous problem are given by Duffin, Peterson, and Zener (1967) where many more examples of geometric programming problems can be found.

Chapter 14

Non-differentiable Optimization

14.1 Introduction

In Volume 1 on unconstrained optimization an early assumption is to exclude all problems for which the objective function $f(\mathbf{x})$ is not smooth. Yet practical problems sometimes arise (*non-differentiable optimization* (NDO) or *non-smooth optimization* problems) which do not meet this requirement; this chapter studies progress which has been made in the practical solution of such problems. Examples of NDO problems occur when solving non-linear equations $c_i(\mathbf{x}) = 0$, $i = 1, 2, \ldots, m$ (see Section 6.1, Volume 1) by minimizing $\|\mathbf{c}(\mathbf{x})\|_1$ or $\|\mathbf{c}(\mathbf{x})\|_\infty$. This arises either when solving simultaneous equations exactly ($m = n$) or when finding best solutions to over-determined systems ($m > n$) as in data fitting applications. Another similar problem is that of finding a feasible point of a system of non-linear inequalities $c_i(\mathbf{x}) \leqslant 0$, $i = 1, 2, \ldots, m$, by minimizing $\|\mathbf{c}(\mathbf{x})^+\|_1$ or $\|\mathbf{c}(\mathbf{x})^+\|_\infty$ where $c_i^+ = \max(c_i, 0)$. A generalization of these problems arises when the equations or inequalities are the constraints in a nonlinear programming problem (Chapter 12). Then a possible approach is to use an exact penalty function and minimize functions like $\nu f(\mathbf{x}) + \|\mathbf{c}(\mathbf{x})\|$ or $\nu f(\mathbf{x}) + \|\mathbf{c}(\mathbf{x})^+\|$, in particular using the L_1 norm. This idea is attracting much research interest and is expanded upon in Section 14.3. Yet another type of problem is to minimize the *max function* $\max_i c_i(\mathbf{x})$ where the max is taken over some finite set. This includes many examples from electrical engineering including microwave network design and digital filter design (Charalambous, 1979). In fact almost all these examples can be considered in a more generalized sort of way as special cases of a certain *composite function* and a major portion of this chapter is devoted to studying this type of function.

Another common source of NDO problems arises when using the decomposition principle. For example the LP problem

$$\underset{\mathbf{x},\, \mathbf{y}}{\text{minimize}} \quad \mathbf{c}^{\mathrm{T}}\mathbf{x} + \mathbf{d}^{\mathrm{T}}\mathbf{y}$$

$$\text{subject to } \mathbf{A}\mathbf{x} + \mathbf{B}\mathbf{y} \leqslant \mathbf{b}$$

can be written as the convex NDO problem

$$\underset{\mathbf{x}}{\text{minimize}}\ f(\mathbf{x}) \triangleq \mathbf{c}^{\mathrm{T}}\mathbf{x} + \underset{\mathbf{y}}{\min}[\mathbf{d}^{\mathrm{T}}\mathbf{y} : \mathbf{B}\mathbf{y} \leqslant \mathbf{b} - \mathbf{A}\mathbf{x}].$$

Application of NDO methods to column generation problems and to a variety of scheduling problems is described by Marsten (1975) and the application to network scheduling problems is described by Fisher, Northup, and Shapiro (1975). Both these papers appear in the Mathematical Programming Study 3 (Balinski and Wolfe, 1975) which is a valuable reference. Another good source is the book by Lemarechal and Mifflin (1978) and their test problems 3 and 4 illustrate further applications.

In so far as I understand them, all these applications could in principle be formulated as max functions and hence as composite functions (see below). However in practice this would be too complicated or require too much storage. Thus a different situation can arise in which the only information available at any point $\mathbf{x}$ is $f(\mathbf{x})$ and a vector $\mathbf{g}$. Usually $\mathbf{g}$ is ∇f, but if f is non-differentiable at $\mathbf{x}$ then $\mathbf{g}$ is an element of the subdifferential ∂f (see Section 14.2) for convex problems, or the generalized gradient in non-convex problems. Since less information about $f(\mathbf{x})$ is available, this type of application is more difficult than the composite function application and is referred to here as *basic NDO*. However both types of application have some structure in common.

In view of this common structure, algorithms for all types of NDO problem are discussed together in Section 14.4. For max functions it is shown that an algorithm with second order convergence can be obtained by making linearizations of the individual functions over which the max is taken, together with an additional (smooth) quadratic approximation which includes curvature information from the functions in the max. Variants of this algorithm have been used in a number of practical applications. The algorithm is closely related to one in which the composite function is approximated in a similar way. This algorithm can be made globally convergent by incorporating the idea of a trust region and details of this result are given in Section 14.5. The algorithm is applicable to both max function and L_1 and L_∞ approximation problems and to nonlinear programming applications of non-differentiable exact penalty functions. Algorithms for basic NDO have not progressed as far because of the difficulties caused by the limited availability of information. One type of method – the *bundle method* – tries to accumulate information locally about the subdifferential of the objective function. Alternatively linearization methods have also been tried and the possibility of introducing second order information into these algorithms is being considered. In fact there is currently much research interest in NDO algorithms of all kinds and further developments can be expected.

A prerequisite for describing NDO problems and algorithms is a study of optimality conditions for non-differentiable functions. This can be done at various levels of generality. I have tried to make the approach here as simple and readable as possible (although this is not easy in view of the inherent difficulty of the material), whilst trying to cover all practical applications. The chapter is largely self-contained and does not rely on any key results from outside sources. One requirement is the extension of the material in Section 9.4 on convex functions to include the non-differentiable case. The concept of a subdifferential is introduced and the resulting definitions of directional derivatives lead to a simple statement of first order conditions. More general applications can be represented by the *composite function*

$$\phi(\mathbf{x}) \triangleq f(\mathbf{x}) + h(\mathbf{c}(\mathbf{x})) \tag{14.1.1}$$

which I shall refer to as *composite NDO*. Here $f(\mathbf{x})$ ($\mathbb{R}^n \to \mathbb{R}^1$) and $\mathbf{c}(\mathbf{x})$ ($\mathbb{R}^n \to \mathbb{R}^m$) are smooth functions ($\mathbb{C}^1$), and $h(\mathbf{c})$ ($\mathbb{R}^m \to \mathbb{R}^1$) is convex but non-smooth ($\mathbb{C}^0$). In some applications the smooth function $f(\mathbf{x})$ may be absent. Important special cases of $h(\mathbf{c})$ are set out in (14.1.3) below. Optimality conditions for (14.1.1) can be obtained as a straightforward extension of those for convex functions. This material is set out in Section 14.2 and both first and second order conditions are studied. There are NDO problems (for example $\phi(\mathbf{x}) = \max(0, \min(x_1, x_2))$) which do not fit into this category and which are not even well modelled by (14.1.1) locally, although I do not know of any which arise in practice. Wider classes of function have been introduced which cover these cases, for example the 'locally Lipschitz' functions of Clarke (1975), but substantial complication in the analysis is introduced. Also these classes do not directly suggest algorithms as the function (14.1.1) does. Furthermore the resulting first order conditions permit descent directions to occur and so are of little practical use (Womersley, 1980). Hence these classes are not studied any further here.

In most cases (but not all – for example the exact penalty function $\phi = \nu f + \|\mathbf{c}\|_2$) a further specialization can be made in which $h(\mathbf{c})$ is restricted to be a *polyhedral convex function*. In this case the graph of $h(\mathbf{c})$ is made up of a finite number of supporting hyperplanes $\mathbf{c}^{\mathrm{T}}\mathbf{h}_i + b_i$, and $h(\mathbf{c})$ is thus defined by

$$h(\mathbf{c}) \triangleq \max_i(\mathbf{c}^{\mathrm{T}}\mathbf{h}_i + b_i) \tag{14.1.2}$$

where the vectors $\mathbf{h}_i$ and scalars b_i are given. Thus the polyhedral convex function is a max function in terms of linear combinations of the elements of $\mathbf{c}$. A simple illustration is given in Figure 14.1.1. Most interest lies in five special cases of (14.1.2), in all of which $b_i = 0$ for all i. These functions are set out below, together with the vectors $\mathbf{h}_i$ which define them given as columns of a matrix $\mathbf{H}$.

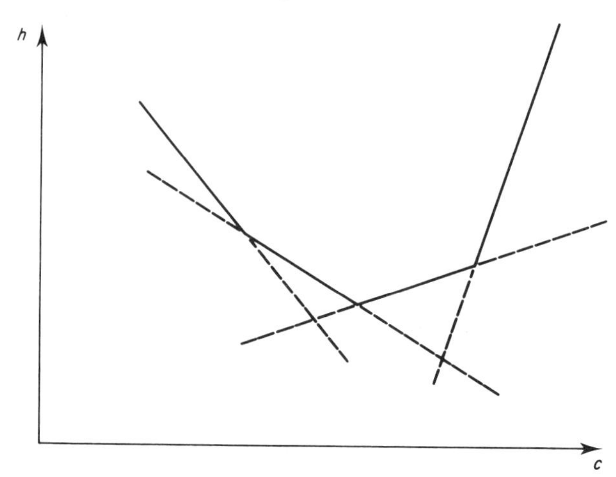

Figure 14.1.1 Polyhedral convex function

Case I	$h(\mathbf{c}) = \max_i c_i$	$\mathbf{H} = \mathbf{I}$ $\quad (m \times m)$	
Case II	$h(\mathbf{c}) = \|\mathbf{c}^+\|_\infty$	$\mathbf{H} = [\mathbf{I} : \mathbf{0}]$ $\quad (m \times (m+1))$	
Case III	$h(\mathbf{c}) = \|\mathbf{c}\|_\infty$	$\mathbf{H} = [\mathbf{I} : -\mathbf{I}]$ $\quad (m \times 2m)$	
Case IV	$h(\mathbf{c}) = \|\mathbf{c}^+\|_1$	columns of **H** are all possible combinations of 1 and 0 $\quad (m \times 2^m)$	(14.1.3)
Case V	$h(\mathbf{c}) = \|\mathbf{c}\|_1$	columns of **H** are all possible combinations of 1 and -1 $\quad (m \times 2^m)$	

For example with $m = 2$ in case (V) the matrix **H** is $\mathbf{H} = \begin{bmatrix} 1 & 1 & -1 & -1 \\ 1 & -1 & 1 & -1 \end{bmatrix}$.

Contours of these functions for $m = 2$ are illustrated in Figure 14.1.2. The broken lines indicate the surfaces along which the different linear pieces join up and along which the derivative is discontinuous. The term *piece* is used to denote that part of the graph of a max function in which one particular function achieves the maximum value, for example the graph illustrated in case (II) of Figure 14.1.2 is made up of

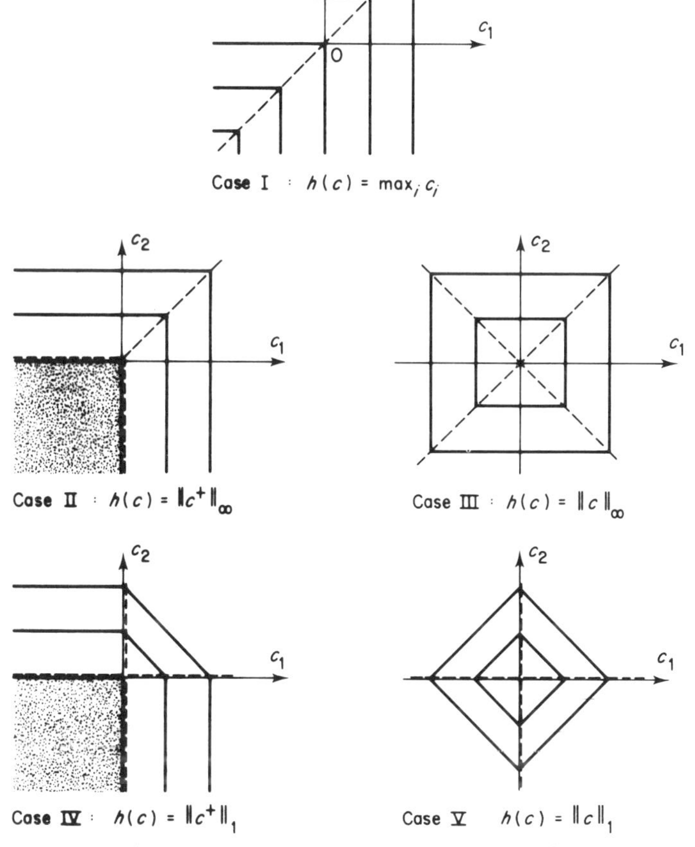

Figure 14.1.2 Contours of polyhedral convex functions

three linear pieces and the active equations are $h(\mathbf{c}) = c_2$ for $c_1 \leqslant c_2$ and $c_2 \geqslant 0$, $h(\mathbf{c}) = c_1$ for $c_2 \leqslant c_1$ and $c_1 \geqslant 0$, and $h(\mathbf{c}) = 0$ for $c_1 \leqslant 0$ and $c_2 \leqslant 0$. The subdifferential $\partial h(\mathbf{c})$ for a polyhedral convex function has the simple form expressed in lemma 14.2.2 as the convex hull of the gradients of the active pieces. The polyhedral nature of $h(\mathbf{c})$ has important consequences in regard to second order conditions, and this is considered later in Section 14.2.

In fact it is possible to derive optimality results for the composite function $\phi(\mathbf{x})$ in (14.1.1) when $h(\mathbf{c})$ is the polyhedral function (14.1.2) without using notions of convexity at all. Clearly $\phi(\mathbf{x})$ can be written

$$\phi(\mathbf{x}) = \max_i (f(\mathbf{x}) + \mathbf{c}(\mathbf{x})^{\mathrm{T}}\mathbf{h}_i + b_i) \tag{14.1.4}$$

which is equivalent to

$$\phi(\mathbf{x}) = \min v : \quad v \geqslant f(\mathbf{x}) + \mathbf{c}(\mathbf{x})^{\mathrm{T}}\mathbf{h}_i + b_i \qquad \forall\, i. \tag{14.1.5}$$

Therefore $\mathbf{x}^*$ minimizes $\phi(\mathbf{x})$ locally iff $\mathbf{x}^*, v^*$ is a local solution of the nonlinear programming problem

$$\begin{aligned} &\underset{\mathbf{x},\, v}{\text{minimize}} \quad v \\ &\text{subject to } v - f(\mathbf{x}) - \mathbf{c}(\mathbf{x})^{\mathrm{T}}\mathbf{h}_i \geqslant b_i \qquad \forall\, i. \end{aligned} \tag{14.1.6}$$

Thus first and second order optimality conditions for (14.1.4) can be obtained by applying the equivalent results in nonlinear programming to (14.1.6), that is theorems 9.1.1, 9.3.1, and 9.3.2. It turns out however that this approach is less general and somewhat clumsy, so it is only sketched out briefly here, and the derivation given in Section 14.2 is preferred. The first difficulty concerns the regularity assumption $\mathscr{F}' = F'$. It is possible (but not trivial – see Question 14.1) to prove that this always holds, and therefore it is important not to make an unnecessary independence assumption. The appropriate Lagrangian function for theorem 9.1.1 is

$$\mathscr{L}(\mathbf{x}, v, \boldsymbol{\mu}) = v - \sum_i \mu_i (v - f(\mathbf{x}) - \mathbf{c}(\mathbf{x})^{\mathrm{T}}\mathbf{h}_i - b_i) \tag{14.1.7}$$

and it is implied at $\mathbf{x}^*, v^*$ that multipliers $\boldsymbol{\mu}^*$ exist such that

$$\begin{aligned} &\frac{\partial}{\partial v}\mathscr{L}(\mathbf{x}^*, v^*, \boldsymbol{\mu}^*) = 0 \qquad \text{or} \qquad \Sigma \mu_i^* = 1 \\ &\nabla_x \mathscr{L}(\mathbf{x}^*, v^*, \boldsymbol{\mu}^*) = \mathbf{0} \qquad \text{or} \qquad \Sigma \mu_i^* (\mathbf{g}^* + \mathbf{A}^*\mathbf{h}_i) = \mathbf{0} \\ &\boldsymbol{\mu}^* \geqslant \mathbf{0} \\ &\mu_i^* > 0 \Rightarrow v^* = f^* + \mathbf{c}^{*\mathrm{T}}\mathbf{h}_i + b_i. \end{aligned} \tag{14.1.8}$$

As in previous chapters the notation $\mathbf{g} = \nabla f$ and $\mathbf{A} = \nabla \mathbf{c}^{\mathrm{T}}$ is used and $f^* = f(\mathbf{x}^*)$, etc. If $\boldsymbol{\lambda} = \mathbf{H}\boldsymbol{\mu}$ is written, then the existence of a vector $\boldsymbol{\mu}^*$ in the above is equivalent to the existence of a vector $\boldsymbol{\lambda}^*$ in the set

$$\partial h^* = \underset{i \in \mathscr{A}^*}{\operatorname{conv}} \mathbf{h}_i \tag{14.1.9}$$

such that $\mathbf{g}^* + \mathbf{A}^*\boldsymbol{\lambda}^* = \mathbf{0}$. In fact ∂h^* is the subdifferential of $h(\mathbf{c}^*)$ as shown in lemma 14.2.2. $\mathscr{A}^*$ is the set of active constraints or equivalently the set of indices at which the max in (14.1.4) is attained, that is

$$\mathscr{A}^* = \{i : \mathbf{c}^{*\mathrm{T}}\mathbf{h}_i + b_i = h(\mathbf{c}^*)\}. \tag{14.1.10}$$

It is possible to use (14.1.9) to obtain equivalent but more convenient expressions for $\partial h(\mathbf{c})$ in cases (I) to (V) in (14.1.3) above as follows:

$$\partial \max_i c_i = \{\boldsymbol{\lambda}: \ \Sigma\lambda_i = 1, \ \boldsymbol{\lambda} \geqslant \mathbf{0}, \ c_i < \max_i c_i \Rightarrow \lambda_i = 0\} \tag{14.1.11}$$

$$\partial \|\mathbf{c}^+\|_\infty = \{\boldsymbol{\lambda}: \ \Sigma\lambda_i \leqslant 1, \boldsymbol{\lambda} \geqslant \mathbf{0}, \ c_i < \|\mathbf{c}^+\|_\infty \Rightarrow \lambda_i = 0\} \tag{14.1.12}$$

$$\begin{aligned}\partial \|\mathbf{c}\|_\infty = \{\boldsymbol{\lambda}: \ \mathbf{c} \neq \mathbf{0} \Rightarrow \ & \Sigma\lambda_i = 1 \\ & |c_i| < \|\mathbf{c}\|_\infty \Rightarrow \lambda_i = 0 \\ & |c_i| = \|\mathbf{c}\|_\infty \Rightarrow \lambda_i c_i \geqslant 0 \\ \mathbf{c} = \mathbf{0} \Rightarrow \ & \Sigma\lambda_i \leqslant 1\}\end{aligned} \tag{14.1.13}$$

$$\begin{aligned}\partial \|\mathbf{c}^+\|_1 = \{\boldsymbol{\lambda}: \ 0 \leqslant \lambda_i \leqslant 1, \ & c_i > 0 \Rightarrow \lambda_i = 1 \\ & c_i < 0 \Rightarrow \lambda_i = 0\}\end{aligned} \tag{14.1.14}$$

$$\partial \|\mathbf{c}\|_1 = \{\boldsymbol{\lambda}: |\lambda_i| \leqslant 1, c_i \neq 0 \Rightarrow \lambda_i = \text{sign } c_i\}. \tag{14.1.15}$$

The equivalence between these sets and those defined by $\text{conv}_{i \in \mathscr{A}}\mathbf{h}_i$ is left as an exercise (Question 14.3). Expressions (14.1.12) to (14.1.15) can also be derived as special cases of (14.3.7) or (14.3.8).

It is also possible to apply the development of (9.3.9) onwards to problem (14.1.6) to obtain the second order conditions of theorems 14.2.2 and 14.2.3, and also the regularity condition (14.2.25) (see lemma 14.2.7). To do this an index $p \in \mathscr{A}_+^*$ is used to eliminate v (or s_{n+1} – see Question 14.1). The analysis is not entirely straightforward so again the more direct approach in Section 14.2 is preferred. The equivalence of the two different approaches is however shown by lemma 14.2.8.

The parameters $\boldsymbol{\lambda}^* \in \partial h^*$ which exist at the solution to an NDO problem are closely related to Lagrange multipliers and indeed they can be given a simple interpretation similar to (9.1.10). Consider a perturbed problem in which $\mathbf{c}(\mathbf{x})$ is replaced by $\mathbf{c}(\mathbf{x}) + \boldsymbol{\epsilon}$ giving a function $\phi_\epsilon(\mathbf{x})$ and assume that the minimizer $\mathbf{x}(\boldsymbol{\epsilon})$ is such that the same constraints are active in (14.1.6). It follows that the right hand side of each active constraint is perturbed by an amount $\boldsymbol{\epsilon}^\mathrm{T}\mathbf{h}_i$. Since μ_i^* is the multiplier of the ith constraint and using (9.1.10) it follows that the change to v is $\Delta v = \Sigma_i \boldsymbol{\epsilon}^\mathrm{T}\mathbf{h}_i\mu_i^* = \boldsymbol{\epsilon}^\mathrm{T}\boldsymbol{\lambda}^*$ and hence that

$$\frac{\mathrm{d}\phi^*}{\mathrm{d}\epsilon_i} = \lambda_i^*. \tag{14.1.16}$$

Thus λ_i^* measures the first order rate of change of the optimum function value consequent on perturbations to c_i. This result can also be used to illustrate the need for the condition $\boldsymbol{\lambda} \geqslant \mathbf{0}$ in (14.1.11). For example if the first order conditions arising

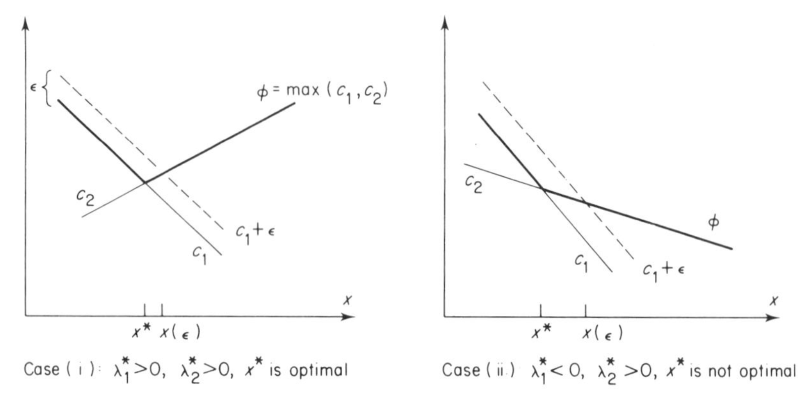

Figure 14.1.3 Interpretation of multipliers in NDO

from (14.1.8) ($\exists\, \boldsymbol{\lambda}^* \in \partial h^*$ such that $\mathbf{g}^* + \mathbf{A}^* \boldsymbol{\lambda}^* = \mathbf{0}$) are satisfied except that $\lambda_i^* < 0$ for some i, then it is possible to show that $\mathbf{x}^*$ is not optimal. Consider a perturbation $\boldsymbol{\epsilon}$ with $\epsilon_i > 0$ and $\epsilon_j = 0$, $j \neq i$. Existence of the solution $\mathbf{x}(\boldsymbol{\epsilon})$ follows under a suitable independence assumption using the implicit function theorem. At $\mathbf{x}(\boldsymbol{\epsilon})$ the max in (14.1.2) is achieved by all $j \in \mathscr{A}^*$ if $\boldsymbol{\epsilon}$ is small enough, so for all $j \in \mathscr{A}^*$, $j \neq i$,

$$c_j(\mathbf{x}(\boldsymbol{\epsilon})) = c_i(\mathbf{x}(\boldsymbol{\epsilon})) + \epsilon_i .$$

Thus $c_i(\mathbf{x}(\boldsymbol{\epsilon})) < c_j(\mathbf{x}(\boldsymbol{\epsilon}))$ and hence $\phi_\epsilon(\mathbf{x}(\boldsymbol{\epsilon})) = \phi_0(\mathbf{x}(\boldsymbol{\epsilon}))$ where ϕ_0 refers to the unperturbed function. It follows from (14.1.16) that $\mathrm{d}\phi_0^*/\mathrm{d}\epsilon_i = \lambda_i^* < 0$ and so $\mathbf{x}^*$ is not a local solution. The situation is illustrated in Figure 14.1.3. Clearly increasing c_1 by ϵ does not reduce $\phi(\mathbf{x})$ in case (i) when $\boldsymbol{\lambda} \geqslant \mathbf{0}$ but does reduce $\phi(\mathbf{x})$ in case (ii). Another example of this is given in Question 14.12.

The requirement that the active constraints in (14.1.6) remain the same is important and is in the nature of an independence assumption. This usually holds for case (I) and (II) problems in (14.1.3) and for case (III) problems when $\mathbf{c}^* \neq \mathbf{0}$. However for the L_1 norm functions of cases (IV) and (V) the assumption is not likely to hold. An alternative interpretation of Lagrange multipliers suitable for the L_1 case is derived at the end of Section 14.3.

Finally it is observed that it is possible to attack the problem of *constrained* NDO, although no details are given here. First order conditions for smooth constraint functions are given by Watson (1978) and for constrained L_∞ approximation by Andreassen and Watson (1976). First and second order conditions for a single non-differentiable constraint function involving a norm are given by Fletcher and Watson (1980). There is no difficulty in extending these ideas to a set of constraints defined in terms of composite functions like (14.1.1).

14.2 Optimality Conditions

In this section a number of results are proved leading to optimality conditions for the composite functions described in the previous section. Since these functions

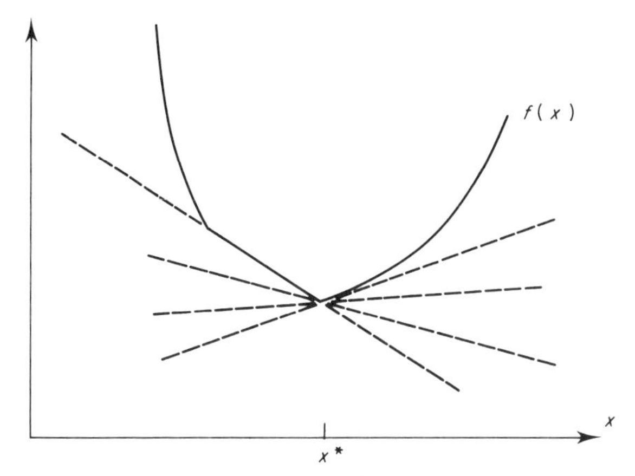

Figure 14.2.1 Supporting hyperplanes to a non-differentiable convex function

are defined in terms of non-differentiable convex functions it is important to study the latter case first. Such a function is illustrated in Figure 14.2.1 and it can be seen that the non-differentiability at $\mathbf{x}^*$ allows the possibility of a number of supporting hyperplanes, in contrast to Figure 9.2.2. (Supporting hyperplanes to the graph of a convex function, which is a convex set, always exist; see Hadley, 1961.) Given a convex function $f(\mathbf{x})$ defined on a convex set $K \subset \mathbb{R}^n$, and given $\mathbf{x} \in$ interior (K), then for each supporting hyperplane at $\mathbf{x}$ it follows that an inequality of the form

$$f(\mathbf{x}+\boldsymbol{\delta}) \geqslant f(\mathbf{x}) + \boldsymbol{\delta}^{\mathrm{T}}\mathbf{g} \qquad \forall\ \mathbf{x}+\boldsymbol{\delta} \in K \tag{14.2.1}$$

holds, and $\mathbf{g}$ is a normal vector of the supporting hyperplane. Such a vector $\mathbf{g}$ is referred to as a *subgradient* at $\mathbf{x}$ and (14.2.1) is known as the *subgradient inequality* and is the generalization of (9.4.4) to non-smooth functions. The set of all subgradients at $\mathbf{x}$ is known as the *subdifferential* at $\mathbf{x}$ and is defined by

$$\partial f(\mathbf{x}) \triangleq \{\mathbf{g}:\ f(\mathbf{x}+\boldsymbol{\delta}) \geqslant f(\mathbf{x}) + \boldsymbol{\delta}^{\mathrm{T}}\mathbf{g} \qquad \forall\ \mathbf{x}+\boldsymbol{\delta} \in K\}. \tag{14.2.2}$$

The notation $\partial f^{(k)} = \partial f(\mathbf{x}^{(k)})$ analogous to $f^{(k)} = f(\mathbf{x}^{(k)})$, etc., is used. It is easy to see that $\partial f(\mathbf{x})$ is a closed convex set, which by lemma 14.2.1 below is also bounded and therefore compact. In fact a more general result can be established.

Lemma 14.2.1

$\partial f(\mathbf{x})$ is bounded for all $\mathbf{x} \in B \subset$ interior (K) where B is compact.

Proof

If the lemma is false, $\exists$ a sequence $\mathbf{g}^{(k)} \in \partial f(\mathbf{x}^{(k)})$, $\mathbf{x}^{(k)} \in B$, such that $\|\mathbf{g}^{(k)}\|_2 \to \infty$. By compactness $\exists$ a subsequence $\mathbf{x}^{(k)} \to \mathbf{x}'$. Define $\boldsymbol{\delta}^{(k)} = \mathbf{g}^{(k)}/\|\mathbf{g}^{(k)}\|_2^2$. Then $\mathbf{x}^{(k)} + \boldsymbol{\delta}^{(k)} \in K$ for k sufficiently large, so by (14.2.1)

$$f(\mathbf{x}^{(k)} + \boldsymbol{\delta}^{(k)}) \geqslant f^{(k)} + \mathbf{g}^{(k)\mathrm{T}}\boldsymbol{\delta}^{(k)} = f^{(k)} + 1.$$

But in the limit, $f^{(k)} \to f'$, $\boldsymbol{\delta}^{(k)} \to \mathbf{0}$, and so $f(\mathbf{x}^{(k)} + \boldsymbol{\delta}^{(k)}) \to f'$, which is a contradiction, so the lemma is true. □

If f is differentiable at $\mathbf{x}$, then

$$f(\mathbf{x} + \boldsymbol{\delta}) = f(\mathbf{x}) + \boldsymbol{\delta}^{\mathrm{T}} \nabla f(\mathbf{x}) + o(\| \boldsymbol{\delta} \|)$$

and subtracting this from (14.2.1) gives

$$\boldsymbol{\delta}^{\mathrm{T}} (\mathbf{g} - \nabla f(\mathbf{x})) \leqslant o(\| \boldsymbol{\delta} \|).$$

Choosing $\boldsymbol{\delta} = \theta(\mathbf{g} - \nabla f(\mathbf{x}))$, $\theta \downarrow 0$, shows that $\mathbf{g} = \nabla f$. Hence in this case $\partial f(\mathbf{x})$ is the single vector $\nabla f(\mathbf{x})$.

In Section 14.1 the particular class of polyhedral convex functions $h(\mathbf{c})$, $\mathrm{IR}^m \to \mathrm{IR}^1$, is defined by

$$h(\mathbf{c}) = \max_i \mathbf{c}^{\mathrm{T}} \mathbf{h}_i + b_i \tag{14.2.3}$$

where $\mathbf{h}_i$ are the columns of a given finite matrix $\mathbf{H}$. Defining

$$\mathscr{A} = \mathscr{A}(\mathbf{c}) \triangleq \{ i: \; \mathbf{c}^{\mathrm{T}} \mathbf{h}_i + b_i = h(\mathbf{c}) \} \tag{14.2.4}$$

as the set of supporting planes which are active at $\mathbf{c}$ and so achieve the maximum, then it is clear that these planes determine the subdifferential $\partial h(\mathbf{c})$. This is proved as follows.

Lemma 14.2.2

$$\partial h(\mathbf{c}) = \operatorname*{conv}_{i \in \mathscr{A}(\mathbf{c})} \mathbf{h}_i \tag{14.2.5}$$

Proof

$\partial h(\mathbf{c})$ is defined by

$$\partial h(\mathbf{c}) = \{ \boldsymbol{\lambda}: \; h(\mathbf{c} + \boldsymbol{\delta}) \geqslant h(\mathbf{c}) + \boldsymbol{\delta}^{\mathrm{T}} \boldsymbol{\lambda} \qquad \forall \; \boldsymbol{\delta} \}. \tag{14.2.6}$$

Let $\boldsymbol{\lambda} = \mathbf{H}\boldsymbol{\mu} \in (14.2.5)$ where $\mu_i \geqslant 0$, $\Sigma \mu_i = 1$. Then for all $\boldsymbol{\delta}$,

$$\begin{aligned} h(\mathbf{c}) + \boldsymbol{\delta}^{\mathrm{T}} \boldsymbol{\lambda} &= \max_i (\mathbf{c}^{\mathrm{T}} \mathbf{h}_i + b_i) + \sum_{i \in \mathscr{A}} \boldsymbol{\delta}^{\mathrm{T}} \mathbf{h}_i \mu_i \\ &\leqslant \max_{i \in \mathscr{A}} (\mathbf{c}^{\mathrm{T}} \mathbf{h}_i + b_i) + \max_{i \in \mathscr{A}} \boldsymbol{\delta}^{\mathrm{T}} \mathbf{h}_i \\ &= \max_{i \in \mathscr{A}} ((\mathbf{c} + \boldsymbol{\delta})^{\mathrm{T}} \mathbf{h}_i + b_i) \leqslant h(\mathbf{c} + \boldsymbol{\delta}). \end{aligned}$$

Thus $\boldsymbol{\lambda} \in (14.2.6)$. Now let $\boldsymbol{\lambda} \in (14.2.6)$ and assume $\boldsymbol{\lambda} \notin (14.2.5)$. Then by lemma 14.2.3 below, $\exists \; \mathbf{s} \neq \mathbf{0}$ such that $\mathbf{s}^{\mathrm{T}} \boldsymbol{\lambda} > \mathbf{s}^{\mathrm{T}} \boldsymbol{\mu} \;\; \forall \; \boldsymbol{\mu} \in (14.2.5)$. Taking $\boldsymbol{\delta} = \alpha \mathbf{s}$, and since $\mathbf{h}_i \in (14.2.5) \; \forall \; i \in \mathscr{A}$,

$$h(\mathbf{c}) + \boldsymbol{\delta}^{\mathrm{T}}\boldsymbol{\lambda} = \max_i(\mathbf{c}^{\mathrm{T}}\mathbf{h}_i + b_i) + \alpha\mathbf{s}^{\mathrm{T}}\boldsymbol{\lambda}$$
$$> \mathbf{c}^{\mathrm{T}}\mathbf{h}_i + b_i + \alpha\mathbf{s}^{\mathrm{T}}\mathbf{h}_i \qquad \forall \;\; i \in \mathscr{A}$$
$$= \max_{i\in\mathscr{A}}((\mathbf{c} + \alpha\mathbf{s})^{\mathrm{T}}\mathbf{h}_i + b_i)$$
$$\geqslant \max_i((\mathbf{c} + \alpha\mathbf{s})^{\mathrm{T}}\mathbf{h}_i + b_i) = h(\mathbf{c} + \boldsymbol{\delta})$$

for α sufficiently small, since the max is then achieved on a subset of $\mathscr{A}$. Thus (14.2.6) is contradicted, proving $\boldsymbol{\lambda} \in$ (14.2.5). Hence the definitions of $\partial h(\mathbf{c})$ in (14.2.5) and (14.2.6) are equivalent. $\square$

Examples of this result for a number of particular cases of $h(\mathbf{c})$ and $\mathbf{H}$ are given in more detail in Section 14.1. For convex functions involving a norm, an alternative description of $\partial h(\mathbf{c})$ is provided by (14.3.7) and (14.3.8). The above lemma makes use of the following important result.

Lemma 14.2.3 (Separating hyperplane lemma for convex sets)

If K is a closed convex set and $\boldsymbol{\lambda} \notin K$ then there exists a hyperplane which separates $\boldsymbol{\lambda}$ and K (see Figure 14.2.2).

Proof

Let $\mathbf{x}_0 \in K$. Then the set $\{\mathbf{x}\colon \|\mathbf{x} - \boldsymbol{\lambda}\|_2 \leqslant \|\mathbf{x}_0 - \boldsymbol{\lambda}\|_2\}$ is bounded and so there exists a minimizer, $\bar{\mathbf{x}}$ say, to the problem: $\min \|\mathbf{x} - \boldsymbol{\lambda}\|_2$, $\mathbf{x} \in K$. Then for any $\mathbf{x} \in K$,

$$\|(1-\theta)\bar{\mathbf{x}} + \theta\mathbf{x} - \boldsymbol{\lambda}\|_2^2 \geqslant \|\bar{\mathbf{x}} - \boldsymbol{\lambda}\|_2^2$$

and in the limit $\theta \downarrow 0$ it follows that

$$(\mathbf{x} - \bar{\mathbf{x}})^{\mathrm{T}}(\boldsymbol{\lambda} - \bar{\mathbf{x}}) \leqslant 0 \qquad \forall \;\; \mathbf{x} \in K.$$

Thus the vector $\mathbf{s} = \boldsymbol{\lambda} - \bar{\mathbf{x}} \neq \mathbf{0}$ satisfies both $\mathbf{s}^{\mathrm{T}}(\boldsymbol{\lambda} - \bar{\mathbf{x}}) > 0$ and $\mathbf{s}^{\mathrm{T}}(\mathbf{x} - \bar{\mathbf{x}}) \leqslant 0$ $\forall\; \mathbf{x} \in K$ and hence

$$\mathbf{s}^{\mathrm{T}}\boldsymbol{\lambda} > \mathbf{s}^{\mathrm{T}}\mathbf{x} \qquad \forall \;\; \mathbf{x} \in K. \tag{14.2.7}$$

The hyperplane $\mathbf{s}^{\mathrm{T}}(\mathbf{x} - \bar{\mathbf{x}}) = 0$ thus separates K and $\boldsymbol{\lambda}$ as illustrated in Figure 14.2.2. $\square$

At any point $\mathbf{x}'$ at which ∇f does not exist, it nonetheless happens that the directional derivative of the convex function in any direction is well defined. It is again assumed that $f(\mathbf{x})$ is defined on a convex set $K \subset \mathbb{R}^n$ and $\mathbf{x}' \in$ interior (K). The following preliminary result is required.

Lemma 14.2.4

Let $\mathbf{x}^{(k)} \to \mathbf{x}'$ and $\mathbf{g}^{(k)} \in \partial f^{(k)}$. Then any accumulation point of $\{\mathbf{g}^{(k)}\}$ is in $\partial f'$.

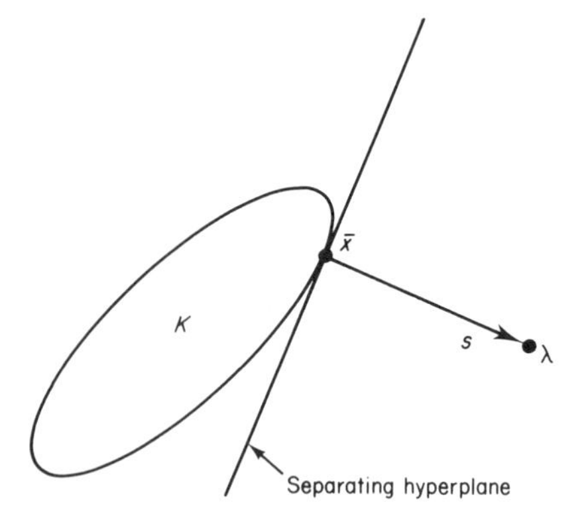

Figure 14.2.2 Existence of a separating hyperplane

Proof

For any $\mathbf{y} \in K$, (14.2.1) can be written as

$$f(\mathbf{y}) \geqslant f^{(k)} + (\mathbf{y} - \mathbf{x}^{(k)})^{\mathrm{T}}\mathbf{g}^{(k)}.$$

Taking any subsequence for which $\mathbf{g}^{(k)} \rightarrow \mathbf{g}'$ it follows that

$$f(\mathbf{y}) \geqslant f' + (\mathbf{y} - \mathbf{x}')^{\mathrm{T}}\mathbf{g}' \qquad \forall \quad \mathbf{y} \in K,$$

that is $\mathbf{g}' \in \partial f'$. $\square$

A result for the directional derivative at $\mathbf{x}'$ in a direction s can now be given in a quite general way for any directional sequence (see Section 9.2).

Lemma 14.2.5

If $\mathbf{x}^{(k)} \rightarrow \mathbf{x}'$ *is any directional sequence with* $\delta^{(k)} \downarrow 0$ *and* $\mathbf{s}^{(k)} \rightarrow \mathbf{s}$ *in (9.2.1) (or more generally* $\delta^{(k)} > 0$ *and* $\delta^{(k)} \rightarrow 0$*), then*

$$\lim_{k\rightarrow\infty} \frac{f^{(k)} - f'}{\delta^{(k)}} = \max_{\mathbf{g} \in \partial f'} \mathbf{s}^{\mathrm{T}}\mathbf{g}. \qquad (14.2.8)$$

Proof

If $\mathbf{g}^{(k)} \in \partial f^{(k)}$ then it follows from (14.2.1) that for k sufficiently large

$$f' \geqslant f^{(k)} - \delta^{(k)}\mathbf{s}^{(k)\mathrm{T}}\mathbf{g}^{(k)}$$

and

$$f^{(k)} \geqslant f' + \delta^{(k)}\mathbf{s}^{(k)\mathrm{T}}\mathbf{g} \qquad \forall \quad \mathbf{g} \in \partial f'$$

both hold, and hence

$$\mathbf{s}^{(k)^\mathrm{T}}\mathbf{g}^{(k)} \geqslant \frac{f^{(k)} - f'}{\delta^{(k)}} \geqslant \max_{\mathbf{g}\in\partial f'} \mathbf{s}^\mathrm{T}\mathbf{g}. \tag{14.2.9}$$

Since $\partial f^{(k)}$ is bounded in a neighbourhood of $\mathbf{x}'$ (lemma 14.2.1), there exists a subsequence for which $\mathbf{g}^{(k)} \to \mathbf{g}'$, and $\mathbf{g}' \in \partial f'$ by lemma 14.2.4. If (14.2.8) is not true then (14.2.9) gives a contradiction in the limit of such a subsequence. □

Thus (14.2.8) shows that the directional derivative is determined by an extreme supporting hyperplane whose subgradient gives the greatest slope: see Figure 14.2.1 for example.

It is possible to deduce from this result that if $\mathbf{x}^*$ is a local minimizer of $f(\mathbf{x})$, then $f^{(k)} \geqslant f^*$ for all k sufficiently large, and hence from (14.2.8) that

$$\max_{\mathbf{g}\in\partial f^*} \mathbf{s}^\mathrm{T}\mathbf{g} \geqslant 0 \qquad \forall\ \mathbf{s}: \|\mathbf{s}\| = 1. \tag{14.2.10}$$

Thus a *first order necessary condition* for a local minimum is that the directional derivative is non-negative in all directions. This can be stated alternatively as

$$\mathbf{0} \in \partial f^* \tag{14.2.11}$$

which generalizes the condition $\mathbf{g}^* = \mathbf{0}$ for smooth functions. Clearly (14.2.11) implies (14.2.10). If $\mathbf{0} \notin \partial f'$ then by lemma 14.2.3 (with $\boldsymbol{\lambda} = \mathbf{0}$, $K = \partial f'$) there exists a vector $\mathbf{s} = -\bar{\mathbf{g}}/\|\bar{\mathbf{g}}\|_2$ for which $\mathbf{s}^\mathrm{T}\mathbf{g} < 0\ \forall\ \mathbf{g} \in \partial f'$, where $\bar{\mathbf{g}}$ is the vector which minimizes $\|\mathbf{g}\|_2\ \forall\ \mathbf{g} \in \partial f'$. Applying this result at $\mathbf{x}^*$ shows that (14.2.10) and (14.2.11) are equivalent. It is now immediate from (14.2.1) that either (14.2.10) or (14.2.11) is also a sufficient condition for a global minimizer at $\mathbf{x}^*$. In fact the vector $\mathbf{s} = -\bar{\mathbf{g}}/\|\bar{\mathbf{g}}\|_2$ defined above is the *steepest descent vector* at $\mathbf{x}'$. Assuming $\mathbf{0} \notin \partial f'$ then from (14.2.8) the direction of least slope is defined by

$$\begin{aligned}\min_{\|\mathbf{s}\|_2=1} \max_{\mathbf{g}\in\partial f'} \mathbf{s}^\mathrm{T}\mathbf{g} &= \max_{\mathbf{g}\in\partial f'} \min_{\|\mathbf{s}\|_2=1} \mathbf{s}^\mathrm{T}\mathbf{g} \\ &= \max_{\mathbf{g}\in\partial f'} -\|\mathbf{g}\|_2 = -\|\bar{\mathbf{g}}\|_2 \end{aligned} \tag{14.2.12}$$

and hence the least slope is achieved when $\mathbf{s} = -\bar{\mathbf{g}}/\|\bar{\mathbf{g}}\|_2$. The justification of interchanging the min and max operations is explored in Question 14.7.

The main aim in introducing the above development is to apply it to composite functions of the form

$$\phi(\mathbf{x}) = f(\mathbf{x}) + h(\mathbf{c}(\mathbf{x})) \tag{14.2.13}$$

where $f(\mathbf{x})$ ($\mathbb{R}^n \to \mathbb{R}^1$) and $\mathbf{c}(\mathbf{x})$ ($\mathbb{R}^n \to \mathbb{R}^m$) are smooth ($\mathbb{C}^1$) functions and $h(\mathbf{c})$ ($\mathbb{R}^m \to \mathbb{R}^1$) is convex but non-smooth ($\mathbb{C}^0$). Let $\mathbf{x}^{(k)} \to \mathbf{x}'$ be a directional sequence with $\delta^{(k)} \downarrow 0$ and $\mathbf{s}^{(k)} \to \mathbf{s}$ in (9.2.1). By Taylor series

$$f^{(k)} = f' + \delta^{(k)}\mathbf{g}'^\mathrm{T}\mathbf{s}^{(k)} + o(\delta^{(k)})$$

so $(f^{(k)} - f')/\delta^{(k)} \to \mathbf{g}'^\mathrm{T}\mathbf{s}$. Likewise

$$\mathbf{c}^{(k)} = \mathbf{c}' + \delta^{(k)}\mathbf{A}'^\mathrm{T}\mathbf{s}^{(k)} + o(\delta^{(k)})$$

so $\mathbf{c}^{(k)} \to \mathbf{c}'$ is a directional sequence in $\mathbb{R}^m$ with $(\mathbf{c}^{(k)} - \mathbf{c}')/\delta^{(k)} \to \mathbf{A}'^{\mathrm{T}}\mathbf{s}$. Thus by applying lemma 14.2.5 to $h(\mathbf{c})$, it follows that

$$\lim_{k\to\infty} \frac{\phi^{(k)} - \phi'}{\delta^{(k)}} = \max_{\boldsymbol{\lambda}\in\partial h'} \mathbf{s}^{\mathrm{T}}(\mathbf{g}' + \mathbf{A}'\boldsymbol{\lambda}) \tag{14.2.14}$$

and this gives the directional derivative at $\mathbf{x}'$ in the direction $\mathbf{s}$ for the function $\phi(\mathbf{x})$ in (14.2.13). Another more general result about directional derivatives is proved in lemma 14.5.1 and its corollary.

It can be deduced from (14.2.14) that if $\mathbf{x}^*$ is a local minimizer of $\phi(\mathbf{x})$, then $\phi^{(k)} \geqslant \phi^*$ for all k sufficiently large, and hence

$$\max_{\boldsymbol{\lambda}\in\partial h^*} \mathbf{s}^{\mathrm{T}}(\mathbf{g}^* + \mathbf{A}^*\boldsymbol{\lambda}) \geqslant 0 \qquad \forall \quad \mathbf{s}: \|\mathbf{s}\| = 1. \tag{14.2.15}$$

This is a first order necessary condition for a local minimizer which like (14.2.10) can be interpreted as a non-negative directional derivative in all directions. Again the result can be stated alternatively as

$$\mathbf{0} \in \partial\phi(\mathbf{x}^*) \triangleq \{\boldsymbol{\gamma}: \boldsymbol{\gamma} = \mathbf{g} + \mathbf{A}\boldsymbol{\lambda} \qquad \forall\ \boldsymbol{\lambda}\in\partial h\}_{\mathbf{x}=\mathbf{x}^*}. \tag{14.2.16}$$

The set $\partial\phi^*$ thus defined, although convex and compact, is not the subdifferential because ϕ may not be a convex function, but it is convenient to use the same notation. (It is in fact the *generalized gradient* of Clarke (1975).) Its definition in (14.2.16) is in the nature of a *generalized chain rule*, by analogy with the expression $\nabla\phi = \mathbf{g} + \mathbf{A}\nabla h$ for smooth functions. The equivalence of (14.2.15) and (14.2.16) is again a consequence of the separating hyperplane lemma 14.2.3. Yet another way to state this condition is to introduce the Lagrangian function

$$\mathscr{L}(\mathbf{x}, \boldsymbol{\lambda}) = f(\mathbf{x}) + \boldsymbol{\lambda}^{\mathrm{T}}\mathbf{c}(\mathbf{x}). \tag{14.2.17}$$

Then an equivalent statement of (14.2.16) is as follows.

Theorem 14.2.1 (First order necessary conditions)

If $\mathbf{x}^*$ minimizes $\phi(\mathbf{x})$ in (14.2.3) then there exists a vector $\boldsymbol{\lambda}^* \in \partial h^*$ such that

$$\nabla\mathscr{L}(\mathbf{x}^*, \boldsymbol{\lambda}^*) = \mathbf{g}^* + \mathbf{A}^*\boldsymbol{\lambda}^* = \mathbf{0}. \tag{14.2.18}$$

Proof

Immediate since $\partial\phi^*$ is the set of vectors $\nabla\mathscr{L}(\mathbf{x}^*, \boldsymbol{\lambda})$ for all $\boldsymbol{\lambda} \in \partial h^*$. □

This form illustrates the close relationship between the vector $\boldsymbol{\lambda}^* \in \partial h^*$ and the Lagrange multipliers in theorem 9.1.1. These conditions are illustrated by Questions 14.8, 14.9, 14.10, and 14.12. In general, since $\phi(\mathbf{x})$ may not be convex, the conditions of theorem 14.2.1 are not sufficient.

In view of this last observation it is important to consider second order conditions for $\mathbf{x}^*$ to be a minimizer of the composite function (14.2.13). These conditions again exhibit a close relationship with the results in Section 9.3. The

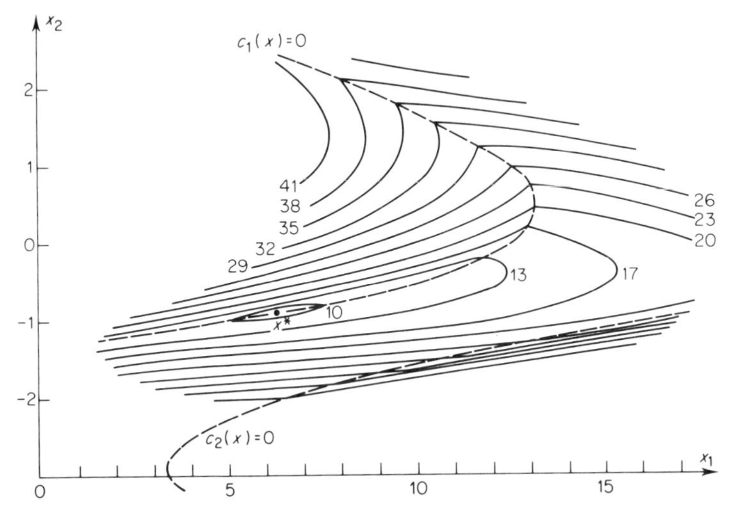

Figure 14.2.3 Contours for the scaled L_1 Freudenstein and Roth problem, Question 14.9

approach is based on that taken by Fletcher and Watson (1980). In considering second order conditions it is necessary to restrict the possible directions to those having zero directional derivative so that second order effects become important. This is illustrated by the contours in Figure 14.2.3. $\mathbf{x}^*$ satisfies first order conditions but there are directions of zero slope along the derivative discontinuity $c_1(\mathbf{x}) = 0$ (broken line), so that first order conditions are not sufficient. However sufficient second order conditions can be derived which imply that $\mathbf{x}^*$ is a local minimizer – see Question 14.9. In general let $\boldsymbol{\lambda}^*$ be any vector which exists in theorem 14.2.1 and consider the set

$$X = \{\mathbf{x}:\ h(\mathbf{c}(\mathbf{x})) = h(\mathbf{c}(\mathbf{x}^*)) + (\mathbf{c}(\mathbf{x}) - \mathbf{c}(\mathbf{x}^*))^{\mathrm{T}} \boldsymbol{\lambda}^*\}. \tag{14.2.19}$$

Define $\mathscr{G}^*$ as the set of normalized feasible directions with respect to X at $\mathbf{x}^*$. (That is $\mathbf{s} \in \mathscr{G}^*$ implies that there exists a directional sequence $\mathbf{x}^{(k)} \to \mathbf{x}^*$, feasible in (14.2.19), such that $\mathbf{s}^{(k)} \to \mathbf{s}$ in (9.2.1) with $\sigma = 1$.) It is possible to show that these directions are closely related to the set G^* of normalized directions of zero slope, that is

$$G^* \triangleq \{\mathbf{s}:\ \max_{\boldsymbol{\lambda} \in \partial h^*} \mathbf{s}^{\mathrm{T}}(\mathbf{g}^* + \mathbf{A}^*\boldsymbol{\lambda}) = 0,\ \|\mathbf{s}\|_2 = 1\}. \tag{14.2.20}$$

The extent to which $\mathscr{G}^*$ and G^* correspond is important and it is first shown that $\mathscr{G}^*$ is a subset of G^*.

Lemma 14.2.6

$$\mathscr{G}^* \subset G^*.$$

Proof

Let $\mathbf{s} \in \mathscr{G}^*$ so a directional sequence in X exists with $\mathbf{s}^{(k)} \to \mathbf{s}$, $\|\mathbf{s}\|_2 = 1$. Using

(14.2.14), (14.2.13), (14.2.19), and a Taylor series it follows that

$$\begin{aligned}\max_{\boldsymbol{\lambda}\in\partial h^*} \mathbf{s}^{\mathrm{T}}(\mathbf{g}^* + \mathbf{A}^*\boldsymbol{\lambda}) &= \lim_{k\to\infty} \frac{\phi^{(k)} - \phi^*}{\delta^{(k)}} \\ &= \lim_{k\to\infty} \frac{f^{(k)} - f^* + h(\mathbf{c}^{(k)}) - h(\mathbf{c}^*)}{\delta^{(k)}} \\ &= \lim_{k\to\infty} \frac{f^{(k)} - f^* + (\mathbf{c}^{(k)} - \mathbf{c}^*)^{\mathrm{T}}\boldsymbol{\lambda}^*}{\delta^{(k)}} \\ &= \mathbf{s}^{\mathrm{T}}(\mathbf{g}^* + \mathbf{A}^*\boldsymbol{\lambda}^*).\end{aligned}$$

It then follows from (14.2.18) that $\mathbf{s}\in G^*$. □

To give a general result going the other way is not always possible although it can be done in important special cases. Further discussion of this is given later in this section: at present the regularity assumption

$$\mathscr{G}^* = G^* \tag{14.2.21}$$

is made (which depends on $\boldsymbol{\lambda}^*$ if more than one such vector exists). It is now possible to state the second order conditions. In doing this it is assumed that f and $\mathbf{c}$ are $\mathbb{C}^2$ functions. Note that as usual the regularity assumption is needed only in the necessary conditions.

Theorem 14.2.2 (Second order necessary conditions)

If $\mathbf{x}^*$ *minimizes* $\phi(\mathbf{x})$ *then theorem 14.2.1 holds; for each vector* $\boldsymbol{\lambda}^*$ *which thus exists, if (14.2.21) holds, then*

$$\mathbf{s}^{\mathrm{T}}\nabla^2\mathscr{L}(\mathbf{x}^*, \boldsymbol{\lambda}^*)\mathbf{s} \geqslant 0 \qquad \forall \quad \mathbf{s}\in G^*. \tag{14.2.22}$$

Proof

For any $\mathbf{s}\in G^*$, $\mathbf{s}\in\mathscr{G}^*$ by (14.2.21) and hence $\exists$ a directional sequence $\mathbf{x}^{(k)}\to\mathbf{x}^*$, feasible in (14.2.19). A Taylor expansion of $\mathscr{L}(\mathbf{x}, \boldsymbol{\lambda}^*)$ about $\mathbf{x}^*$ yields

$$\begin{aligned}\mathscr{L}(\mathbf{x}^{(k)}, \boldsymbol{\lambda}^*) = f^* + \mathbf{c}^{*\mathrm{T}}\boldsymbol{\lambda}^* + \mathbf{e}^{(k)\mathrm{T}}\nabla\mathscr{L}(\mathbf{x}^*, \boldsymbol{\lambda}^*) \\ + \tfrac{1}{2}\mathbf{e}^{(k)\mathrm{T}}\nabla^2\mathscr{L}(\mathbf{x}^*, \boldsymbol{\lambda}^*)\mathbf{e}^{(k)} + o(\|\mathbf{e}^{(k)}\|^2)\end{aligned}$$

where $\mathbf{e}^{(k)} = \mathbf{x}^{(k)} - \mathbf{x}^*$, and using (14.2.18) and (9.2.1),

$$\mathscr{L}(\mathbf{x}^{(k)}, \boldsymbol{\lambda}^*) = f^* + \mathbf{c}^{*\mathrm{T}}\boldsymbol{\lambda}^* + \tfrac{1}{2}\delta^{(k)^2}\mathbf{s}^{(k)\mathrm{T}}\nabla^2\mathscr{L}(\mathbf{x}^*, \boldsymbol{\lambda}^*)\mathbf{s}^{(k)} + o(\delta^{(k)^2}). \tag{14.2.23}$$

Since $\mathbf{x}^{(k)}$ is feasible in (14.2.19) it follows from (14.2.13) that

$$\phi^{(k)} = \phi^* + \tfrac{1}{2}\delta^{(k)^2}\mathbf{s}^{(k)\mathrm{T}}\nabla^2\mathscr{L}(\mathbf{x}^*, \boldsymbol{\lambda}^*)\mathbf{s}^{(k)} + o(\delta^{(k)^2}).$$

Since $\mathbf{x}^*$ is a local minimizer, $\phi^{(k)} \geqslant \phi^*$ for all k sufficiently large, so dividing by $\frac{1}{2}\delta^{(k)^2}$ and taking the limit yields (14.2.22). □

For sufficient conditions, it is firstly observed that if the directional derivative is positive in all directions, that is

$$\max_{\boldsymbol{\lambda} \in \partial h^*} \mathbf{s}^{\mathrm{T}}(\mathbf{g}^* + \mathbf{A}^*\boldsymbol{\lambda}) > 0 \qquad \forall \quad \mathbf{s}: \|\mathbf{s}\| = 1,$$

or equivalently G^* is empty, then the conditions of theorem 14.2.1 imply that $\mathbf{x}^*$ is an isolated local minimizer (see Question 14.12). This result is in fact a special case of theorem 14.2.3 below. When G^* is non-empty, second order effects come into play and this is expressed in the following.

Theorem 14.2.3 (Second order sufficient conditions)

If there exists $\boldsymbol{\lambda}^* \in \partial h^*$ such that (14.2.18) holds, and if

$$\mathbf{s}^{\mathrm{T}}\nabla^2\mathscr{L}(\mathbf{x}^*, \boldsymbol{\lambda}^*)\mathbf{s} > 0 \qquad \forall \quad \mathbf{s} \in G^* \tag{14.2.24}$$

then $\mathbf{x}^*$ is an isolated local minimizer of $\phi(\mathbf{x})$.

Proof

Assume the contrary, that $\exists$ a sequence and hence a directional sequence $\mathbf{x}^{(k)} \to \mathbf{x}^*$ such that $\phi^{(k)} \leqslant \phi^*$. By (14.2.18),

$$0 \leqslant \max_{\boldsymbol{\lambda} \in \partial h^*} \mathbf{s}^{\mathrm{T}}(\mathbf{g}^* + \mathbf{A}^*\boldsymbol{\lambda}) = \mu$$

say. If $\mu > 0$ then $\lim(\phi^{(k)} - \phi^*)/\delta^{(k)} = \mu$ (see (14.2.14)) which contradicts $\phi^{(k)} \leqslant \phi^*$. Thus $\mu = 0$ and hence $\mathbf{s} \in G^*$. Now from (14.2.17) and (14.2.13)

$$\begin{aligned}\mathscr{L}(\mathbf{x}^{(k)}, \boldsymbol{\lambda}^*) - \mathscr{L}(\mathbf{x}^*, \boldsymbol{\lambda}^*) &= f^{(k)} + \mathbf{c}^{(k)^{\mathrm{T}}}\boldsymbol{\lambda}^* - f^* - \mathbf{c}^{*\mathrm{T}}\boldsymbol{\lambda}^* \\ &= \phi^{(k)} - \phi^* - (h(\mathbf{c}^{(k)}) - h(\mathbf{c}^*) - (\mathbf{c}^{(k)} - \mathbf{c}^*)^{\mathrm{T}}\boldsymbol{\lambda}^*) \\ &\leqslant \phi^{(k)} - \phi^*\end{aligned}$$

using the subgradient inequality. Hence from (14.2.23)

$$0 \geqslant \phi^{(k)} - \phi^* \geqslant \tfrac{1}{2}\delta^{(k)^2}\mathbf{s}^{(k)^{\mathrm{T}}}\nabla^2\mathscr{L}(\mathbf{x}^*, \boldsymbol{\lambda}^*)\mathbf{s}^{(k)} + o(\delta^{(k)^2}).$$

Dividing by $\frac{1}{2}\delta^{(k)^2}$ and taking the limit contradicts (14.2.24) and establishes the theorem. □

These conditions (with G^* non-empty) are illustrated by Questions 14.9 and 14.10.

The second order conditions for nonlinear programming given in Section 9.3 are almost necessary and sufficient because the regularity assumption is mild and there is a gap only in the case of zero curvature. The same is *not* true here and it is important to realize that there are well-behaved cases in which $\mathscr{G}^* \neq G^*$. However there are also two important cases in which $\mathscr{G}^* = G^*$ and the conditions are therefore almost equivalent. One case is when $h(\mathbf{c})$ is $\|\mathbf{c}\|$ or $\|\mathbf{c}^+\|$, and when $\mathbf{c}^* = \mathbf{0}$ or $\mathbf{c}^{*+} = \mathbf{0}$. This is described in Section 14.3 in the context of applying an exact penalty function to solve a nonlinear programming problem. Then an assumption that the vectors $\mathbf{a}_i^*$ of the active constraints are independent yields $\mathscr{G}^* = G^*$ (lemma 14.3.1). Another case is when $h(\mathbf{c})$ is a polyhedral convex function in which case a different independence assumption also ensures that $\mathscr{G}^* = G^*$, as follows.

Lemma 14.2.7

If $h(\mathbf{c})$ is defined by (14.2.3), if $\boldsymbol{\mu}^*$ are the multipliers which exist in (14.1.8), if p is an index with $\mu_p^* > 0$, and if the vectors

$$\mathbf{A}^*(\mathbf{h}_p - \mathbf{h}_i), \qquad i \in \mathscr{A}^* - p \tag{14.2.25}$$

are linearly independent, then $\mathscr{G}^* = G^*$.

Proof

Let $\boldsymbol{\lambda}^* = \Sigma \mu_i^* \mathbf{h}_i$ be the corresponding vector in ∂h^* and let $\mathscr{A}_+^* \triangleq \{i: \mu_i^* > 0\}$. Let $\mathbf{s} \in G^*$ so that

$$\max_{\boldsymbol{\lambda} \in \partial h^*} \mathbf{s}^{\mathrm{T}}(\mathbf{g}^* + \mathbf{A}^*\boldsymbol{\lambda}) = 0.$$

Let $\mathscr{A}_s^* \triangleq \{i: \mathbf{s}^{\mathrm{T}}(\mathbf{g}^* + \mathbf{A}^*\mathbf{h}_i) = 0,\ i \in \mathscr{A}^*\}$ denote the set of vectors $\mathbf{h}_i$ (dependent on $\mathbf{s}$) which achieve the max. By (14.2.18) $\boldsymbol{\lambda}^*$ also achieves the max so $\mu_i^* > 0$ implies $i \in \mathscr{A}_s^*$ and hence $\mathscr{A}_+^* \subset \mathscr{A}_s^* \subset \mathscr{A}^*$. Thus for any $p \in \mathscr{A}_+^*$,

$$\begin{aligned} \mathbf{s}^{\mathrm{T}}\mathbf{A}^*(\mathbf{h}_p - \mathbf{h}_i) &= 0, \qquad i \in \mathscr{A}_s^* - p \\ \mathbf{s}^{\mathrm{T}}\mathbf{A}^*(\mathbf{h}_p - \mathbf{h}_i) &> 0, \qquad i \in \mathscr{A}^* - \mathscr{A}_s^*. \end{aligned}$$

If (14.2.5) holds, $\mathscr{A}_s^* - p$ contains fewer than n elements (else $\|\mathbf{s}\| = 1$ is contradicted) and so (as in the proof of lemma 9.2.2) there exists an arc $\mathbf{x}(\theta)$ with $\mathbf{x}^* = \mathbf{x}(0)$ and $\dot{\mathbf{x}}(0) = \mathbf{s}$ defined for $\theta \geqslant 0$ sufficiently small such that for $\theta > 0$

$$\begin{aligned} \mathbf{h}_p^{\mathrm{T}}\mathbf{c}(\mathbf{x}(\theta)) + b_p &= \mathbf{h}_i^{\mathrm{T}}\mathbf{c}(\mathbf{x}(\theta)) + b_i, \qquad i \in \mathscr{A}_s^* - p \\ \mathbf{h}_p^{\mathrm{T}}\mathbf{c}(\mathbf{x}(\theta)) + b_p &> \mathbf{h}_i^{\mathrm{T}}\mathbf{c}(\mathbf{x}(\theta)) + b_i, \qquad i \in \mathscr{A}^* - \mathscr{A}_s^*. \end{aligned}$$

Thus for θ sufficiently small, $h(\mathbf{c}(\mathbf{x}(\theta))) = \mathbf{h}_i^{\mathrm{T}}\mathbf{c}(\mathbf{x}(\theta)) + b_i$ and hence $h(\mathbf{c}(\mathbf{x}(\theta))) - h^* = \mathbf{h}_i^{\mathrm{T}}(\mathbf{c}(\mathbf{x}(\theta)) - \mathbf{c}^*)\ \ \forall\ i \in \mathscr{A}_s^*$. An inner product with μ_i^* shows that $\mathbf{x}(\theta)$ is feasible in (14.2.9) and hence $\mathbf{s} \in \mathscr{G}^*$. □

Corollary

If in addition $\mathscr{A}^*$ contains $n+1$ elements and $\mathscr{A}_+^* = \mathscr{A}^*$ then G^* (and hence $\mathscr{G}^*$ by lemma 14.2.6) is empty.

Proof

$\mathscr{A}_s^* - p$ must now have n elements so as above $\|\mathbf{s}\| = 1$ is contradicted. □

Conditions (14.2.25) are realistic in cases (I) and (II) (see (14.1.3)) and in case (III) when $\mathbf{c}^* \neq \mathbf{0}$. However in cases (IV) and (V) involving $\|\cdot\|_1$, the set $\mathscr{A}^*$ contains 2^q elements where q is the number of zero components of $\mathbf{c}^*$, and so independence of the vectors (14.2.25) is unlikely. In this case however the following device can be used. In the neighbourhood of a solution $\mathbf{x}^*$, the function

$$\phi(\mathbf{x}) = f(\mathbf{x}) + \|\mathbf{c}(\mathbf{x})\|_1 \tag{14.2.26}$$

can equivalently be written as

$$\phi(\mathbf{x}) = f(\mathbf{x}) + \sum_{i \notin Z} \lambda_i^* c_i(\mathbf{x}) + \sum_{i \in Z} |c_i(\mathbf{x})| \tag{14.2.27}$$

where $Z = \{i: c_i^* = 0\}$, and using (14.1.15). Essentially the smooth part of (14.2.26) is extended to include terms $\lambda_i^* c_i(\mathbf{x})$, $i \notin Z$, and the $\|\mathbf{c}(\mathbf{x})\|_1$ part is reduced correspondingly. Then lemma 14.3.1 can be invoked to show that independence of the vectors $\mathbf{a}_i^*$, $i \in Z$, yields $\mathscr{G}^* = G^*$. A slightly more general result is given by Fletcher and Watson (1980). A similar result holds good when $\|\mathbf{c}^+\|_1$ is used in (14.2.26).

A final observation about second order conditions is contained in the following.

Lemma 14.2.8

For given $\mathbf{x}^*$ let $\boldsymbol{\mu}^*$ satisfy (14.1.8) and $\boldsymbol{\lambda}^* = \mathbf{H}\boldsymbol{\mu}^*$. Then the second order conditions of theorem 9.3.2 for problem (14.1.6) are equivalent to those in theorem 14.2.3.

Proof

In both cases the first order conditions hold. Let $\bar{\mathbf{s}}^{\mathrm{T}} = (\mathbf{s}^{\mathrm{T}}, s_{n+1})$. The second order conditions of theorem 9.3.2 for (14.1.6) involve the set

$$\begin{aligned} \{\bar{\mathbf{s}}: \bar{\mathbf{s}} \neq \mathbf{0}, \quad & s_{n+1} = \mathbf{s}^{\mathrm{T}}(\mathbf{g}^* + \mathbf{A}^*\mathbf{h}_i), \ i \in \mathscr{A}_+^* \\ & s_{n+1} \geqslant \mathbf{s}^{\mathrm{T}}(\mathbf{g}^* + \mathbf{A}^*\mathbf{h}_i), \ i \in \mathscr{A}^* - \mathscr{A}_+^*\}. \end{aligned} \tag{14.2.28}$$

Let $\bar{\mathbf{s}} \in (14.2.28)$. An inner product with μ_i^* implies $s_{n+1} = 0$. For any $\boldsymbol{\lambda} \in \partial h^*$ let $\boldsymbol{\lambda} = \mathbf{H}\boldsymbol{\mu}$; then an inner product with μ_i gives $\mathbf{s}^{\mathrm{T}}(\mathbf{g}^* + \mathbf{A}^*\boldsymbol{\lambda}) \leqslant 0$ and hence $\mathbf{s}/\|\mathbf{s}\|_2 \in G^*$ in (14.2.20). Now let $\mathbf{s} \in G^*$, let $p \in \mathscr{A}_+^*$, and define $s_{n+1} = \mathbf{s}^{\mathrm{T}}(\mathbf{g}^* + \mathbf{A}^*\mathbf{h}_p)$. Then as in lemma 14.2.7,

$$\begin{aligned} s_{n+1} &= \mathbf{s}^{\mathrm{T}}(\mathbf{g}^* + \mathbf{A}^*\mathbf{h}_i), \qquad i \in \mathscr{A}_s^* \\ s_{n+1} &\geqslant \mathbf{s}^{\mathrm{T}}(\mathbf{g}^* + \mathbf{A}^*\mathbf{h}_i), \qquad i \in \mathscr{A}^* - \mathscr{A}_s^*. \end{aligned}$$

Since $\mathscr{A}_s^* \supset \mathscr{A}_+^*$ it follows that $\bar{\mathbf{s}} \in (14.2.28)$. Thus apart from normalization the vectors $\mathbf{s}$ in (14.2.28) and in G^* are equivalent. The second order conditions that $\mathbf{s}^{\mathrm{T}}\mathbf{W}^*\mathbf{s} > 0$ for all $\mathbf{s}$ in either (14.2.28) or G^* are also equivalent so the lemma is proved. □

14.3 Exact Penalty Functions

One of the most important applications of NDO is in the area of nonlinear programming through the use of an exact penalty function and is described in this section. To simplify the presentation, two basic types of nonlinear programming problem are considered, the equality constraint problem

$$\begin{aligned}&\underset{\mathbf{x}}{\text{minimize}}\ f(\mathbf{x})\\&\text{subject to } \mathbf{c}(\mathbf{x})=\mathbf{0}\end{aligned} \tag{14.3.1}$$

for which the corresponding exact penalty function problem is

$$\underset{\mathbf{x}}{\text{minimize}}\ \phi(\mathbf{x}) \triangleq \nu f(\mathbf{x}) + \|\mathbf{c}(\mathbf{x})\|, \tag{14.3.2}$$

and the inequality constraint problem

$$\begin{aligned}&\underset{\mathbf{x}}{\text{minimize}}\ f(\mathbf{x})\\&\text{subject to } \mathbf{c}(\mathbf{x})\leqslant\mathbf{0}\end{aligned} \tag{14.3.3}$$

for which the corresponding exact penalty function problem is

$$\underset{\mathbf{x}}{\text{minimize}}\ \phi(\mathbf{x}) \triangleq \nu f(\mathbf{x}) + \|\mathbf{c}(\mathbf{x})^+\| \tag{14.3.4}$$

where c_i^+ denotes max $(c_i, 0)$. There is no difficulty in generalizing to the mixed problem (7.1.1) and this can be done within the framework of (14.3.3) by replacing $c_i = 0$ by $c_i \leqslant 0$ and $-c_i \leqslant 0$. Note that in (14.3.3) the inequality is in the reverse direction from that considered in Chapter 9. This is equivalent to a change of sign in the constraint residuals $\mathbf{c}(\mathbf{x})$ which must be kept in mind. These penalty functions are exact in the sense of Section 12.5, so that (14.3.1) and (14.3.2) (or (14.3.3) and (14.3.4)) are equivalent in that for sufficiently small ν a local solution of (14.3.1) is a local minimizer of (14.3.2), and vice versa. This result is made precise in theorems 14.3.1 and 14.3.2 below. Practical considerations in choosing ν are discussed later in the section. The conditions under which equivalence holds are usually satisfied in practice but there are examples when the problems are not equivalent. Examples where $\mathbf{x}^*$ solves (14.3.3) and not (14.3.4) are given at the end of theorem 14.3.1. However these examples are either pathological or limiting cases and need not greatly concern the user. The alternative possibility is that a local minimizer $\mathbf{x}^*$ of (14.3.2) may not be feasible in (14.3.1), even though the latter may have a solution. This is the same situation as that illustrated in Figure 6.2.1 and described in Section 6.2 (Volume 1): it is an inevitable consequence of the use of any penalty approach as a means of inducing global convergence. To circumvent the difficulty is a global minimization problem and hence generally impracticable. Corresponding advantages are that *best* solutions can be determined when no feasible point exists in (14.3.1) and that the difficulty of finding an initial feasible point is avoided. In practice the most likely unfavourable situation which arises when applying an exact penalty function is that a sequence $\{\mathbf{x}^{(k)}\}$ is calculated such that $\phi^{(k)} \to -\infty$, which is an indication that the calculation should be repeated with a smaller value of ν. It may be necessary to pre-scale the $c_i(\mathbf{x})$ to be of comparable magnitude so that the use of a single penalty parameter ν is reasonable (see also (14.3.17)). As an illustration the non-linear programming problem: min $-x_1 - x_2$ subject to $x_1^2 + x_2^2 \leqslant 1$ with solution $x_1^* = x_2^* = \lambda^* = 1/\sqrt{2}$ is considered. The corresponding exact penalty function is

$$\phi(\mathbf{x}) = \nu(-x_1 - x_2) + \max(x_1^2 + x_2^2 - 1, 0) \tag{14.3.5}$$

for which the threshold value of ν in theorem 14.3.1 is $\nu < 1/\lambda^* = \sqrt{2}$. The contours of $\phi(\mathbf{x})$ for $\nu = 1$ are shown in Figure 14.3.1 and it is clear that $\mathbf{x}^*$ minimizes $\phi(\mathbf{x})$. The non-differentiable nature of $\phi(\mathbf{x})$ on the unit circle (broken line) can also be seen.

The main attraction in using (14.3.2) or (14.3.4) is that it holds out the possibility of avoiding the sequential nature of the penalty functions in Sections 12.1 and 12.2 so that only a single unconstrained minimization calculation is required. Unfortunately (even if $\|\cdot\|_2$ is used) (14.3.2) and (14.3.4) are not smooth ($\mathbb{C}^1$) functions so that the many effective techniques for smooth minimization described in Volume 1 cannot adequately be used. The study of algorithms for non-smooth problems is a relatively recent development and is described in some detail in Section 14.4 and some of the algorithms can be used here. Another approach (Han, 1977; Coleman and Conn, 1980; Mayne, 1980) is to use an algorithm for nonlinear programming as a means of generating a direction of search, and to use the exact penalty function as the criterion function to be minimized (approximately) in the line search. This approach often works well in practice but unfortunately can fail (see Fletcher, 1980b) and I think it is important to take into account the non-smooth nature of the penalty function when choosing the direction in which to search. The discussion of algorithms in Section 14.4 leads to a globally convergent algorithm for composite NDO problems (Fletcher, 1980a) which works well when

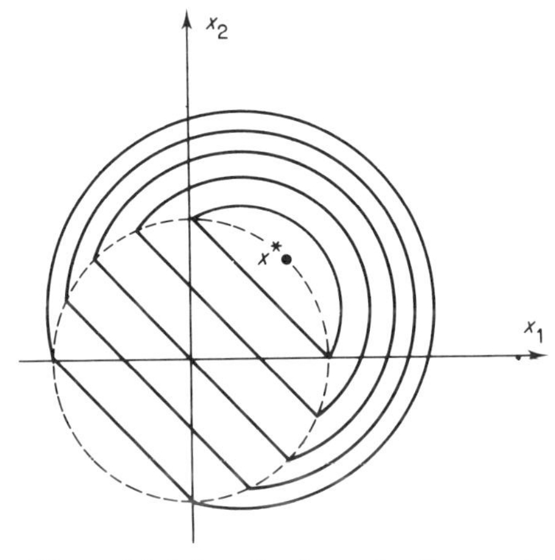

Figure 14.3.1 Contours for the exact penalty function (14.3.5)

applied to minimize exact penalty functions (Fletcher, 1980b), and it is this type of algorithm which I currently favour. Use of the $\|\cdot\|_1$ is the most convenient for reasons given later in this section. An example of the algorithm applied to minimize (14.3.5) is shown in Section 14.4. The general algorithm is described in more detail in Section 14.5; it usually converges at a second order rate and is closely related to the SOLVER method (Section 12.3) when applied to exact penalty functions. In some cases however a second order rate of convergence is not always obtained (the Maratos effect – see Section 14.4) and the best way of dealing with this is currently the subject of research.

The theory for exact penalty functions can be set out in a very concise and general way. The definition in (14.3.2) allows the use of any norm and in (14.3.4) any *monotonic* norm ($|\mathbf{x}| \leqslant |\mathbf{y}| \Rightarrow \|\mathbf{x}\| \leqslant \|\mathbf{y}\|$). This latter condition ensures that $\|\mathbf{c}^+\|$ is a convex function (see Question 14.2) and includes all L_p and scaled L_p norms for $1 \leqslant p \leqslant \infty$. Thus the results of this section are not just restricted to polyhedral norms. It is convenient to introduce the concept of a *dual norm*

$$\|\mathbf{u}\|_D \triangleq \max_{\|\mathbf{v}\| \leqslant 1} \mathbf{u}^{\mathrm{T}}\mathbf{v}. \tag{14.3.6}$$

The dual of $\|\cdot\|_1$ is $\|\cdot\|_\infty$, and vice versa, and the $\|\cdot\|_2$ is self-dual. Expressions for the subdifferential of the functions $\|\mathbf{c}\|$ and $\|\mathbf{c}^+\|$ are given by

$$\partial\|\mathbf{c}\| = \{\boldsymbol{\lambda}: \boldsymbol{\lambda}^{\mathrm{T}}\mathbf{c} = \|\mathbf{c}\|,\ \|\boldsymbol{\lambda}\|_D \leqslant 1\} \tag{14.3.7}$$

and

$$\partial\|\mathbf{c}^+\| = \{\boldsymbol{\lambda}: \boldsymbol{\lambda}^{\mathrm{T}}\mathbf{c} = \|\mathbf{c}^+\|,\ \boldsymbol{\lambda} \geqslant \mathbf{0},\ \|\boldsymbol{\lambda}\|_D \leqslant 1\}. \tag{14.3.8}$$

The proof of these expressions is sketched out in some detail in Questions 14.4 and 14.5. Expressions (14.1.12) to (14.1.15) are special cases of (14.3.7) and (14.3.8). The main results giving the equivalence between local solutions of (14.3.1) and (14.3.2) (or between (14.3.3) and (14.3.4)) can now be given. In fact the latter result only is given; the former case is similar (but easier). A point to bear in mind is that in these theorems $\boldsymbol{\lambda}^*$ refers to a multiplier vector for the constrained problem (14.3.3) and $\nu\boldsymbol{\lambda}^*$ is the equivalent multiplier vector for the exact penalty function problem (14.3.4).

Theorem 14.3.1

If $\nu < 1/\|\boldsymbol{\lambda}^*\|_D$ *and* $\mathbf{c}^{*+} = \mathbf{0}$ *then the second order sufficient conditions at* $\mathbf{x}^*$ *for problems (14.3.3) and (14.3.4) are equivalent. Therefore if they hold, the fact that* $\mathbf{x}^*$ *solves (14.3.3) implies that* $\mathbf{x}^*$ *solves (14.3.4), and vice versa.*

Proof

Second order conditions are given for problem (14.3.3) in theorem 9.3.2 and for problem (14.3.4) in theorem 14.2.3. The first order requirements are that $\mathbf{g}^* + \mathbf{A}^*\boldsymbol{\lambda}^* = \mathbf{0}$, $\boldsymbol{\lambda}^* \geqslant \mathbf{0}$, $\boldsymbol{\lambda}^{*\mathrm{T}}\mathbf{c}^* = \|\mathbf{c}^{*+}\| = 0$ (with a suitable sign change), and by

$\nu\mathbf{g}^* + \mathbf{A}^*(\nu\boldsymbol{\lambda}^*) = \mathbf{0}$, $\nu\boldsymbol{\lambda}^* \in \partial h^*$ respectively, which are clearly equivalent from (14.3.8) if $\nu < 1/\|\boldsymbol{\lambda}^*\|_D$. Next the sets G^* defined in (9.3.11) (with $\|\mathbf{s}\| = 1$) and (14.2.20) are shown to be equivalent. For convenience refer to these as G_9^* and G_{14}^* respectively. Let $\mathbf{s} \in G_{14}^*$. From (14.2.20) and the above it follows that

$$\mathbf{s}^{\mathrm{T}}\mathbf{A}^*(\boldsymbol{\lambda} - \nu\boldsymbol{\lambda}^*) \leqslant 0 \qquad \forall \quad \boldsymbol{\lambda} \in \partial \|\mathbf{c}^{*+}\|. \tag{14.3.9}$$

Let $\mathscr{A}^*$ denote active constraints at $\mathbf{x}^*$ in (14.3.3). For $i \in \mathscr{A}^*$ and small ϵ it follows using $\|\nu\boldsymbol{\lambda}^*\|_D < 1$ that $\boldsymbol{\lambda} = \nu\boldsymbol{\lambda}^* + \epsilon\mathbf{e}_i \in \partial\|\mathbf{c}^{*+}\|$ either if $\lambda_i^* = 0$ and $\epsilon > 0$ or if $\lambda_i^* > 0$ and $\pm\epsilon > 0$. Hence from (14.3.9),

$$\begin{aligned} \mathbf{s}^{\mathrm{T}}\mathbf{a}_i^* &\leqslant 0 \qquad \text{if } \lambda_i^* = 0, \\ \mathbf{s}^{\mathrm{T}}\mathbf{a}_i^* &= 0 \qquad \text{if } \lambda_i^* > 0, \end{aligned} \qquad i \in \mathscr{A}^*. \tag{14.3.10}$$

Thus $\mathbf{s} \in G_9^*$. Conversely let $\mathbf{s} \in G_9^*$. It follows from (14.3.10) that $\mathbf{s}^{\mathrm{T}}\mathbf{a}_i^*\lambda_i^* = 0$ which also implies that $\mathbf{s}^{\mathrm{T}}\mathbf{g}^* = 0$ from the first order conditions. Let $\boldsymbol{\lambda} \in \partial h^*$; then $\boldsymbol{\lambda}^{\mathrm{T}}\mathbf{c}^* = 0$ and $\boldsymbol{\lambda} \geqslant \mathbf{0}$ imply that if $c_i^* < 0$ then $\lambda_i = 0$ so constraints $i \notin \mathscr{A}^*$ can be ignored. Otherwise $\mathbf{s}^{\mathrm{T}}\mathbf{a}_i^*\lambda_i \leqslant 0$ follows from (14.3.10) and hence $\max_{\boldsymbol{\lambda} \in \partial h^*} \mathbf{s}^{\mathrm{T}}(\mathbf{g}^* + \mathbf{A}^*\boldsymbol{\lambda}) = 0$. Thus $\mathbf{s} \in G_{14}^*$ and the equivalence of G_9^* and G_{14}^* is shown. Finally conditions (9.3.15) and (14.2.24) are clearly equivalent so the equivalence of the second order conditions is demonstrated. □

Similar theorems relating the solutions of (14.3.3) and (14.3.4) are given by Charalambous (1979) and Han and Mangasarian (1979).

The requirement that second order conditions hold and that $\nu < 1/\|\boldsymbol{\lambda}^*\|_D$ in theorem 14.3.1 cannot easily be relaxed, as the following simple examples show. In all of these, $x^* = 0$ solves (14.3.3) but does not solve (14.3.4) using $\|\cdot\|_1$, for the reasons given. For the problem: $\min x$ subject to $x^2 \leqslant 0$, x^* is not a KT point. For the problem: $\min x^3$ subject to $x^5 \geqslant 0$, x^* is a KT point and $\lambda^* = 0$ but the curvature condition (9.3.14) is not strict. For the problem: $\min x - \frac{1}{2}x^2$ subject to $0 \leqslant x \leqslant 1$, x^* is a KT point with $\lambda^* = 1$. Then if $\nu = 1$ the condition $\nu < 1/\|\boldsymbol{\lambda}^*\|_D$ is not strict for L_1 norm. The last example illustrates another fact. The proof of theorem 14.3.1 shows that $G_9^* \subset G_{14}^*$ without requiring that $\nu < 1/\|\boldsymbol{\lambda}^*\|_D$. (In the last example G_9^* is empty whilst G_{14}^* is the direction $\mathbf{s} = -1$.) Thus if $\mathbf{x}^*$ satisfies second order sufficient conditions for (14.3.4) it follows that both $\nu \leqslant 1/\|\boldsymbol{\lambda}^*\|_D$ and $\mathbf{x}^*$ satisfies second order sufficient conditions for (14.3.3). Finally if $\nu > 1/\|\boldsymbol{\lambda}^*\|_D$ then a result going the other way can be proved.

Theorem 14.3.2

If first order conditions $\mathbf{g}^* + \mathbf{A}^*\boldsymbol{\lambda}^* = \mathbf{0}$ *hold for (14.3.3) and if* $\nu > 1/\|\boldsymbol{\lambda}^*\|_D$ *then* $\mathbf{x}^*$ *is not a local minimizer of (14.3.4).*

Proof

Since $\|\nu\boldsymbol{\lambda}^*\|_D > 1$, the vector $\nu\boldsymbol{\lambda}^*$ is not in $\partial\|\mathbf{c}^{*+}\|$ and so the vector $\nu\mathbf{g}^* + \mathbf{A}^*(\nu\boldsymbol{\lambda}^*) = \mathbf{0}$ is not in $\partial\phi^*$. Hence by (14.2.6) $\mathbf{x}^*$ is not a local minimizer of $\phi(\mathbf{x})$. □

Another important result concerns the regularity assumption in (14.2.21). If $\mathbf{x}^*$, $\boldsymbol{\lambda}^*$ satisfy first order conditions for either (14.3.1) or (14.3.3) (including feasibility) then under mild assumptions (14.2.21) holds, even though the norm may not be polyhedral. Again the result is proved only in the more difficult case.

Lemma 14.3.1

If $\mathbf{x}^*$, $\boldsymbol{\lambda}^*$ is a KT point for (14.3.3), if the vectors $\mathbf{a}_i^*$, $i \in \mathscr{A}^*$, are linearly independent, and if $\|\nu\boldsymbol{\lambda}^*\|_D < 1$, then $\mathscr{G}^* = G^*$.

Proof

Let $\mathbf{s} \in G^*$. Then as in the proof of theorem 14.3.1, (14.3.10) holds. By the independence assumption there exists an arc $\mathbf{x}(\theta)$ with $\mathbf{x}(0) = \mathbf{x}^*$ and $\dot{\mathbf{x}}(0) = \mathbf{s}$, for which

$$\left.\begin{array}{ll} c_i(\mathbf{x}(\theta)) = 0 & \text{if } \lambda_i^* > 0, \\ c_i(\mathbf{x}(\theta)) \leqslant 0 & \text{if } \lambda_i^* = 0, \end{array}\right\} \quad i \in \mathscr{A}^*$$

for $\theta \geqslant 0$ and sufficiently small (see the proof of lemma 9.2.2). It follows that $\|\mathbf{c}(\mathbf{x}(\theta))^+\| = \mathbf{c}(\mathbf{x}(\theta))^{\mathrm{T}}\boldsymbol{\lambda}^* = 0$ so $\mathbf{x}(\theta)$ is feasible in (14.2.19) and hence $\mathbf{s} \in \mathscr{G}^*$. $\square$

In practice there are considerable advantages in using the L_1 norm to define (14.3.1) or (14.3.3). This choice has been researched widely, for example by Pietrzykowski (1969), Conn (1973), Han (1977), Coleman and Conn (1980), Fletcher (1980b), and Mayne (1980) in nonlinear programming applications and by Barrodale (1970) in L_1 data fitting problems. The general nonlinear programming problem (7.1.1) can be written

$$\begin{array}{lll} \underset{\mathbf{x}}{\text{minimize}} & f(\mathbf{x}) & \\ \text{subject to} & c_i(\mathbf{x}) = 0, & i \in E \\ & c_i(\mathbf{x}) \leqslant 0, & i \in I \end{array} \tag{14.3.11}$$

with the sign convention of this chapter. The corresponding L_1 exact penalty

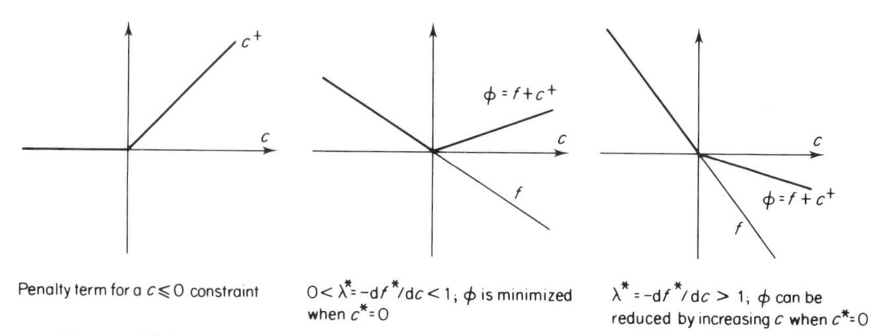

Figure 14.3.2 Optimality conditions for an L_1 exact penalty function

function is

$$\phi(\mathbf{x}) = \nu f(\mathbf{x}) + \sum_{i\in E} |c_i(\mathbf{x})| + \sum_{i\in I} \max(c_i(\mathbf{x}), 0). \tag{14.3.12}$$

From theorem 14.2.1 and a combination of (14.1.14) and (14.1.15) it follows that first order necessary conditions for (14.3.12) are that there exists a Lagrange multiplier vector $\boldsymbol{\lambda}^*$ such that

$$\begin{aligned} &\nu\mathbf{g}^* + \mathbf{A}^*\boldsymbol{\lambda}^* = \mathbf{0} \\ &\left.\begin{aligned} &|\lambda_i^*| \leqslant 1 \\ &\lambda_i^* = \operatorname{sign} c_i^*, \text{ if } c_i^* \neq 0 \end{aligned}\right\} i \in E \\ &\left.\begin{aligned} &0 \leqslant \lambda_i^* \leqslant 1 \\ &\lambda_i^* = 1, \quad \text{if } c_i^* > 0 \\ &\lambda_i^* = 0, \quad \text{if } c_i^* < 0 \end{aligned}\right\} i \in I. \end{aligned} \tag{14.3.13}$$

The differences between this system and the first order conditions for (14.3.11) (see (9.1.16)) are not great. Infeasible points in (14.3.11) are permitted in (14.3.13) in which case the corresponding Lagrange multipliers must take the value $\lambda_i^* = \operatorname{sign} c_i^*$. If $\mathbf{x}^*$ is feasible in (14.3.11) however, this difference does not arise. The parameter ν occurs, which is equivalent to saying that $\boldsymbol{\lambda}^*$ is the multiplier of the scaled problem in which $\nu f(\mathbf{x})$ replaces $f(\mathbf{x})$ in (14.3.11). The only other significant difference between (9.1.16) and (14.3.13) is the requirement that $|\lambda_i^*| \leqslant 1$ in the latter. A simple interpretation of this is the following. Define $Z^* = \{i: c_i^* = 0\}$ and perturb the constraints $i \in Z^*$ as in (9.1.9). Now $\mathbf{x}^*$ also solves

$$\begin{aligned} &\underset{\mathbf{x}}{\text{minimize}} \quad \nu f(\mathbf{x}) + \sum_{j\notin Z^*} \lambda_j^* c_j(\mathbf{x}) \\ &\text{subject to } c_i(\mathbf{x}) = 0, \qquad i \in Z^* \end{aligned} \tag{14.3.14}$$

so using (9.1.10) it follows that

$$\frac{\mathrm{d}(\nu f + \sum_{j\notin Z^*} \lambda_j^* c_j)}{\mathrm{d}\epsilon_i} = -\lambda_i^*, \qquad i \in Z^* \tag{14.3.15}$$

(allowing for the sign change). If in addition $i \in I$, then for increasing ϵ_i the derivative of the remaining terms which make up $\phi(\mathbf{x})$ is just unity. Thus

$$\frac{\mathrm{d}\phi^*}{\mathrm{d}\epsilon_i} = -\lambda_i^* + 1, \qquad i \in Z^* \cap I. \tag{14.3.16}$$

The necessity for $\mathbf{x}^*$ to be minimal thus ensures that $\lambda_i^* \leqslant 1$. A similar development gives $|\lambda_i^*| \leqslant 1$ for $i \in Z^* \cap E$. Thus for example if (14.3.13) holds except that $\lambda_i^* > 1$ for some $i \in Z^* \cap I$, then there exists a perturbation with $\epsilon_i > 0$, $\epsilon_j = 0$, $j \neq i$, such that the rate of decrease $-\lambda_i^*$ in the terms $\nu f + \sum_{j\notin Z^*} \lambda_j^* c_j$ is greater than the unit rate of increase in the penalty term $\max(c_i, 0)$. The situation is illustrated in Figure 14.3.2 for a single constraint and $\nu = 1$. Also the requirement $\nu < 1/\|\boldsymbol{\lambda}^*\|_D$ in theorem 14.3.1 is seen largely as a need to ensure that the scaled multipliers $\nu\boldsymbol{\lambda}^*$ satisfy the conditions $|\nu\lambda_i^*| \leqslant 1$ (since $\|\cdot\|_D = \|\cdot\|_\infty$ in this case). A further feature of the multipliers in (14.3.13) is that they can be interpreted as the optimum variables in a dual problem in the case that $f(\mathbf{x})$ is convex. This result is set out in Question 14.14.

An important advantage of using the L_1 norm is that the discontinuities in derivative occur along surfaces $c_i(\mathbf{x}) = 0$ and the NDO problem can therefore be considered as the equivalent smooth problem (14.3.14) which has a very simple structure. Also the subproblem (14.5.2) which is solved in algorithm (14.5.6) is closely related to the regular QP problem (12.3.13) which arises in the SOLVER method. It is readily solved without adding extra variables as described in (10.3.5) and (10.3.6).

Finally some remarks about the practical choice of the parameter ν are made. The smaller ν is, the more f is damped out relative to $\|\mathbf{c}\|$, and the less accurately is the solution located in the tangent plane of the zeroed constraints (cf. Figure 12.1.2, $\sigma = 100$). Also when following a 'curved valley' along the discontinuity $c_i(\mathbf{x}) = 0$ by a sequence of line searches, then as ν decreases the length of the correction which can be made decreases and hence the total number of line searches increases. Thus it is advantageous to keep ν reasonably large, yet less than the threshold value $1/\|\boldsymbol{\lambda}^*\|_\infty$. Too small a value of ν is indicated by the multipliers $\boldsymbol{\lambda}^*$ in (14.3.14) being uniformly small. Then it can be advantageous to increase ν (by a power of 10 say) so that the active λ_i^* are approximately in the range (0.1, 1). On the other hand, a sequence $\phi^{(k)} \to -\infty$ may occur which can require a smaller value of ν or a different starting point to be chosen. If the active λ_i^* differ widely in magnitude this can be an indication that the constraints should be rescaled. Some algorithms allow this to be done automatically, in which case it can be better (for example Powell, 1978a) to rewrite (14.3.12) as

$$\phi(\mathbf{x}) = f(\mathbf{x}) + \sum_{i \in E} \mu_i |c_i(\mathbf{x})| + \sum_{i \in I} \mu_i \max(c_i(\mathbf{x}), 0) \tag{14.3.17}$$

so that the weighting parameter multiplies the constraint functions and their relative weight can be adjusted.

14.4 Algorithms

This section reviews progress in the development of algorithms for many different classes of NDO problem. It happens that similar ideas have been tried in different situations and it is convenient to discuss them in a unified way. At present there is a considerable amount of interest in these developments and it is not possible to say yet what the best approaches are. This is particularly true when curvature estimates are updated by quasi-Newton like schemes. However some important common features are emerging which this review tries to bring out.

Many methods are line search methods in which on each iteration a direction of search $\mathbf{s}^{(k)}$ is determined from some model situation, and $\mathbf{x}^{(k+1)} = \mathbf{x}^{(k)} + \alpha^{(k)}\mathbf{s}^{(k)}$ is obtained by choosing $\alpha^{(k)}$ to minimize approximately the objective function $\phi(\mathbf{x}^{(k)} + \alpha\mathbf{s}^{(k)})$ along the line (see Section 2.3, Volume 1). A typical line search algorithm (various have been suggested) operates under similar principles to those described in Section 2.6 and uses a combination of sectioning and interpolation, but there are some new features. One is that since $\phi(\mathbf{x})$ may contain a polyhedral component $h(\mathbf{c}(\mathbf{x}))$, the interpolating function must also have the same type of structure. The simplest possibility is to interpolate a one variable function of the form

$$\psi(\alpha) = q(\alpha) + h(\ell(\alpha)) \tag{14.4.1}$$

where $q(\alpha)$ is quadratic and the functions $\ell(\alpha)$ are linear. For composite NDO applications q and ℓ can be estimated from information about f and $\mathbf{c}$, and $\alpha^{(k)}$ is then determined by minimizing (14.4.1). For basic NDO, only the values of ϕ and $\mathrm{d}\phi/\mathrm{d}\alpha$ are known at any point so it must be assumed that $\psi(\alpha)$ has a more simple structure, for example the max of just two linear functions $\ell_i(\alpha)$. Many other possibilities exist. Another new feature concerns what acceptability tests to use, analogous to (2.4.2), (2.4.3), (2.4.7), and (2.4.8) for smooth functions. If the line search minimum is non-smooth (see Figure 14.4.1) then it is not appropriate to try to make the directional derivative small, as (2.4.8) does, since such a point may not exist. Many line searches use a combination of (2.4.2) and (2.4.7). For this choice the range of acceptable α-values is the interval $[a, b]$ in Figure 14.4.1, and this should be compared with Figure 2.4.1. It can be seen that this line search has the effect that the acceptable point *always overshoots the minimizing value of* α, and in fact this may occur by a substantial amount, so as to considerably slow down the rate of convergence of the algorithm in which it is embedded. I have found it valuable to use a different test (Fletcher, 1980b) that the line search is terminated when the predicted reduction based on (14.4.1) is sufficiently small, and in particular when the predicted reduction on any subinterval is no greater than 0.1 times the total reduction in $\phi(\mathbf{x})$ so far achieved in the search. This condition has been found to work well in ensuring that $\alpha^{(k)}$ is close to the minimizing value of α. Also when using a Newton-like method and when $\mathbf{x}^{(k)}$ is close to the solution, then the unit step can be accepted without further searching.

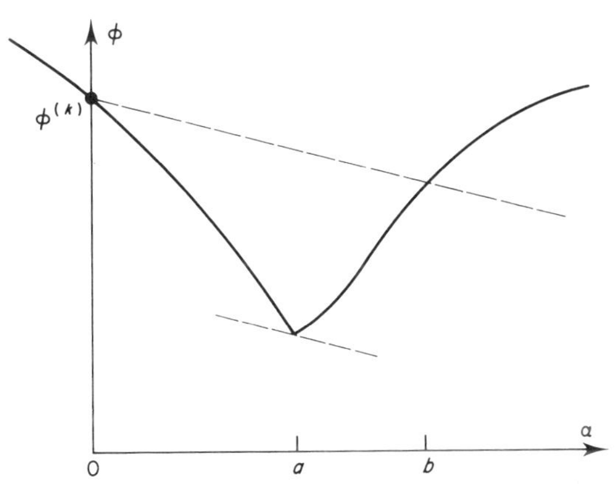

Figure 14.4.1 Line search for non-differentiable functions

It is now possible to examine algorithms for NDO problems in many variables. In basic NDO only a limited amount of information is available at any given point $\mathbf{x}$, namely $\phi(\mathbf{x})$ and one element $\mathbf{g} \in \partial\phi(\mathbf{x})$ (usually $\nabla\phi(\mathbf{x})$ since $\phi(\mathbf{x})$ is almost everywhere differentiable). This makes the basic NDO problem more difficult than composite NDO in which values of f and $\mathbf{c}$ (and their derivatives) are given in (14.1.1). The simplest method for basic NDO is an analogue of the steepest descent method in which $\mathbf{s}^{(k)} = -\mathbf{g}^{(k)}$ is used as a search direction; this is known as *subgradient optimization*. Because of its simplicity the method has received much attention both in theory and in practice (see the references given in Lemarechal and Mifflin (1978)), but it is at best linearly convergent, as illustrated by the results for smooth functions (see Section 2.3). In fact the situation is worse in the non-smooth case because the convergence result (theorem 2.4.1) no longer holds. In fact examples of non-convergence are easily constructed. Assume that the line search is exact and that the subgradient obtained at $\mathbf{x}^{(k+1)}$ corresponds to the piece which is active for $\alpha \geqslant \alpha^{(k)}$. Then the example

$$\begin{aligned} \phi(\mathbf{x}) &= \max_{i=1,2,3} c_i(\mathbf{x}) \\ c_1(\mathbf{x}) &= -5x_1 + x_2 \\ c_2(\mathbf{x}) &= x_1^2 + x_2^2 + 4x_2 \\ c_3(\mathbf{x}) &= 5x_1 + x_2 \end{aligned} \tag{14.4.2}$$

due to Demyanov and Malozemov (1971) illustrates convergence from the initial point $\mathbf{x}^{(1)}$ (see Figure 14.4.2) to the non-optimal point $\mathbf{x}^{\infty} = \mathbf{0}$. The solution is at $\mathbf{x}^* = (0, -3)^{\mathrm{T}}$. At $\mathbf{x}^{(1)}, \mathbf{x}^{(3)}, \ldots$ the subgradient corresponding to c_1 is used and at $\mathbf{x}^{(2)}, \mathbf{x}^{(4)}, \ldots$ the subgradient corresponding to c_3. It is easily seen from Figure 14.4.2 that the sequence $\{\mathbf{x}^{(k)}\}$ oscillates between the two curved surfaces of non-differentiability and converges to $\mathbf{x}^{\infty}$ which is not optimal. In fact it is not necessary for the surfaces of non-differentiability to be curved, and a similar polyhedral (piecewise linear) example can easily be constructed for which the algorithm also fails.

It can be argued that a closer analogue to the steepest descent method at a point of non-differentiability is to search along the steepest descent direction $\mathbf{s}^{(k)} = -\bar{\mathbf{g}}^{(k)}$ where $\bar{\mathbf{g}}^{(k)}$ minimizes $\|\mathbf{g}\|_2$ for $\mathbf{g} \in \partial\phi^{(k)}$. This interpretation is justified for convex functions in (14.2.12) and a similar result holds when the composite function $\phi(\mathbf{x})$ is in use. Let $\partial\phi^{(k)}$ be defined by the convex hull of its extreme points, $\partial\phi^{(k)} = \operatorname{conv} \mathbf{g}_i^{(k)}$, say. (For composite functions $\mathbf{g}_i^{(k)} = \mathbf{g}^{(k)} + \mathbf{A}^{(k)}\mathbf{h}_i$, $i \in \mathscr{A}^{(k)}$, see (14.2.16).) Then $\bar{\mathbf{g}}^{(k)}$ is defined by solving the problem

$$\begin{aligned} &\underset{\mathbf{g},\,\boldsymbol{\mu}}{\text{minimize}} \quad \mathbf{g}^{\mathrm{T}}\mathbf{g} \\ &\text{subject to } \mathbf{g} = \sum_i \mu_i \mathbf{g}_i^{(k)}, \\ &\qquad \sum_i \mu_i = 1, \qquad \boldsymbol{\mu} \geqslant \mathbf{0}, \end{aligned} \tag{14.4.3}$$

and $\mathbf{s}^{(k)} = -\bar{\mathbf{g}}^{(k)}$. This problem is similar to a least distance QP problem and is

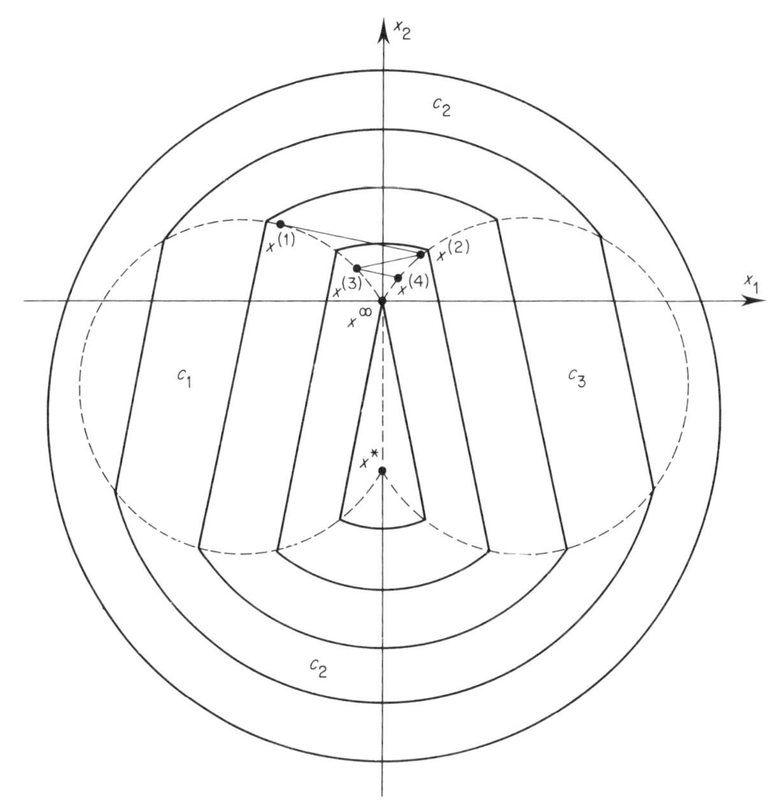

Figure 14.4.2 False convergence of steepest-descent-like methods for problem (14.4.2)

readily solved by the methods of Section 10.5. The resulting method terminates finitely when $\phi(\mathbf{x})$ is polyhedral (piecewise linear) and exact line searches are used. This latter condition ensures that the $\mathbf{x}^{(k)}$ are on the surfaces of non-differentiability for which $\partial\phi^{(k)}$ has more than one element. The algorithm is idealized however in that exact line searches are not generally possible in practice. Also the whole subdifferential $\partial\phi^{(k)}$ is not usually available, and even if it were the non-exact line search would cause $\partial\phi^{(k)} = \nabla\phi^{(k)}$ to hold and the method would revert to subgradient optimization.

The spirit of this type of method however is preserved in *bundle* methods (Lemarechal, 1978). A bundle method is a line search method which solves subproblem (14.4.3) to define $\mathbf{s}^{(k)}$, except that the vectors $\mathbf{g}_i^{(k)}$ are elements of a 'bundle' B rather than the extreme points of $\partial\phi^{(k)}$. Initially B is set to $B = \mathbf{g}^{(1)} \in \partial\phi^{(1)}$, and in the simplest form of the algorithm, subgradients $\mathbf{g}^{(2)}$, $\mathbf{g}^{(3)}, \ldots$ are added to B on successive iterations. The method continues in this way until $\mathbf{0} \in B$. Then the bundle B is reset, for instance to the current $\mathbf{g}^{(k)}$, and the iteration is continued. With careful manipulation of B (see for example the *conjugate subgradient method* of Wolfe (1975)) a convergence result for the algorithm can be proved and a suitable termination test obtained. However Wolfe's method does not terminate finitely at the solution for polyhedral $\phi(\mathbf{x})$. An example (due to Powell) has the pentagonal contours illustrated in Figure 14.4.3. At $\mathbf{x}^{(1)}$, B is $\mathbf{g}_1$, and at $\mathbf{x}^{(2)}$, B is $\{\mathbf{g}_1, \mathbf{g}_3\}$. At $\mathbf{x}^{(3)}$, B is $\{\mathbf{g}_1, \mathbf{g}_3, \mathbf{g}_4\}$ and $\mathbf{0} \in B$ so the process is restarted with $B = \mathbf{g}_3$ (or $B = \mathbf{g}_4$ with the same conclusions). The whole sequence is repeated and the points $\mathbf{x}^{(k)}$ cycle round the pentagon, approaching the centre but without ever terminating there.

This non-termination can be corrected by including more elements in B when restarting (for example both $\mathbf{g}_3$ and $\mathbf{g}_4$) or alternatively deleting old elements like $\mathbf{g}_1$. This can be regarded as providing a better estimate of $\partial\phi^{(k)}$ by B. However it is by no means certain that this would give a better method: in fact the idealized steepest descent method itself can fail to converge. Problem (14.4.2) again illustrates this; this time the search directions are tangential to the surfaces of non-differentiability and oscillatory convergence to $\mathbf{x}^{\infty}$ occurs. This example illustrates the need to have information about the subderivative at $\partial\phi^{\infty}$ (which is lacking in this example) to avoid the false convergence at $\mathbf{x}^{\infty}$. Generalizations of bundle methods have therefore been suggested which attempt to enlarge $\partial\phi^{(k)}$ so as to contain subgradient information at neighbouring points. One possibility is to use the ϵ-subdifferential

$$\partial_{\epsilon}\phi(\mathbf{x}) = \{\mathbf{g}:\ \phi(\mathbf{x}+\boldsymbol{\delta}) \geqslant \phi(\mathbf{x}) + \mathbf{g}^{\mathrm{T}}\boldsymbol{\delta} - \epsilon \quad \forall\, \boldsymbol{\delta}\}$$

which contains $\partial\phi(\mathbf{x})$. Bundle methods in which B approximates this set are discussed by Lemarechal (1978) and are currently being researched.

A quite different type of method for basic NDO is to use the information $\mathbf{x}^{(k)}, \phi^{(k)}, \mathbf{g}^{(k)}$ obtained at any point to define the linear function

$$\phi^{(k)} + (\mathbf{x} - \mathbf{x}^{(k)})^{\mathrm{T}}\mathbf{g}^{(k)} \tag{14.4.4}$$

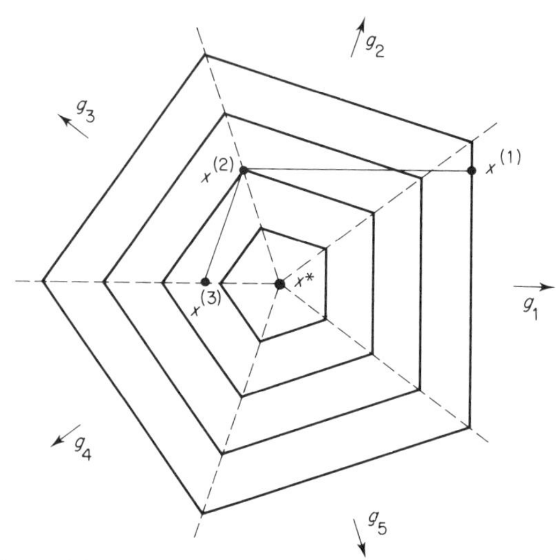

Figure 14.4.3 Non-termination of a bundle method

and to use these linearizations in an attempt to model $\phi(\mathbf{x})$. If $\phi(\mathbf{x})$ is convex then this function is a supporting hyperplane and this fact is exploited in the *cutting plane method*. The linearizations are used to define a model polyhedral convex function and the minimizer of this function determines the next iterate. Specifically the linear program

$$\begin{aligned} &\underset{\mathbf{x},\, v}{\text{minimize}} \quad v \\ &\text{subject to } v \geqslant \phi^{(i)} + (\mathbf{x} - \mathbf{x}^{(i)})^{\mathrm{T}}\mathbf{g}^{(i)}, \qquad i = 1, 2, \ldots, k \end{aligned} \tag{14.4.5}$$

is solved to determine $\mathbf{x}^{(k+1)}$. Then the linearization determined by $\mathbf{x}^{(k+1)}$, $\phi^{(k+1)}$, and $\mathbf{g}^{(k+1)}$ is added to the set and the process is repeated. On the early iterations a step restriction $\|\mathbf{x} - \mathbf{x}^{(k)}\|_\infty \leqslant h$ is required to ensure that (14.4.5) is not unbounded. A line search can also be added to the method. A similar method can be used on non-convex problems if old linearizations are deleted in a systematic way.

More sophisticated algorithms for basic NDO are discussed at the end of this section and attention is now turned to algorithms for minimizing the composite function (14.1.1) (which includes max functions, etc., when $h(\mathbf{c})$ is the polyhedral function (14.1.2) – see the special cases in (14.1.3)). Linear approximations have also been used widely and the simplest method is to replace $\mathbf{c}(\mathbf{x})$ by the first order Taylor series approximation

$$\mathbf{c}(\mathbf{x}^{(k)} + \boldsymbol{\delta}) \approx \boldsymbol{\ell}^{(k)}(\boldsymbol{\delta}) = \mathbf{c}^{(k)} + \mathbf{A}^{(k)^{\mathrm{T}}}\boldsymbol{\delta} \tag{14.4.6}$$

and $f(\mathbf{x})$ (if present) by

$$f(\mathbf{x}^{(k)} + \boldsymbol{\delta}) \approx f^{(k)} + \mathbf{g}^{(k)^{\mathrm{T}}}\boldsymbol{\delta}, \tag{14.4.7}$$

and to substitute these approximations into (14.1.6) which is an equivalent form of the original problem. The linear program

$$\begin{aligned} &\underset{\boldsymbol{\delta},\, v}{\text{minimize}} \quad v \\ &\text{subject to } v - (\mathbf{g}^{(k)} + \mathbf{A}^{(k)}\mathbf{h}_i)^{\mathrm{T}}\boldsymbol{\delta} \geqslant f^{(k)} + \mathbf{c}^{(k)^{\mathrm{T}}}\mathbf{h}_i + b_i \qquad \forall\ i \end{aligned} \tag{14.4.8}$$

is solved to determine $\boldsymbol{\delta}^{(k)}$ and hence $\mathbf{x}^{(k+1)} = \mathbf{x}^{(k)} + \boldsymbol{\delta}^{(k)}$. The only difference between this and (14.4.5) is that here there is sufficient information available at $\mathbf{x}^{(k)}$ to determine linear approximations to all the pieces, whereas in basic NDO this information has to be accumulated on a sequence of iterations. When applied to $\min \|\mathbf{c}(\mathbf{x})\|_\infty$ or $\min \|\mathbf{c}(\mathbf{x})\|_1$ the iteration based on (14.4.8) can be considered as a *Gauss–Newton method* by analogy with the same method for solving $\min \|\mathbf{c}(\mathbf{x})\|_2^2$ (see Section 6.1, Volume 1) which is based on the same linear approximations. An early study of this type of method is by Osborne and Watson (1969) and a more elaborate recent method is given by Charalambous and Conn (1978). As with Gauss–Newton methods, convergence is not guaranteed but this can likewise be rectified by going to a restricted step type of method. In this case $\|\mathbf{x} - \mathbf{x}^{(k)}\|_\infty \leqslant h^{(k)}$ is a suitable choice since it preserves a linear programming subproblem. This approach has been investigated by Madsen (1975).

Linear approximation methods are most successful when the linearizations of the active pieces at $\mathbf{x}^*$ fully determine $\mathbf{x}^*$ (that is to say when the first order conditions are sufficient, which occurs when the set G^* in (14.2.20) corresponding to directions of zero slope is empty and curvature effects are negligible). A necessary condition for this is that there are $n + 1$ (or more) active pieces at $\mathbf{x}^*$. In these circumstances the rate of convergence of the method based on (14.4.8) is second order (to prove this, show that the method is ultimately equivalent to the Newton–Raphson method for solving the active constraints in (14.4.8)). This situation is most likely to occur in problems such as over-determined L_1 or L_∞ data fitting, where these methods can be very successful.

In general however it is not possible to exclude curvature effects, and when these are important then methods based only on linear information converge slowly and become unreliable. However the problem min $\phi(\mathbf{x})$ is equivalent to the non-linear program (14.1.6), and it is known that the SOLVER method of Section 12.3 has a second order rate of convergence. Following (12.3.13), and given an approximation $\boldsymbol{\mu}^{(k)}$ to the optimum multipliers $\boldsymbol{\mu}^*$ in (14.1.8), the QP problem

$$\begin{aligned}&\underset{\boldsymbol{\delta},\, v}{\text{minimize}} \quad v + \tfrac{1}{2}\boldsymbol{\delta}^{\mathrm{T}}\Big(\sum_i \mu_i^{(k)}(\nabla^2 f^{(k)} + \nabla^2(\mathbf{c}^{\mathrm{T}}\mathbf{h}_i)^{(k)}\Big)\boldsymbol{\delta} \\ &\text{subject to } v - (\mathbf{g}^{(k)} + \mathbf{A}^{(k)}\mathbf{h}_i)^{\mathrm{T}}\boldsymbol{\delta} \geqslant f^{(k)} + \mathbf{c}^{(k)\mathrm{T}}\mathbf{h}_i + b_i \qquad \forall\ i\end{aligned} \tag{14.4.9}$$

is set up. Using $\Sigma\mu_i^{(k)} = 1$, $\boldsymbol{\lambda}^{(k)} = \mathbf{H}\boldsymbol{\mu}^{(k)}$, and writing the matrix $\nabla^2 f^{(k)} + \Sigma\lambda_i^{(k)}\nabla^2 c_i^{(k)} = \nabla_x^2\mathscr{L}(\mathbf{x}^{(k)}, \boldsymbol{\lambda}^{(k)})$ as $\mathbf{W}^{(k)}$, (14.4.9) can be written as

$$\begin{aligned}&\underset{\boldsymbol{\delta},\, v}{\text{minimize}} \quad v + \tfrac{1}{2}\boldsymbol{\delta}^{\mathrm{T}}\mathbf{W}^{(k)}\boldsymbol{\delta} \\ &\text{subject to } v - (\mathbf{g}^{(k)} + \mathbf{A}^{(k)}\mathbf{h}_i)^{\mathrm{T}}\boldsymbol{\delta} \geqslant f^{(k)} + \mathbf{c}^{(k)\mathrm{T}}\mathbf{h}_i + b_i \qquad \forall\ i.\end{aligned} \tag{14.4.10}$$

This is similar to (14.4.8) but contains an extra quadratic term in the objective function which accounts for curvature in the functions f and $\mathbf{c}$. Thus the analogue of the SOLVER method is to solve (14.4.10) to determine a correction $\boldsymbol{\delta}^{(k)}$ and then set $\mathbf{x}^{(k+1)} = \mathbf{x}^{(k)} + \boldsymbol{\delta}^{(k)}$. In addition the multipliers of the constraints in (14.4.10) become the next approximation $\boldsymbol{\mu}^{(k+1)}$ to be used in constructing $\mathbf{W}^{(k+1)}$. This method is mentioned briefly by Pshenichnyi (1978) in the context of minimizing max functions.

An important observation is that the above algorithm is closely related to one in which $\boldsymbol{\delta}^{(k)}$ is chosen to solve

$$\underset{\boldsymbol{\delta}}{\text{minimize}} \quad \psi^{(k)}(\boldsymbol{\delta}) \triangleq q^{(k)}(\boldsymbol{\delta}) + h(\boldsymbol{\ell}^{(k)}(\boldsymbol{\delta})) \tag{14.4.11}$$

where

$$q^{(k)}(\boldsymbol{\delta}) \triangleq \tfrac{1}{2}\boldsymbol{\delta}^{\mathrm{T}}\mathbf{W}^{(k)}\boldsymbol{\delta} + \mathbf{g}^{(k)\mathrm{T}}\boldsymbol{\delta} + f^{(k)} \tag{14.4.12}$$

and where $\boldsymbol{\ell}^{(k)}(\boldsymbol{\delta})$ is defined as in (14.4.6). Then $\mathbf{x}^{(k+1)} = \mathbf{x}^{(k)} + \boldsymbol{\delta}^{(k)}$ and $\boldsymbol{\lambda}^{(k+1)}$ is chosen as the multipliers which exist at the solution to (14.4.11). This method is referred to here as the *QL method* since it makes both quadratic and linear approximations. It is of wider application than (14.4.10) since the latter requires that $h(\mathbf{c})$

is polyhedral. However when this occurs then the QL method and that based on (14.4.10) are equivalent. This is readily shown by setting $w = v + \frac{1}{2}\boldsymbol{\delta}^T \mathbf{W}^{(k)}\boldsymbol{\delta}$ so that (14.4.10) becomes

$$\begin{aligned}&\underset{\boldsymbol{\delta},\, w}{\text{minimize}}\quad w\\&\text{subject to } w \geqslant q^{(k)}(\boldsymbol{\delta}) + \mathbf{h}_i^T \boldsymbol{\ell}^{(k)}(\boldsymbol{\delta}) + b_i \qquad \forall\ i\end{aligned}$$

which is equivalent to (14.4.11) (Fletcher, 1980a). The QL method is seen to have a nice interpretation. The function $\psi^{(k)}(\boldsymbol{\delta})$ defined in (14.4.11) approximates the composite function $\phi(\mathbf{x})$ $(= \phi(\mathbf{x}^{(k)} + \boldsymbol{\delta}))$ local to $\mathbf{x}^{(k)}$. The approximation is constructed by replacing $\mathbf{c}(\mathbf{x})$ in (14.1.1) by the linear approximation $\boldsymbol{\ell}^{(k)}(\boldsymbol{\delta})$ and $f(\mathbf{x})$ by the quadratic approximation $q^{(k)}(\boldsymbol{\delta})$. This contains the matrix $\Sigma \lambda_i^{(k)} \nabla^2 c_i^{(k)}$ which accounts for curvature in the functions $c_i(\mathbf{x})$. This is exactly the way in which the SOLVER method itself is obtained from a nonlinear programming problem (Section 12.3).

An illustration of the QL method applied to solve problem (14.3.5) with $\nu = 1$ is given. Initial approximations $\mathbf{x}^{(1)} = (2, 0)^T$ and $\lambda^{(1)} = 1$ are taken and the progress of the iterations is given in Table 14.4.1. It can be seen that the rate of convergence is second order with $\|\boldsymbol{\delta}^{(k+1)}\|_2 / \|\boldsymbol{\delta}^{(k)}\|_2^2 \approx 0.5$. The contours of the approximating function $\psi^{(k)}$ for $k = 1$ and $k = 2$ are illustrated in Figure 14.4.4. Each function has quadratic pieces with a discontinuous derivative on a linear surface (partly dotted) which is the linearization (through (14.4.6)) of the unit circle on which the discontinuity in Figure 14.3.1 occurs.

It is possible under mild conditions to prove that the QL method converges locally at a second order rate in a variety of circumstances. The most straightforward case uses the equivalence between (14.4.10) and (14.4.11). It is assumed that second order sufficient conditions hold at $\mathbf{x}^*$, $\boldsymbol{\mu}^*$ for problem (14.1.6) (or equivalently at $\mathbf{x}^*$, $\boldsymbol{\lambda}^*$ for problem (14.1.1) – see lemma 14.2.8), that $\mathscr{A}^* = \mathscr{A}_+^*$, and that the independence assumption (14.2.25) holds. Using the result of Question 14.11, these conditions imply the local convergence at a second order rate of the SOLVER method applied to (14.1.6), that is method (14.4.10), by virtue of theorem 12.3.1 and Question 12.18. Thus the QL method also converges locally at a second order rate. The second order sufficient conditions are almost equivalent to the necessary conditions (see lemma 14.2.7) so this assumption is likely to be valid in practice. The independence assumption is likely to hold

Table 14.4.1 Application of the QL method to problem (14.3.5)

k	1	2	3	4	5	6
$\mathbf{x}^{(k)}$	2	1.25	0.804310	0.712691	0.707133	0.707107
	0	0.5	0.801724	0.712981	0.707127	0.707107
$\lambda^{(k)}$	1	0.625	0.622844	0.692599	0.706968	0.707107
$\phi^{(k)}$	1	−0.9375	−1.316358	−1.409402	−1.414194	−1.414214
$\boldsymbol{\delta}^{(k)}$	−0.75	−0.445690	−0.091619	−0.005558	−0.000026	$2_{10^{-10}}$
	0.5	0.301724	−0.088744	−0.005854	−0.000020	$-9_{10^{-10}}$

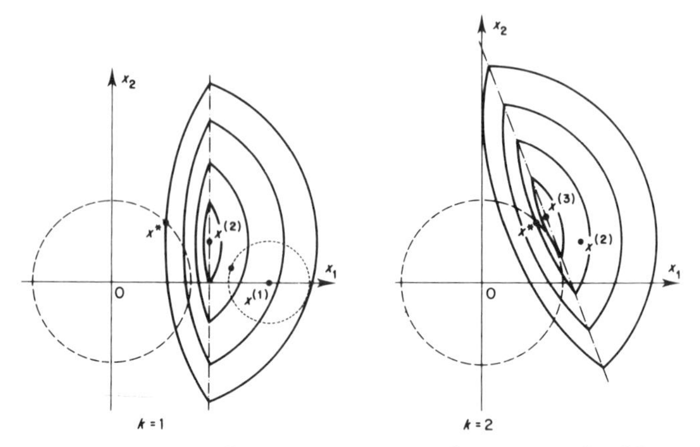

Figure 14.4.4 Contours of the approximating function in the QL method

when there are at most $n + 1$ elements in $\mathscr{A}^*$, that is in case (I) and (II) situations in (14.1.3) and in case (III) when $\mathbf{c}^* \neq \mathbf{0}$. However case (IV) and (V) situations are not included – see the comments after lemma 14.2.7.

Another situation which can be handled is when $h(\mathbf{c})$ is $\|\mathbf{c}\|$ (or $\|\mathbf{c}^+\|$) and $\mathbf{c}^* = \mathbf{0}$ (or $\mathbf{c}^{*+} = \mathbf{0}$). The norm need not be polyhedral; this is essentially the situation of Section 14.3 that an exact penalty function is being used to solve the nonlinear programming problem (14.3.1) (or (14.3.3)). Let $\mathscr{A}^*$ refer here to the active constraints in (14.3.1) or (14.3.3). It is assumed that $\nu\boldsymbol{\lambda}^* \in \text{interior}(\partial h^*)$ (that is $\|\nu\boldsymbol{\lambda}^*\|_D < 1$ and $\lambda_i^* > 0$, $i \in \mathscr{A}^*$, when $h(\mathbf{c}) = \|\mathbf{c}^+\|$) and that the vectors $\mathbf{a}_i^*$, $i \in \mathscr{A}^*$, are linearly independent, which are mild assumptions. The argument is given in the more difficult case that $h(\mathbf{c}) = \|\mathbf{c}^+\|$. It follows from theorem 14.3.1 that second order conditions for (14.3.3) and (14.3.4) are equivalent and these are assumed to hold. Let $\mathbf{A}^*$ have columns $\mathbf{a}_i^*$, $i \in \mathscr{A}^*$, and let $\mathbf{Z}^*$ be any basis for the null column space of $\mathbf{A}^*$ as in Section 10.1. Then the second order conditions for (14.3.3) are equivalent to the condition that $\mathbf{Z}^{*\mathrm{T}}\mathbf{W}^*\mathbf{Z}^*$ is positive definite (Question 11.1). Define $\mathbf{Z}^*$ by (10.1.22) (direct elimination) assuming without loss of generality that $\mathbf{A}_1^*$ is non-singular. Then $\mathbf{Z}$ is a continuous function of $\mathbf{A}$ in a neighbourhood of $\mathbf{A}^*$. Thus for $\mathbf{x}^{(k)}$, $\boldsymbol{\lambda}^{(k)}$ in a neighbourhood of $\mathbf{x}^*$, $\boldsymbol{\lambda}^*$, $\mathbf{Z}^{(k)\mathrm{T}}\mathbf{W}^{(k)}\mathbf{Z}^{(k)}$ is positive definite (continuity of eigenvalues) and hence as in Questions 11.1 and 12.18, second order sufficient conditions hold for the problem

$$\begin{aligned} &\underset{\boldsymbol{\delta}}{\text{minimize}} \quad q^{(k)}(\boldsymbol{\delta}) \\ &\text{subject to } \boldsymbol{\ell}^{(k)}(\boldsymbol{\delta}) \leqslant \mathbf{0}. \end{aligned}$$

This is the subproblem which is solved when the SOLVER method is applied to (14.3.3). It follows by theorem 14.3.1 that this subproblem is equivalent to (14.4.11) which is the QL method applied to (14.3.4). Hence the QL method is equivalent to SOLVER in this case also. Under the same assumptions the SOLVER method converges locally at a second order rate so the same is true of the QL method. Also the assumptions imply that lemma 14.3.1 holds so that the sufficient conditions are almost necessary and hence likely to be achieved in practice.

This result can also be used to tackle the more difficult case (IV) and (V) situations in (14.1.3) involving $\|\cdot\|_1$, when $\mathbf{c}^{*+} \neq \mathbf{0}$ or $\mathbf{c}^* \neq \mathbf{0}$. Then using a splitting like (14.2.27) the problem is reduced to that in the previous paragraph, and the same conclusions can be deduced under the mild assumption that the vectors $\mathbf{a}_i^*$, $i \in Z$, are linearly independent. Thus in most practical cases the QL method converges locally at a second order rate. A final remark about rates of convergence is that when $h(\mathbf{c}) = \|\mathbf{c}\|$ is a smooth norm (for example $\|\cdot\|_2$) and $\mathbf{c}^* \neq \mathbf{0}$ then $\mathscr{G}^* \neq G^*$ and the above development is not appropriate. In this case the most appropriate second order sufficient conditions are those for a smooth problem that $\nabla\phi^* = \mathbf{0}$ and $\nabla^2\phi^*$ is positive definite. However the QL method does not reduce to Newton's method so it is an open question as to whether a second order rate of convergence can be deduced.

These results on second order convergence make it clear that the QL method is attractive as a starting point for developing a general purpose algorithm for composite NDO problems. However the basic method itself is not robust and can fail to converge (when $m = 0$ it is just the basic Newton's method – see Section 3.1, Volume 1). An obvious precaution to prevent divergence is to require that the sequence $\{\phi^{(k)}\}$ is non-increasing. One possibility is to use the correction from solving (14.4.11) as a direction of search $\mathbf{s}^{(k)}$ along which to approximately minimize the objective function $\phi(\mathbf{x}^{(k)} + \alpha\mathbf{s}^{(k)})$. Both the basic method and the line search modification can fail in another way however, in that remote from the solution the function $\psi^{(k)}(\boldsymbol{\delta})$ may not have a minimizer. This is analogous to the case that $\mathbf{G}^{(k)}$ is indefinite in smooth unconstrained minimization. An effective modification in the latter case is the restricted step or trust region approach (Chapter 5, Volume 1) and it is fruitful to consider modifying the QL method in this way. Since (14.4.11) contains no side conditions the incorporation of a step restriction causes no difficulty. This is illustrated in Figure 14.4.4 ($k = 1$) in which the dotted circle with centre $\mathbf{x}^{(1)}$ is a possible trust region. Clearly there is no difficulty in minimizing $\psi^{(1)}(\boldsymbol{\delta})$ within this region: the solution is the point on the periphery of the circle. Contrast this situation with Section 12.3 where a trust region cannot be used in conjunction with the SOLVER method because there may not exist feasible points in the resulting problem. Fletcher (1980a) describes a model restricted step method for solving composite NDO problems which is globally convergent without the need to assume that vectors $\mathbf{a}_i^\infty$ are independent or multipliers $\boldsymbol{\lambda}^{(k)}$ are bounded. This algorithm is described in more detail in Section 14.5 where the global convergence property is justified. Good practical experience with the method in L_1 exact penalty function applications is reported by Fletcher (1980b), including the solution of test problems on which other methods fail.

There is one adverse feature of the QL method with line search or trust region modifications, and also of other second order algorithms for NDO, which differs from experience with similar algorithms for smooth unconstrained optimization. In the latter case when $\mathbf{x}^{(k)}$ is close to $\mathbf{x}^*$ then the unit step of the basic method reduces the objective function so that the line search or trust region modification is not brought into play and the second order rate of the basic method is observed (for example theorem 5.1.2). The same situation does *not* hold in NDO as Maratos (1978) observes (see also Mayne, 1980, Chamberlain *et al.*, 1980, and Question 14.13). Thus in some well-behaved NDO problems in which second order effects are significant at the solution, $\mathbf{x}^{(k)}$ can be arbitrarily close to $\mathbf{x}^*$ and the unit step of the basic algorithm can fail to reduce $\phi(\mathbf{x})$. The *Maratos effect*, as it might be called, thus causes the unit step to be rejected and may obviate a second order rate of convergence result. The effect is most likely to occur when the jump discontinuity in derivative is large. In my practical experience with L_1 exact penalty functions however the effect has not been noticeable: this may be due to keeping the weighting parameter ν relatively large to reduce ill-conditioning and hence not typical of all types of application. Further modifications of second order methods are currently being researched to circumvent this difficulty.

Another disadvantage of the QL method is that it requires second derivative matrices $\nabla^2 f$ and $\nabla^2 c_i$ to be supplied by the user. However there is no difficulty in approximating $\mathbf{W}^{(k)}$, for instance by finite differences or a quasi-Newton update, so that only first derivative information is required. One possibility is to use the modified BFGS formula given by Powell (1978a) to update an approximation $\mathbf{B}^{(k)}$ to $\mathbf{W}^{(k)}$ as described earlier in (12.3.18). In view of the superlinear convergence result for nonlinear programming applications the same is likely to hold here, assuming that ultimately the unit step $\mathbf{x}^{(k+1)} = \mathbf{x}^{(k)} + \boldsymbol{\delta}^{(k)}$ is taken on every iteration. However the feature of forcing $\mathbf{B}^{(k)}$ to be positive definite when $\mathbf{W}^*$ may not be, does seem artificial. Alternative approaches which avoid this, and which may require much less storage on large problems, keep an approximation $\mathbf{B}^{(k)}$ to the *reduced* curvature matrix $\mathbf{Z}^T\mathbf{W}^{(k)}\mathbf{Z}$ (see (11.1.7) and (10.1.3)) appropriate to the problem (14.4.11). Womersley (1981) uses the BFGS formula (unmodified) to update $\mathbf{B}^{(k)}$ and Conn (1979) obtains $\mathbf{B}^{(k)}$ by a finite difference process. Methods which use curvature information appear to be much more efficient in practice than those which do not, when second order effects are important. Womersley (1981) gives a comparison on a number of test problems. Typically for the exact penalty function determined by the TP2 problem (Section 12.3), Charalambous and Conn (1978) require 281 gradient evaluations whereas Womersley (1981) requires 48. Because of such results there is currently considerable research interest in second order methods, as shown by the number of presentations at the recent X Symposium on Mathematical Programming (August 1979) in Montreal.

Techniques based on the QL method are only appropriate for composite NDO problems, and it is also important to consider applications to basic NDO in which second order information has been used. Womersley (1978) describes a method which requires $\phi^{(i)}$, $\nabla\phi^{(i)}$, and $\nabla^2\phi^{(i)}$ to be available for any $\mathbf{x}^{(i)}$, $i = 1, 2, \ldots, k$, and each such set of information is assumed to arise from an active piece of $\phi(\mathbf{x})$. This information is used to build up a linear approximation for each piece valid at $\mathbf{x}^{(k)}$, and also to give the matrix $\mathbf{W}^{(k)}$. This is then used as in (14.4.10). Information about pieces may be rejected if the Lagrange multipliers $\mu_i^{(k+1)}$ determine that the piece is no longer active. Also a Levenberg–Marquardt modification (Section 5.2 of Volume 1) is made to $\mathbf{W}^{(k)}$ to ensure positive definiteness. A difficulty with the algorithm is that repeated approximations to the same piece tend to be collected, which cause degeneracy in the QP solver close to the solution. A modification is proposed in which an extra item of information is supplied at each $\mathbf{x}^{(i)}$ which is an integer *label* of the active piece. Such information can be supplied in most basic NDO applications. This labelling enables a single most recent approximation to each piece to be maintained and circumvents the degeneracy problem. Again this approach is considerably more effective than the subgradient or cutting plane methods as numerical evidence given by Womersley suggests (~20 iterations as against ~300 on a typical problem). A quasi-Newton version of the algorithm is currently being researched as described in Womersley (1981).

14.5 A Globally Convergent Model Algorithm

The aim of this section is to show that the QL method based on solving the subproblem (14.4.11) can be incorporated readily with the idea of a step restriction which is known to give good numerical results in smooth unconstrained optimization (Chapter 5, Volume 1). Subproblem (14.4.11) contains curvature information and potentially allows convergence at a second order rate, whereas the step length restriction is shown to ensure global convergence. The resulting method (Fletcher, 1980a) is applicable to the solution of all composite NDO problems, including exact penalty functions, best L_1 and L_∞ approximation, etc., but excluding basic NDO problems. The term 'model algorithm' indicates that the algorithm is presented in a simple format as a means of making its convergence properties clear. It admits of the algorithm being modified to improve its practical performance whilst not detracting from these properties. The motivation for using a step restriction is that it defines a *trust region* for the correction $\boldsymbol{\delta}$ by

$$\|\boldsymbol{\delta}\| \leqslant h^{(k)} \tag{14.5.1}$$

in which the Taylor series approximations (14.4.6) and (14.4.12) are assumed to be adequate. The norm in (14.5.1) is arbitrary but either the $\|\cdot\|_\infty$ or the $\|\cdot\|_2$ is the most likely choice, especially the former since (14.5.2) below can then be solved by QP-like methods (see (10.3.5)). On each iteration the subproblem is to minimize (14.4.11) subject to this restriction, that is

$$\begin{aligned} &\underset{\boldsymbol{\delta}}{\text{minimize}} \quad \psi^{(k)}(\boldsymbol{\delta}) \\ &\text{subject to } \|\boldsymbol{\delta}\| \leqslant h^{(k)}. \end{aligned} \tag{14.5.2}$$

The radius $h^{(k)}$ of the trust region is adjusted adaptively to be as large as possible subject to adequate agreement between $\phi(\mathbf{x}^{(k)} + \boldsymbol{\delta})$ and $\psi^{(k)}(\boldsymbol{\delta})$ being maintained. This can be quantified by defining the *actual reduction*

$$\Delta\phi^{(k)} = \phi(\mathbf{x}^{(k)}) - \phi(\mathbf{x}^{(k)} + \boldsymbol{\delta}^{(k)}) \tag{14.5.3}$$

and the *predicted reduction*

$$\Delta\psi^{(k)} = \phi(\mathbf{x}^{(k)}) - \psi^{(k)}(\boldsymbol{\delta}^{(k)}). \tag{14.5.4}$$

Then the ratio

$$r^{(k)} = \frac{\Delta\phi^{(k)}}{\Delta\psi^{(k)}} \tag{14.5.5}$$

measures the extent to which ϕ and $\psi^{(k)}$ agree local to $\mathbf{x}^{(k)}$. The rules for changing $h^{(k)}$ in the model algorithm are those given in Section 5.1 and are not elaborated on further, except to emphasize that in practice the rule for reducing $h^{(k)}$ can be more elaborate, based perhaps on some sort of interpolation (Fletcher, 1980b).

The kth iteration of the model algorithm is as follows.

$$\begin{aligned}
&\text{(i) Given } \mathbf{x}^{(k)}, \boldsymbol{\lambda}^{(k)}, \text{ and } h^{(k)}, \text{ calculate } f^{(k)}, \mathbf{g}^{(k)}, \mathbf{c}^{(k)}, \mathbf{A}^{(k)}, \text{ and}\\
&\qquad \mathbf{W}^{(k)} \text{ which determine } \phi^{(k)} \text{ and } \psi^{(k)}(\boldsymbol{\delta}).\\
&\text{(ii) Find a global solution } \boldsymbol{\delta}^{(k)} \text{ to (14.5.2).}\\
&\text{(iii) Evaluate } \phi(x^{(k)} + \boldsymbol{\delta}^{(k)}) \text{ and calculate } \Delta\phi^{(k)}, \Delta\psi^{(k)}, \text{ and } r^{(k)}.\\
&\text{(iv) If } r^{(k)} < 0.25 \text{ set } h^{(k+1)} = \|\boldsymbol{\delta}^{(k)}\|/4,\\
&\qquad \text{if } r^{(k)} > 0.75 \text{ and } \|\boldsymbol{\delta}^{(k)}\| = h^{(k)} \text{ set } h^{(k+1)} = 2h^{(k)},\\
&\qquad \text{otherwise set } h^{(k+1)} = h^{(k)}.\\
&\text{(v) If } r^{(k)} \leqslant 0 \text{ set } \mathbf{x}^{(k+1)} = \mathbf{x}^{(k)}, \boldsymbol{\lambda}^{(k+1)} = \boldsymbol{\lambda}^{(k)},\\
&\qquad\qquad \text{else } \mathbf{x}^{(k+1)} = \mathbf{x}^{(k)} + \boldsymbol{\delta}^{(k)}, \boldsymbol{\lambda}^{(k+1)} = \text{multipliers from (14.5.2).}
\end{aligned} \tag{14.5.6}$$

The parameters 0.25, 0.75, etc., which arise are arbitrary and are not very sensitive but the values given here are typical. In fact the solution of problems like (14.5.2) has not been considered in this chapter but it is a straightforward extension of these ideas (see Watson, 1978, Fletcher and Watson, 1980) and multipliers $\boldsymbol{\lambda}^{(k+1)} \in \partial h(\boldsymbol{\ell}^{(k)}(\boldsymbol{\delta}^{(k)}))$ exist at the solution in an analogous way to theorem 14.2.1. It is these multipliers that are used in step (v).

In proving global convergence, a result is used relating the directional derivative (14.2.14) of the composite function (14.2.13) and the difference quotient between two points $\mathbf{x}^{(k)}$ and $\mathbf{x}^{(k)} + \epsilon^{(k)}\mathbf{s}$ in a common direction $\mathbf{s}$ $(\neq \mathbf{0})$, as both points approach a fixed point $\mathbf{x}'$. The result is a special case of one due to Clarke (1975) but is proved here directly for completeness.

Lemma 14.5.1

Let S be the set of all sequences $\mathbf{x}^{(k)} \to \mathbf{x}'$, $\epsilon^{(k)} \to 0$, and let $\mathbf{c}^{(k)} \triangleq \mathbf{c}(\mathbf{x}^{(k)})$, etc., and $\mathbf{c}_\epsilon^{(k)} \triangleq \mathbf{c}(\mathbf{x}^{(k)} + \epsilon^{(k)}\mathbf{s})$; then

$$\limsup_{S} \frac{h(\mathbf{c}_\epsilon^{(k)}) - h(\mathbf{c}^{(k)})}{\epsilon^{(k)}} = \max_{\boldsymbol{\lambda} \in \partial h'} \mathbf{s}^{\mathrm{T}}\mathbf{A}'\boldsymbol{\lambda} \tag{14.5.7}$$

in the sense that the difference quotient is bounded above and the sup of all accumulation points of the quotient taken over all sequences in S is the directional derivative $\max_{\boldsymbol{\lambda} \in \partial h'} \mathbf{s}^{\mathrm{T}}\mathbf{A}'$.

Proof

By the integral form of the Taylor series

$$\mathbf{c}_\epsilon^{(k)} = \mathbf{c}^{(k)} + \epsilon^{(k)} \int_0^1 [\mathbf{A}(\mathbf{x}^{(k)} + \theta\epsilon^{(k)}\mathbf{s})]^{\mathrm{T}}\mathbf{s}\, d\theta = \mathbf{c}^{(k)} + \epsilon^{(k)}\mathbf{d}^{(k)} \tag{14.5.8}$$

say, where $\mathbf{d}^{(k)} \to \mathbf{A}'^{\mathrm{T}}\mathbf{s}$ as $k \to \infty$. Let $\boldsymbol{\lambda}_\epsilon^{(k)} \in \partial h_\epsilon^{(k)} \triangleq \partial h(\mathbf{c}_\epsilon^{(k)})$. Using the subgradient inequality and (14.5.8),

$$h(\mathbf{c}^{(k)}) \geqslant h(\mathbf{c}_\epsilon^{(k)}) + (\mathbf{c}^{(k)} - \mathbf{c}_\epsilon^{(k)})^{\mathrm{T}}\boldsymbol{\lambda}_\epsilon^{(k)} = h(\mathbf{c}_\epsilon^{(k)}) - \epsilon^{(k)}\mathbf{d}^{(k)^{\mathrm{T}}}\boldsymbol{\lambda}_\epsilon^{(k)}$$

or

$$\mathbf{d}^{(k)^{\mathrm{T}}}\boldsymbol{\lambda}_\epsilon^{(k)} \geqslant \frac{h(\mathbf{c}_\epsilon^{(k)}) - h(\mathbf{c}^{(k)})}{\epsilon^{(k)}}. \tag{14.5.9}$$

Since $\partial h_\epsilon^{(k)}$ is bounded in a neighbourhood of $\mathbf{x}'$ (lemma 14.2.1) the difference quotient is bounded above. Now consider a subsequence for which the difference quotient accumulates, and let

$$\lim \frac{h(\mathbf{c}_\epsilon^{(k)}) - h(\mathbf{c}^{(k)})}{\epsilon^{(k)}} > \max_{\boldsymbol{\lambda} \in \partial h'} \mathbf{s}^{\mathrm{T}}\mathbf{A}'\boldsymbol{\lambda}. \tag{14.5.10}$$

Since $\boldsymbol{\lambda}_\epsilon^{(k)}$ is bounded there exists a thinner subsequence for which $\boldsymbol{\lambda}_\epsilon^{(k)} \to \boldsymbol{\lambda}'$ and since $\mathbf{c}_\epsilon^{(k)} \to \mathbf{c}'$ it follows by lemma 14.2.4 that $\boldsymbol{\lambda}' \in \partial h'$. Thus from (14.5.9),

$$\lim \frac{h(\mathbf{c}_\epsilon^{(k)}) - h(\mathbf{c}^{(k)})}{\epsilon^{(k)}} \leqslant \mathbf{s}^{\mathrm{T}}\mathbf{A}'\boldsymbol{\lambda}' \leqslant \max_{\boldsymbol{\lambda} \in \partial h'} \mathbf{s}^{\mathrm{T}}\mathbf{A}'\boldsymbol{\lambda} \tag{14.5.11}$$

which contradicts (14.5.10), so that the reverse inequality ($\leqslant$) in (14.5.10) is true. Finally by taking $\mathbf{x}^{(k)} = \mathbf{x}'$, (14.2.14) shows that there is a sequence in S which achieves equality with the directional derivative. Thus (14.5.7) is established. □

Corollary

Define $\phi(\mathbf{x})$ by (14.2.13) where $f \in \mathbb{C}^1$. Then

$$\limsup_{S} \frac{\phi(\mathbf{x}^{(k)} + \epsilon^{(k)}\mathbf{s}) - \phi(\mathbf{x}^{(k)})}{\epsilon^{(k)}} = \max_{\boldsymbol{\lambda} \in \partial h'} \mathbf{s}^{\mathrm{T}}(\mathbf{g}' + \mathbf{A}'\boldsymbol{\lambda}). \tag{14.5.12}$$

Proof

The result follows by using an analogous Taylor series for $f(\mathbf{x})$ as in the proof of the lemma. □

It is now possible to state the main result of this section.

Theorem 14.5.1

Let $\mathbf{x}^{(k)} \in B \subset \mathbb{R}^n$ $\forall$ k where B is bounded and let f, $\mathbf{c}$ be $\mathbb{C}^2$ functions whose second derivative matrices are bounded on B. Then there exists an accumulation point $\mathbf{x}^\infty$ of algorithm (14.5.6) at which first order conditions hold, that is

$$\max_{\boldsymbol{\lambda} \in \partial h^\infty} \mathbf{s}^{\mathrm{T}}(\mathbf{g}^\infty + \mathbf{A}^\infty\boldsymbol{\lambda}) \geqslant 0 \qquad \forall \ \mathbf{s}. \tag{14.5.13}$$

Proof

There exists a convergent subsequence $\mathbf{x}^{(k)} \to \mathbf{x}^\infty$ for which either

(i) $r^{(k)} < 0.25, h^{(k+1)} \to 0$, and hence $\| \boldsymbol{\delta}^{(k)} \| \to 0$ or
(ii) $r^{(k)} \geqslant 0.25$ and $\inf h^{(k)} > 0$.

In either case (14.5.13) is shown to hold. In case (i) let $\exists$ a descent direction $\mathbf{s}$ ($\| \mathbf{s} \| = 1$) at $\mathbf{x}^\infty$, that is

$$\max_{\boldsymbol{\lambda} \in \partial h^\infty} \mathbf{s}^{\mathrm{T}}(\mathbf{g}^\infty + \mathbf{A}^\infty \boldsymbol{\lambda}) = -d, \qquad d > 0. \tag{14.5.14}$$

By Taylor series

$$\begin{aligned} f(\mathbf{x}^{(k)} + \epsilon^{(k)}\mathbf{s}) &= f^{(k)} + \epsilon^{(k)}\mathbf{s}^{\mathrm{T}}\mathbf{g}^{(k)} + o(\epsilon^{(k)}) \\ &= q^{(k)}(\epsilon^{(k)}\mathbf{s}) + o(\epsilon^{(k)}) \end{aligned} \tag{14.5.15}$$

by (14.4.12), since $\boldsymbol{\lambda}^{(k)}$ is bounded by lemma 14.2.1 and $\nabla^2 f$, $\nabla^2 c_i$ are bounded by assumption. Likewise by (14.4.6),

$$\mathbf{c}(\mathbf{x}^{(k)} + \epsilon^{(k)}\mathbf{s}) = \boldsymbol{\ell}^{(k)}(\epsilon^{(k)}\mathbf{s}) + o(\epsilon^{(k)}) \tag{14.5.16}$$

and hence by (14.1.2), the boundedness of ∂h, and (14.4.11), it follows that

$$\begin{aligned} \phi(\mathbf{x}^{(k)} + \epsilon^{(k)}\mathbf{s}) &= q^{(k)}(\epsilon^{(k)}\mathbf{s}) + h(\boldsymbol{\ell}^{(k)}(\epsilon^{(k)}\mathbf{s}) + o(\epsilon^{(k)})) + o(\epsilon^{(k)}) \\ &= q^{(k)}(\epsilon^{(k)}\mathbf{s}) + h(\boldsymbol{\ell}^{(k)}(\epsilon^{(k)}\mathbf{s})) + o(\epsilon^{(k)}) \\ &= \psi^{(k)}(\epsilon^{(k)}\mathbf{s}) + o(\epsilon^{(k)}). \end{aligned} \tag{14.5.17}$$

Writing $\epsilon^{(k)} = \| \boldsymbol{\delta}^{(k)} \|$ and considering a step along $\mathbf{s}$ in the subproblem, it follows by the optimality of $\boldsymbol{\delta}^{(k)}$ that

$$\begin{aligned} \Delta\psi^{(k)} &\geqslant \phi^{(k)} - \psi^{(k)}(\epsilon^{(k)}\mathbf{s}) \\ &= \phi^{(k)} - \phi(\mathbf{x}^{(k)} + \epsilon^{(k)}\mathbf{s}) + o(\epsilon^{(k)}) \\ &\geqslant \epsilon^{(k)}(d + o(1)) + o(\epsilon^{(k)}) = d\epsilon^{(k)} + o(\epsilon^{(k)}) \end{aligned} \tag{14.5.18}$$

by the corollary to lemma 14.5.1 and (14.5.14). But (14.5.17) implies that

$$\Delta\phi^{(k)} = \Delta\psi^{(k)} + o(\epsilon^{(k)})$$

and hence $r^{(k)} = \Delta\phi^{(k)}/\Delta\psi^{(k)} = 1 + o(\epsilon^{(k)})/\Delta\psi^{(k)} = 1 + o(1)$ from (14.5.8) since $d > 0$, which contradicts $r^{(k)} < 0.25$. Thus $d \leqslant 0$ for all $\mathbf{s}$ and hence (15.5.13) holds at $\mathbf{x}^\infty$.

In case (ii) the argument of case (ii) of theorem 5.1.1 (Volume 1) is largely followed (see also Fletcher, 1980a). The only extra step is to deduce from $\mathbf{x}^{(k)} \in B$ that $\partial h^{(k)}$ is uniformly bounded for all k, so that the parameters $\boldsymbol{\lambda}^{(k)}$ are bounded. Thus a thinner subsequence can be chosen such that $\boldsymbol{\lambda}^{(k)} \to \boldsymbol{\lambda}^\infty$ and hence $\mathbf{W}^{(k)} \to \mathbf{W}^\infty$. Then functions $q^\infty(\boldsymbol{\delta})$, $\boldsymbol{\ell}^\infty(\boldsymbol{\delta})$, and $\psi^\infty(\boldsymbol{\delta})$ are defined and it is concluded as in theorem 5.1.1 that $\boldsymbol{\delta} = \mathbf{0}$ minimizes $\psi^\infty(\boldsymbol{\delta})$. It follows that the first order conditions (14.5.13) hold at $\mathbf{x}^\infty$. In case (ii) it is also possible to conclude that second order conditions hold. $\square$

Note that the existence of a bounded region B which the theorem requires is implied if any level set $\{\mathbf{x}: \phi(\mathbf{x}) \leqslant \phi^{(k)}\}$ is bounded. Also the theorem assumes that the sequence $\{\mathbf{x}^{(k)}\}$ is infinite; if not then $\Delta\psi^{(k)} = 0$ for some k, the iteration terminates, and first order conditions are satisfied.

One point to emphasize about the theorem is that there are no hidden assumptions that certain vectors $\mathbf{a}_i^\infty$ are linearly independent or that the multipliers $\boldsymbol{\lambda}^{(k)}$ are bounded. Methods for NDO or nonlinear programming can often be proved to be convergent under such assumptions, yet can fail in practice. Thus it is important that this theorem avoids such assumptions. Another point is that $\mathbf{W}^{(k)}$ does not need to be defined as in (14.4.10) but can be any bounded matrix. Thus the theorem indicates that a corresponding quasi-Newton method, using for example $\mathbf{B}^{(k)}$ as defined in (12.3.18) in place of $\mathbf{W}^{(k)}$, can only fail if $\mathbf{B}^{(k)}$ becomes unbounded. The theorem also subsumes a result due to Madsen (1975) for L_∞ approximation in a first order method like (14.4.8) for which $\mathbf{W}^{(k)} = \mathbf{0}$.

For smooth unconstrained optimization a stronger result (theorem 5.1.1) can be proved that the accumulation point also satisfies second order necessary conditions. To prove this for algorithm (14.5.6) would require the expression

$$\phi(\mathbf{x}^{(k)} + \epsilon^{(k)}\mathbf{s}) = \psi^{(k)}(\epsilon^{(k)}\mathbf{s}) + o(\epsilon^{(k)2})$$

which however does not hold in NDO applications. Nonetheless when $\inf h^{(k)} > 0$, then second order necessary conditions do hold (see the proof of theorem 14.5.1). Also for smooth problems, and when second order sufficient conditions hold, the rate of convergence can be shown to be second order (theorem 5.1.2). This result is unlikely to hold in general for algorithm (14.5.6) in an NDO application because of the possibility of the Maratos effect (Section 14.4). However if an assumption that $\inf h^{(k)} > 0$ is made, then the unit step of the QL method is taken for k sufficiently large, in which case the second order rate of convergence results described in Section 14.4 are valid. Current research is aimed at modifications of algorithm (14.5.6) which avoid the Maratos effect and enable the second order rate to be established more generally (see the *Note in proof* on p. 214).

Questions for Chapter 14

1. For the nonlinear program (14.1.6) prove that $\mathscr{F}' = F'$ (defined in Section 9.2). Let $\bar{\mathbf{s}} = (\mathbf{s}^T, s_{n+1})^T$ be a feasible direction, and use the linearized constraint equations to show that

$$s_{n+1} \geqslant \max_{i \in \mathscr{A}'} \mathbf{s}^T(\mathbf{g}' + \mathbf{A}'\mathbf{h}_i). \tag{a}$$

   Let $\delta^{(k)} \downarrow 0$ be any sequence, and define $\mathbf{x}^{(k)} = \mathbf{x}' + \delta^{(k)}\mathbf{s}$. If equality holds in (a), and if $v^{(k)} = \phi(\mathbf{x}^{(k)})$, show that $\mathbf{x}^{(k)}, v^{(k)}$ gives a feasible directional sequence in (14.1.6) for sufficiently large k. If strict inequality holds in (a), and if $v^{(k)} = v^* + \delta^{(k)}s_{n+1}$, again show that $\mathbf{x}^{(k)}, v^{(k)}$ is a feasible directional sequence.
2. Prove that $\|\mathbf{c}^+\|$ is a convex function of $\mathbf{c}$ when the norm is monotonic

($|\mathbf{x}| \leqslant |\mathbf{y}| \Rightarrow \|\mathbf{x}\| \leqslant \|\mathbf{y}\|$). Deduce that $\mathbf{c}_\theta^+ \leqslant (1-\theta)\mathbf{c}_0^+ + \theta\mathbf{c}_1^+$ as an intermediate stage.

3. Establish the equivalence between each of (14.1.11) to (14.1.15) and the general expression $\partial h(\mathbf{c}) = \text{conv}_{i\in} \mathbf{h}_i$. In (14.1.12) let $\lambda_i = \mu_i$, $i \leqslant m$, and μ_{m+1} acts as the slack variable for $\Sigma_i \lambda_i \leqslant 1$. In (14.1.13) define $\lambda_i = \mu_i - \mu_{m+i}$ or $\mu_i = \max(\lambda_i, 0)$, $\mu_{m+i} = \max(-\lambda_i, 0)$. In (14.1.14) use the fact that the cube $0 \leqslant \lambda_i \leqslant 1$ has extreme points which are all combinations of 1 and 0, and similarly for (14.1.15).
4. Establish the equivalence between the subdifferential expression

$$\partial \|\mathbf{c}\| = \{\boldsymbol{\lambda}: \|\mathbf{c}+\mathbf{h}\| \geqslant \|\mathbf{c}\| + \boldsymbol{\lambda}^{\mathrm{T}}\mathbf{h} \quad \forall\, \mathbf{h}\} \tag{b}$$

and (14.3.7). Use the generalized Cauchy inequality $\mathbf{a}^{\mathrm{T}}\mathbf{b} \leqslant \|\mathbf{a}\| \|\mathbf{b}\|_D$ on $(\mathbf{c}+\mathbf{h})^{\mathrm{T}}\boldsymbol{\lambda}$ to show that $\boldsymbol{\lambda} \in$ (14.3.7) implies $\boldsymbol{\lambda} \in$ (b). If $\boldsymbol{\lambda} \in$ (b), use the triangle inequality to show that $\mathbf{h}^{\mathrm{T}}\boldsymbol{\lambda} \leqslant \|\mathbf{h}\|\ \forall\, \mathbf{h}$ and (14.3.4) to show $\|\boldsymbol{\lambda}\|_D \leqslant 1$. Hence $\boldsymbol{\lambda}^{\mathrm{T}}\mathbf{c} \leqslant \|\mathbf{c}\|$. Then with $\mathbf{h} = -\mathbf{c}$ in (b) show that $\boldsymbol{\lambda}^{\mathrm{T}}\mathbf{c} \geqslant \|\mathbf{c}\|$ and hence $\boldsymbol{\lambda} \in$ (14.3.7).
5. If the norm is monotonic (see Question 14.2) establish the equivalence between

$$\partial \|\mathbf{c}^+\| = \{\boldsymbol{\lambda}: \|(\mathbf{c}+\mathbf{h})^+\| \geqslant \|\mathbf{c}^+\| + \boldsymbol{\lambda}^{\mathrm{T}}\mathbf{h} \quad \forall\, \mathbf{h}\} \tag{c}$$

and (14.3.8). The proof is similar to that in Question 14.4. In the first part also use the fact that $\boldsymbol{\lambda} \geqslant \mathbf{0}$ implies $(\mathbf{c}+\mathbf{h})^{+\mathrm{T}}\boldsymbol{\lambda} \geqslant (\mathbf{c}+\mathbf{h})^{\mathrm{T}}\boldsymbol{\lambda}$. In the second part also use the monotonic norm property to establish $\|\mathbf{h}^+\| \geqslant \mathbf{h}^{\mathrm{T}}\boldsymbol{\lambda}\ \forall\, \mathbf{h}$. Then $\mathbf{h} = -\mathbf{e}_i$ yields $\lambda_i \geqslant 0$. Use $\mathbf{h}^+ \leqslant |\mathbf{h}|$ to show $\|\boldsymbol{\lambda}\|_D \leqslant 1$. Then proceed much as in Question 14.4.
6. Show that the subdifferential in (14.2.2) is a closed convex set.
7. Justify equation (14.2.12) when $\mathbf{0} \notin \partial f'$. Show by straightforward arguments that

$$\min_{\|\mathbf{s}\|_2=1}\ \max_{\mathbf{g}\in\partial f'} \mathbf{s}^{\mathrm{T}}\mathbf{g} \geqslant \max_{\mathbf{g}\in\partial f'}\ \min_{\|\mathbf{s}\|_2=1} \mathbf{s}^{\mathrm{T}}\mathbf{g} = -\|\bar{\mathbf{g}}\|_2.$$

Then use the separating hyperplane result (lemma 14.2.3) to show that equality is achieved when $\mathbf{s} = -\bar{\mathbf{g}}/\|\bar{\mathbf{g}}\|_2$. Show by considering $f = |x|$ that the result is not true when $\mathbf{0} \in \partial f'$. The result can be generalized to include the case $\mathbf{0} \in \partial f'$ by writing $\|\mathbf{s}\|_2 \leqslant 1$ in place of $\|\mathbf{s}\|_2 = 1$.
8. Consider the Freudenstein and Roth equations

$$c_1(\mathbf{x}) = x_1 - x_2^3 + 5x_2^2 - 2x_2 - 13$$
$$c_2(\mathbf{x}) = x_1 + x_2^3 + x_2^2 - 14x_2 - 29$$

(see Chapter 6, Volume 1) and consider minimizing $\|\mathbf{c}(\mathbf{x})\|_p$ for $1 \leqslant p \leqslant \infty$. A local solution in all cases is $\mathbf{x}^* = (11.4128, -0.8968)^{\mathrm{T}}$. Find the sets $\partial \|\mathbf{c}^*\|_p$ (that is ∂h^*) for $p = 1, 2, \infty$ and the vector $\boldsymbol{\lambda}^*$ which satisfies theorem 14.2.1. For $p = 1$ the local solution is not unique and any $x_1^* \in [6.4638, 11.4128]$ gives a solution. Find $\partial \|\mathbf{c}^*\|_1$ when $x_1^* = 6.4638$.
9. Consider minimizing $\|\mathbf{c}\|_1$ when the equation $c_1(\mathbf{x})$ in Question 14.8 is scaled by multiplying the right hand side by 2. Show that $\mathbf{x}^* = (6.4638, -0.8968)^{\mathrm{T}}$

satisfies the second order sufficient conditions of theorem 14.2.3 and find $\boldsymbol{\lambda}^*$. Find a trajectory which is feasible in (14.2.19) (see Figure 14.2.3) and hence find the sets $\mathscr{G}^*$ and G^* and verify that they are non-empty and equal. Verify that the vectors $\mathbf{a}_i^*$, $i \in Z$, are linearly independent which implies $\mathscr{G}^* = G^*$ (see 14.2.27) ff.).

10. Consider minimizing $\|\mathbf{c}\|_\infty$ for the system which results on adding an equation $c_3(\mathbf{x}) = \alpha x_1$ to those in Question 14.8. If $\alpha = 0.4336$ show that $\mathbf{x}^* = (11.4128, -0.8968)^{\mathrm{T}}$ satisfies the second order sufficient conditions of theorem 14.2.3 and find $\boldsymbol{\lambda}^*$. Find a trajectory which is feasible in (14.2.19) and hence find the sets $\mathscr{G}^*$ and G^* and verify that they are non-empty and equal. Verify that (14.2.25) is satisfied which implies $\mathscr{G}^* = G^*$.
11. Show that the vectors in (14.2.25) are independent iff the vectors $\begin{pmatrix} -\mathbf{g}^* - \mathbf{A}^*\mathbf{h}_i \\ 1 \end{pmatrix}$ are independent. To do this let the vectors be the columns of matrices $\mathbf{B}$ and $\begin{bmatrix} \mathbf{C} \\ \mathbf{e}^{\mathrm{T}} \end{bmatrix}$ respectively, where

$$\begin{bmatrix} \mathbf{C} \\ \mathbf{e}^{\mathrm{T}} \end{bmatrix} = \begin{bmatrix} \mathbf{B} & \mathbf{0} \\ \mathbf{0}^{\mathrm{T}} & 0 \end{bmatrix} - \begin{pmatrix} \mathbf{g}^* + \mathbf{A}^*\mathbf{h}_p \\ 1 \end{pmatrix} \mathbf{e}^{\mathrm{T}}.$$

Clearly if $\begin{bmatrix} \mathbf{C} \\ \mathbf{e}^{\mathrm{T}} \end{bmatrix}$ has full rank then so has $\mathbf{B}$. If $\begin{bmatrix} \mathbf{C} \\ \mathbf{e}^{\mathrm{T}} \end{bmatrix}$ has not full rank then there exists $\bar{\mathbf{u}} = \begin{pmatrix} \mathbf{u} \\ u_{n+1} \end{pmatrix} \neq \mathbf{0}$ such that $\begin{bmatrix} \mathbf{C} \\ \mathbf{e}^{\mathrm{T}} \end{bmatrix}^{\mathrm{T}} \bar{\mathbf{u}} = \mathbf{0}$. Show that this implies that $\mathbf{B}$ does not have full rank.
12. Consider the problem defined in (14.4.2). At $\mathbf{x}^\infty$ find the multipliers $\boldsymbol{\lambda}^\infty$ for which $\mathbf{g}^\infty + \mathbf{A}^\infty \boldsymbol{\lambda}^\infty = \mathbf{0}$ and show that $\lambda_2^\infty < 0$ which implies that $\phi(\mathbf{x})$ can be reduced by making the inequality $v \geqslant c_2(\mathbf{x})$ in (14.1.6) inactive (see end of Section 14.1). Find the multipliers $\boldsymbol{\lambda}^*$ at $\mathbf{x}^*$ and show that the conditions of theorem 14.2.1 hold. Apply the method of (14.4.8) from $\mathbf{x}^{(1)} = (0, -4)^{\mathrm{T}}$ and verify that a second order rate of convergence is obtained. Why does this happen in the absence of curvature information?
13. In problem (14.3.5) let $\mathbf{x}^{(1)}$ lie on the unit circle arbitrarily close to $\mathbf{x}^*$, and let $\boldsymbol{\lambda}^{(1)} = \boldsymbol{\lambda}^*$. Show that there exists a range of values of ν with $\nu < 1/\lambda^*$ for which the unit step determined by solving (14.4.11) fails to reduce $\phi(\mathbf{x})$.
14. Consider the unconstrained NDO problem: $\min f(\mathbf{x}) + \sum_{i=1}^m |r_i(\mathbf{x})|$, where $\mathbf{r}(\mathbf{x}) = \mathbf{A}^{\mathrm{T}}\mathbf{x} + \mathbf{b}$ ($\mathbf{x} \in \mathbb{R}^n$) and $f(\mathbf{x})$ is convex. By introducing variables

$$r_i^+ = \max(r_i, 0), \qquad r_i^- = \max(-r_i, 0),$$

show that $r_i^+ - r_i \geqslant 0$, $r_i^- + r_i \geqslant 0$, and $r_i^+ + r_i^- = |r_i|$. Hence show that the unconstrained problem can be restated as

$$\begin{aligned} &\underset{\mathbf{x}, \mathbf{r}^+, \mathbf{r}^-}{\text{minimize}} \quad f(\mathbf{x}) + \mathbf{e}^{\mathrm{T}}(\mathbf{r}^+ + \mathbf{r}^-) \\ &\text{subject to } \mathbf{r}^+ - \mathbf{A}^{\mathrm{T}}\mathbf{x} - \mathbf{b} \geqslant \mathbf{0} \\ &\qquad\qquad \mathbf{r}^- + \mathbf{A}^{\mathrm{T}}\mathbf{x} + \mathbf{b} \geqslant \mathbf{0}, \qquad \mathbf{r}^+ \geqslant \mathbf{0},\ \mathbf{r}^- \geqslant \mathbf{0}, \end{aligned}$$

where $\mathbf{e} = (1, 1, \ldots, 1)^{\mathrm{T}}$. Show that this is a convex programming problem.

Write down the dual of this problem, denoting the multipliers of the constraints by $\boldsymbol{\lambda}^+$, $\boldsymbol{\lambda}^-$, $\boldsymbol{\mu}^+$, $\boldsymbol{\mu}^-$ respectively. By eliminating $\boldsymbol{\mu}^+$ and $\boldsymbol{\mu}^-$, and writing $\boldsymbol{\lambda}^+ - \boldsymbol{\lambda}^- = \boldsymbol{\lambda}$, show that the dual can be restated more simply as

$$\underset{\mathbf{x},\, \boldsymbol{\lambda}}{\text{maximize}} \;\; f(\mathbf{x}) + \boldsymbol{\lambda}^{\mathrm{T}}(\mathbf{A}^{\mathrm{T}}\mathbf{x} + \mathbf{b})$$

$$\text{subject to} \;\; \mathbf{g}(\mathbf{x}) + \mathbf{A}\boldsymbol{\lambda} = \mathbf{0}, \qquad -\mathbf{e} \leqslant \boldsymbol{\lambda} \leqslant \mathbf{e},$$

where $\mathbf{g} = \nabla_x f$ (see also Watson, 1978).

Note added in proof: some recent suggestions for averting the Maratos effect are given in the paper 'Second order corrections for nondifferentiable optimization' by R. Fletcher in *Numerical Analysis – Dundee 1981* (Ed. G. A. Watson), Springer-Verlag, Berlin (to be published).

References

Abadie, J. and Carpentier, J. (1969). 'Generalization of the Wolfe reduced gradient method to the case of nonlinear constraints', in *Optimization* (Ed. R. Fletcher), Academic Press, London.

Andreassen, D. O. and Watson, G. A. (1976). 'Nonlinear Chebyshev approximation subject to constraints', *J. Approx. Theo.*, **18**, 241–250.

Apostol, T. M. (1957). *Mathematical Analysis*, Addison-Wesley, Reading, Mass.

Appelgren, L. (1971). 'Integer programming methods for a vessel scheduling problem', *Transportation Science*, **5**, 64–78.

Balinski, M. L. and Wolfe, P. (1975). *Nondifferentiable Optimization*, Mathematical Programming Study 3, North Holland, Amsterdam.

Bandler, J. W. and Charalambous, C. (1972). 'Practical least p-th optimization of networks', *IEEE Trans. Microwave Theo. Tech. (1972 Symposium Issue)*, **20**, 834–840.

Barrodale, I. (1970). 'On computing best L_1 approximations', in *Approximation Theory* (Ed. A. Talbot), Academic Press, London.

Barrodale, I. and Roberts, F. D. K. (1973). 'An improved algorithm for discrete ℓ_1 linear approximation', *SIAM J. Num. Anal.*, **10**, 839–848.

Bartels, R. H. (1971). 'A stabilization of the simplex method', *Num. Math.*, **16**, 414–434.

Beale, E. M. L. (1959). 'On quadratic programming', *Naval. Res. Log. Quart.*, **6**, 227–244.

Beale, E. M. L. (1967). 'Numerical methods', in *Nonlinear Programming* (Ed. J. Abadie), North Holland, Amsterdam.

Beale, E. M. L. (1970). 'Advanced algorithmic features for general mathematical programming systems', in *Integer and Nonlinear Programming* (Ed. J. Abadie), North Holland, Amsterdam.

Beale, E. M. L. (1978). 'Integer programming', in *The State of the Art in Numerical Analysis* (Ed. D. A. H. Jacobs), Academic Press, London.

Benveniste, R. (1979). 'A quadratic programming algorithm using conjugate search directions', *Math. Prog.*, **16**, 63–80.

Biggs, M. C. (1975). 'Constrained minimization using recursive quadratic programming: some alternative subproblem formulations', in *Towards Global Optimization* (Eds L. C. W. Dixon and G. P. Szego), North Holland, Amsterdam.

Biggs, M. C. (1978). 'On the convergence of some constrained minimization algorithms based on recursive quadratic programming', *J. Inst. Maths Applns*, **21**, 67–81.

Bradley, J. and Clyne, H. M. (1976). 'Applications of geometric programming to building design problems', in *Optimization in Action* (Ed. L. C. W. Dixon), Academic Press, London.

Breu, R. and Burdet, C.-A. (1974). 'Branch and bound experiments in zero–one programming', in *Approaches to Integer Programming* (Ed. M. L. Balinski), Mathematical Programming Study 2, North Holland, Amsterdam.

Buckley, A. (1975). 'Constrained minimization using Powell's conjugacy approach', AERE Harwell report CSS 22.

Bunch, J. R. and Parlett, B. N. (1971). 'Direct methods for solving symmetric indefinite systems of linear equations', *SIAM J. Num. Anal.*, **8**, 639–655.

Carroll, C. W. (1961). 'The created response surface technique for optimizing nonlinear restrained systems', *Operations Res.*, **9**, 169–184.

Chamberlain, R. M. (1979). 'Some examples of cycling in variable metric methods for constrained minimization', *Math. Prog.*, **16**, 378–383.

Chamberlain, R. M., Lemarechal, C., Pedersen, H. C., and Powell, M. J. D. (1980). 'The watchdog technique for forcing convergence in algorithms for constrained optimization', University of Cambridge DAMTP Report 80/NA1.

Charalambous, C. (1977). 'Nonlinear least p-th optimization and nonlinear programming', *Math. Prog.*, **12**, 195–225.

Charalambous, C. (1979). 'Acceleration of the least p-th algorithm for minimax optimization with engineering applications', *Math. Prog.*, **17**, 270–297.

Charalambous, C. and Conn, A. R. (1978). 'An efficient method to solve the minimax problem directly', *SIAM J. Num. Anal.*, **15**, 162–187.

Charnes, A. (1952). 'Optimality and degeneracy in linear programming', *Econometrica*, **20**, 160–170.

Clarke, F. H. (1975). 'Generalized gradients and applications', *Trans. Amer. Math. Soc.*, **205**, 247–262.

Coleman, T. F. and Conn, A. R. (1980). 'Nonlinear programming via an exact penalty function method: global analysis', University of Waterloo Comp. Sci. Tech. Report CS-80-31.

Colville, A. R. (1968). 'A comparative study on non-linear programming codes', IBM NY Scientific Center Report 320-2949.

Conn, A. R. (1973). 'Constrained optimization using a nondifferentiable penalty function', *SIAM J. Num. Anal.*, **13**, 145–154.

Conn, A. R. (1979). 'An efficient second order method to solve the (constrained) minimax problem', University of Waterloo Dept of Combinatorics and Optimization Res. Report CORR-79-5.

Coope, I. D. and Fletcher, R. (1979). 'Some numerical experience with a globally convergent algorithm for nonlinearly constrained optimization', University of Dundee Maths Dept Report NA/30; also in *J. Opt. Theo. Applns* (to be published).

Cottle, R. W. and Dantzig, G. B. (1968). 'Complementary pivot theory of mathematical programming', *J. Linear Algebra Applns*, **1**, 103–125.

Courant, R. (1943). 'Variational methods for the solution of problems of equilibrium and vibration', *Bull. Amer. Math. Soc.*, **49**, 1–23.

Dantzig, G. B. (1963). *Linear Programming and Extensions*, Princeton University Press, Princeton, N.J.

Dantzig, G. B., Orden, A., and Wolfe, P. (1955). 'The generalized simplex method for minimizing a linear form under linear inequality restraints', *Pacific J. Maths*, **5**, 183–195.

Dantzig, G. B. and Wolfe, P. (1960). 'A decomposition principle for linear programs', *Operations Res.*, **8**, 101–111.

Dembo, R. S. (1976). 'A set of geometric programming test problems and their solutions', *Math. Prog.*, **10**, 192–213.

Dembo, R. S. (1979). 'Second order algorithms for the posynomial geometric programming dual, Part I: Analysis', *Math. Prog.*, **17**, 156–175.

Dembo, R. S. and Avriel, M. (1978). 'Optimal design of a membrane separation process using geometric programming', *Math. Prog.*, **15**, 12–25.

Demyanov, V. F. and Malozemov, V. N. (1971). 'The theory of nonlinear minimax problems', *Uspekhi Matematcheski Nauk*, **26**, 53–104.
Duffin, R. J., Peterson, E. L., and Zener, C. (1967). *Geometric Programming – Theory and Application*, John Wiley, New York.
Eriksson, J. (1980). 'A note on solution of large sparse maximum entropy problems with linear equality constraints', *Math. Prog.*, **18**, 146–154.
Fiacco, A. V. and McCormick, G. P. (1968). *Nonlinear Programming*, John Wiley, New York.
Fisher, M. L., Northup, W. D., and Shapiro, J. F. (1975). 'Using duality to solve discrete optimization problems: theory and computational experience', in Balinski and Wolfe (1975).
Fletcher, R. (1970). 'The calculation of feasible points for linearly constrained optimization problems', AERE Harwell Report AERE-R6354.
Fletcher, R. (1971). 'A general quadratic programming algorithm', *J. Inst. Maths Applns*, **7**, 76–91.
Fletcher, R. (1972a). 'Minimizing general functions subject to linear constraints', in *Numerical Methods for Nonlinear Optimization* (Ed. F. A. Lootsma), Academic Press, London.
Fletcher, R. (1972b). 'An algorithm for solving linearly constrained optimization problems', *Math. Prog.*, **2**, 133–165.
Fletcher, R. (1973). 'An exact penalty function for nonlinear programming with inequalities', *Math. Prog.*, **5**, 129–150.
Fletcher, R. (1975). 'An ideal penalty function for constrained optimization', *J. Inst. Maths Applns*, **15**, 319–342.
Fletcher, R. (1980a). 'A model algorithm for composite NDO problems', Workshop on Numerical Techniques in Systems Engineering, University of Kentucky, June 1980, in Mathematical Programming Studies, North Holland, Amsterdam (to be published).
Fletcher, R. (1980b). 'Numerical experiments with an L_1 exact penalty function method', Nonlinear Programming 4 Symposium, Madison, July 1980, Academic Press (to be published).
Fletcher, R. and Jackson, M. P. (1974). 'Minimization of a quadratic function of many variables subject only to lower and upper bounds', *J. Inst. Maths Applns*, **14**, 159–174.
Fletcher, R. and McCann, A. P. (1969). 'Acceleration techniques for nonlinear programming', in *Optimization* (Ed. R. Fletcher), Academic Press, London.
Fletcher, R. and Powell, M. J. D. (1975). 'On the modification of LDL^{T} factorizations', *Math. Comp.*, **29**, 1067–1087.
Fletcher, R. and Watson, G. A. (1980). 'First and second order conditions for a class of nondifferentiable optimization problems', *Math. Prog.*, **18**, 291–307; abridged from a University of Dundee Dept of Mathematics Report NA/28 (1978).
Frisch, K. R. (1955). 'The logarithmic potential method of convex programming', Memorandum, University Institute of Economics, Oslo, May 1955.
Gill, P. E. and Murray, W. (1973). 'A numerically stable form of the simplex algorithm', *J. Linear Algebra Applns*, **7**, 99–138.
Gill, P. E. and Murray, W. (1974a). 'Newton type methods for linearly constrained optimization', in *Numerical Methods for Constrained Optimization* (Eds P. E. Gill and W. Murray), Academic Press, London.
Gill, P. E. and Murray, W. (1974b). 'Methods for large-scale linearly constrained problems', in *Numerical Methods for Constrained Optimization* (Eds P. E. Gill and W. Murray), Academic Press, London.
Gill, P. E. and Murray, W. (1974c). 'Quasi-Newton methods for linearly constrained optimization', in *Numerical Methods for Constrained Optimization* (Eds P. E. Gill and W. Murray), Academic Press, London.

Gill, P. E. and Murray, W. (1976a). 'Nonlinear least squares and nonlinearly constrained optimization', in *Numerical Analysis, Dundee 1975* (Ed. G. A. Watson), Lecture Notes in Mathematics 506, Springer-Verlag, Berlin.

Gill, P. E. and Murray, W. (1976b). 'Minimization of a nonlinear function subject to bounds on the variables', NPL Report NAC 72.

Gill, P. E. and Murray, W. (1978a). 'Modification of matrix factorizations after a rank-one change', in *The State of the Art in Numerical Analysis* (Ed. D. A. H. Jacobs), Academic Press, London.

Gill, P. E. and Murray, W. (1978b). 'Numerically stable methods for quadratic programming', *Math. Prog.*, **14**, 349–372.

Goldfarb, D. (1969). Extension of Davidon's variable metric method to maximization under linear inequality and equality constraints', *SIAM J. Appl. Math.*, **17**, 739–764.

Goldfarb, D. (1972). 'Extensions of Newton's method and simplex methods for solving quadratic programs', in *Numerical Methods for Nonlinear Optimization* (Ed. F. A. Lootsma), Academic Press, London.

Goldfarb, D. and Reid, J. K. (1977). 'A practicable steepest-edge simplex algorithm', *Math. Prog.*, **12**, 361–371.

Graves, G. W. and Brown, G. G. (1979). 'Computational implications of degeneracy in large scale mathematical programming', X Symposium on Mathematical Programming, Montreal, August 1979.

Hadley, G. (1961). *Linear Algebra*, Addison-Wesley, Reading, Mass.

Hadley, G. (1962). *Linear Programming*, Addison-Wesley, Reading, Mass.

Han, S. P. (1976). 'Superlinearly convergent variable metric algorithms for general nonlinear programming problems', *Math. Prog.*, **11**, 263–282.

Han, S. P. (1977). 'A globally convergent method for nonlinear programming', *J. Opt. Theo. Applns*, **22**, 297–309.

Han, S. P. and Mangasarian, O. L. (1979). 'Exact penalty functions in nonlinear programming', *Math. Prog.*, **17**, 251–269.

Hestenes, M. R. (1969). 'Multiplier and gradient methods', *J. Opt. Theo. Applns*, **4**, 303–320.

Holt, J. N. and Fletcher, R. (1979). 'An algorithm for constrained non-linear least squares', *J. Inst. Maths Applns*, **23**, 449–464.

Kuhn, H. W. and Tucker, A. W. (1951). 'Nonlinear programming', in *Proceedings of the Second Berkeley Symposium on Mathematical Statistics and Probability* (Ed. J. Neyman), University of California Press.

Lancaster, P. (1969). *Theory of Matrices*, Academic Press, New York.

Lemarechal, C. (1978). 'Bundle methods in nonsmooth optimization', in Lemarechal and Mifflin (1978).

Lemarechal, C. and Mifflin, R. (1978). *Nonsmooth Optimization*, IIASA Proceedings 3, Pergamon, Oxford.

Madsen, K. (1975). 'An algorithm for minimax solution of overdetermined systems of nonlinear equations', *J. Inst. Maths Applns*, **16**, 321–328.

Maratos, N. (1978). 'Exact penalty function algorithms for finite dimensional and control optimization problems', Ph.D. thesis, University of London.

Marsten, R. E. (1975). 'The use of the Boxstep method in discrete optimization', in Balinski and Wolfe (1975).

Mayne, D. Q. (1980). 'On the use of exact penalty functions to determine step length in optimization algorithms', in *Numerical Analysis, Dundee 1979* (Ed. G. A. Watson), Lecture Notes in Mathematics 773, Springer-Verlag, Berlin.

Mifflin, R. (1978). 'A feasible descent algorithm for linearly constrained least squares problems', in Lemarechal and Mifflin (1978).

Morrison, D. D. (1968). 'Optimization by least squares', *SIAM J. Num. Anal.*, **5**, 83–88.

Murray, W. (1969). 'An algorithm for constrained minimization', in *Optimization* (Ed. R. Fletcher), Academic Press, London.
Murray, W. (1971). 'An algorithm for finding a local minimum of an indefinite quadratic program', NPL Report NAC 1.
Murtagh, B. A. and Sargent, R. W. H. (1969). 'A constrained minimization method with quadratic convergence', in *Optimization* (Ed. R. Fletcher), Academic Press, London.
Osborne, M. R. (1972). 'Topics in optimization', Stanford University Report STAN-CS-72-279.
Osborne, M. R. and Watson, G. A. (1969). 'An algorithm for minimax approximation in the nonlinear case', *Computer J.*, **12**, 63–68.
Pietrzykowski, T. (1969). 'An exact potential method for constrained maxima', *SIAM J. Num. Anal.*, **6**, 217–238.
Powell, M. J. D. (1969). 'A method for nonlinear constraints in minimization problems', in *Optimization* (Ed. R. Fletcher), Academic Press, London.
Powell, M. J. D. (1972). 'Quadratic termination properties of minimization algorithms, I and II', *J. Inst. Maths Applns*, **10**, 333–342 and 343–357.
Powell, M. J. D. (1977). 'Constrained optimization by a variable metric method', Cambridge University DAMTP Report 77/NA6.
Powell, M. J. D. (1978a). 'A fast algorithm for nonlinearly constrained optimization calculations', in *Numerical Analysis, Dundee 1977* (Ed. G. A. Watson), Lecture Notes in Mathematics 630, Springer-Verlag, Berlin.
Powell, M. J. D. (1978b). 'The convergence of variable metric methods for nonlinearly constrained optimization calculations', in *Nonlinear Programming 3* (Eds O. L. Mangasarian, R. R. Meyer, and S. M. Robinson), Academic Press, New York.
Pshenichnyi, B. N. (1978). 'Nonsmooth optimization and nonlinear programming', in Lemarechal and Mifflin (1978).
Reid, J. K. (Ed.) (1971). *Large Sparse Sets of Linear Equations*, Academic Press, London.
Reid, J. K. (1975). 'A sparsity-exploiting version of the Bartels–Golub decomposition for linear programming bases', AERE Harwell Report CSS 20.
Rockafellar, R. T. (1974). 'Augmented Lagrange multiplier functions and duality in non-convex programming', *SIAM J. Control*, **12**, 268–285.
Rosen, J. B. (1960). 'The gradient projection method for non-linear programming, Part I: Linear constraints', *J. SIAM*, **8**, 181–217.
Rosen, J. B. (1961). 'The gradient projection method for nonlinear programming, Part II: Nonlinear constraints', *J. SIAM*, **9**, 514–532.
Sargent, R. W. H. (1974). 'Reduced gradient and projection methods for nonlinear programming', in *Numerical Methods for Constrained Optimization* (Eds P. E. Gill and W. Murray), Academic Press, London.
Saunders, M. A. (1972). 'Product form of the Cholesky factorization for large-scale linear programming', Stanford University Report STAN-CS-72-301.
Swann, W. H. (1974). 'Constrained optimization by direct search', in *Numerical Methods for Constrained Optimization* (Eds P. E. Gill and W. Murray), Academic Press, London.
Watson, G. A. (1978). 'A class of programming problems whose objective function contains a norm', *J. Approx. Theo.*, **23**, 401–411.
Wilson, R. B. (1963). 'A simplicial algorithm for concave programming', Ph.D. dissertation, Harvard University Graduate School of Business Administration.
Wolfe, P. (1961). 'A duality theorem for non-linear programming', *Quart. Appl. Math.*, **19**, 239–244.
Wolfe, P. (1963a). 'Methods of nonlinear programming', in *Recent Advances in Mathematical Programming* (Eds R. L. Graves and P. Wolfe), McGraw-Hill, New York.

Wolfe, P. (1963b). 'A technique for resolving degeneracy in linear programming', *J. SIAM*, **11**, 205–211.

Wolfe, P. (1965). 'The composite simplex algorithm', *SIAM Rev.*, **7**, 42–54.

Wolfe, P. (1972). 'On the convergence of gradient methods under constraint', *IBM J. Res. and Dev.*, **16**, 407–411.

Wolfe, P. (1975). 'A method of conjugate subgradients', in Balinski and Wolfe (1975).

Womersley, R. S. (1978). 'An approach to nondifferentiable optimization', M.Sc. thesis, Mathematics Dept, University of Dundee.

Womersley, R. S. (1980). 'Optimality conditions for piecewise smooth functions', Workshop on Numerical Techniques in Systems Engineering, University of Kentucky, June 1980, in Mathematical Programming Studies, North Holland, Amsterdam (to be published).

Womersley, R. S. (1981). 'Numerical methods for structured problems in non-smooth optimization', Ph.D. thesis, Mathematics Dept, University of Dundee.

Zoutendijk, G. (1960). *Methods of Feasible Directions*, Elsevier, Amsterdam.

Subject Index